Amorphous and Nanocrystalline Semiconductors: Selected Papers from ICANS 29

Amorphous and Nanocrystalline Semiconductors: Selected Papers from ICANS 29

Editors

Kunji Chen
Shunri Oda
Linwei Yu

 Basel • Beijing • Wuhan • Barcelona • Belgrade • Novi Sad • Cluj • Manchester

Editors

Kunji Chen
School of Electronic Science
and Engineering
Nanjing University
Nanjing
China

Shunri Oda
Quantum Nanoelectronics
Research Center
Tokyo Institute of Technology
Tokyo
Japan

Linwei Yu
School of Electronic Science
and Engineering
Nanjing University
Nanjing
China

Editorial Office
MDPI
St. Alban-Anlage 66
4052 Basel, Switzerland

This is a reprint of articles from the Special Issue published online in the open access journal *Nanomaterials* (ISSN 2079-4991) (available at: www.mdpi.com/journal/nanomaterials/special_issues/22K7M6019H).

For citation purposes, cite each article independently as indicated on the article page online and as indicated below:

Lastname, A.A.; Lastname, B.B. Article Title. *Journal Name* **Year**, *Volume Number*, Page Range.

ISBN 978-3-7258-0120-6 (Hbk)
ISBN 978-3-7258-0119-0 (PDF)
doi.org/10.3390/books978-3-7258-0119-0

© 2024 by the authors. Articles in this book are Open Access and distributed under the Creative Commons Attribution (CC BY) license. The book as a whole is distributed by MDPI under the terms and conditions of the Creative Commons Attribution-NonCommercial-NoDerivs (CC BY-NC-ND) license.

Contents

About the Editors vii

Preface ix

Kunji Chen, Shunri Oda and Linwei Yu
Editorial for the Special Issue "Amorphous and Nanocrystalline Semiconductors: Selected Papers from ICANS 29"
Reprinted from: *Nanomaterials* 2023, 13, 2594, doi:10.3390/nano13182594 1

Yinzi Cheng, Xin Gan, Zongguang Liu, Junzhuan Wang, Jun Xu and Kunji Chen et al.
Nanostripe-Confined Catalyst Formation for Uniform Growth of Ultrathin Silicon Nanowires
Reprinted from: *Nanomaterials* 2022, 13, 121, doi:10.3390/nano13010121 5

Teng Sun, Dongke Li, Jiaming Chen, Yuhao Wang, Junnan Han and Ting Zhu et al.
Enhanced Electroluminescence from a Silicon Nanocrystal/Silicon Carbide Multilayer Light-Emitting Diode
Reprinted from: *Nanomaterials* 2023, 13, 1109, doi:10.3390/nano13061109 14

Kun Wang, Qiang He, Deren Yang and Xiaodong Pi
Highly Efficient Energy Transfer from Silicon to Erbium in Erbium-Hyperdoped Silicon Quantum Dots
Reprinted from: *Nanomaterials* 2023, 13, 277, doi:10.3390/nano13020277 25

Katsunori Makihara, Yuji Yamamoto, Yuki Imai, Noriyuki Taoka, Markus Andreas Schubert and Bernd Tillack et al.
Room Temperature Light Emission from Superatom-like Ge–Core/Si–Shell Quantum Dots
Reprinted from: *Nanomaterials* 2023, 13, 1475, doi:10.3390/nano13091475 35

Mingyue Shao, Yang Qiao, Yuan Xue, Sannian Song, Zhitang Song and Xiaodan Li
Advantages of Ta-Doped Sb_3Te_1 Materials for Phase Change Memory Applications
Reprinted from: *Nanomaterials* 2023, 13, 633, doi:10.3390/nano13040633 43

Jie Zhang, Ningning Rong, Peng Xu, Yuchen Xiao, Aijiang Lu and Wenxiong Song et al.
The Effect of Carbon Doping on the Crystal Structure and Electrical Properties of Sb_2Te_3
Reprinted from: *Nanomaterials* 2023, 13, 671, doi:10.3390/nano13040671 51

Cheng Liu, Yonghui Zheng, Tianjiao Xin, Yunzhe Zheng, Rui Wang and Yan Cheng
The Relationship between Electron Transport and Microstructure in $Ge_2Sb_2Te_5$ Alloy
Reprinted from: *Nanomaterials* 2023, 13, 582, doi:10.3390/nano13030582 63

Renjie Wu, Yuting Sun, Shuhao Zhang, Zihao Zhao and Zhitang Song
Great Potential of Si-Te Ovonic Threshold Selector in Electrical Performance and Scalability
Reprinted from: *Nanomaterials* 2023, 13, 1114, doi:10.3390/nano13061114 70

Vladimir G. Kuznetsov, Anton A. Gavrikov, Milos Krbal, Vladimir A. Trepakov and Alexander V. Kolobov
Amorphous As_2S_3 Doped with Transition Metals: An *Ab Initio* Study of Electronic Structure and Magnetic Properties
Reprinted from: *Nanomaterials* 2023, 13, 896, doi:10.3390/nano13050896 78

Yang Yang, Xu Zhu, Zhongyuan Ma, Hongsheng Hu, Tong Chen and Wei Li et al.
Artificial HfO_2/TiO_x Synapses with Controllable Memory Window and High Uniformity for Brain-Inspired Computing
Reprinted from: *Nanomaterials* 2023, 13, 605, doi:10.3390/nano13030605 95

Tong Chen, Kangmin Leng, Zhongyuan Ma, Xiaofan Jiang, Kunji Chen and Wei Li et al.
Tracing the Si Dangling Bond Nanopathway Evolution in a-SiN$_x$:H Resistive Switching Memory by the Transient Current
Reprinted from: *Nanomaterials* 2022, 13, 85, doi:10.3390/nano13010085 106

Hongsheng Hu, Zhongyuan Ma, Xinyue Yu, Tong Chen, Chengfeng Zhou and Wei Li et al.
Controlling the Carrier Injection Efficiency in 3D Nanocrystalline Silicon Floating Gate Memory by Novel Design of Control Layer
Reprinted from: *Nanomaterials* 2023, 13, 962, doi:10.3390/nano13060962 115

Yang Yang, Haiyan Pei, Zejun Ye and Jiaming Sun
Enhancement of the Electroluminescence from Amorphous Er-Doped Al$_2$O$_3$ Nanolaminate Films by Y$_2$O$_3$ Cladding Layers Using Atomic Layer Deposition
Reprinted from: *Nanomaterials* 2023, 13, 849, doi:10.3390/nano13050849 128

Zhimin Yu, Yang Yang and Jiaming Sun
Narrow UVB-Emitted YBO$_3$ Phosphor Activated by Bi^{3+} and Gd^{3+} Co-Doping
Reprinted from: *Nanomaterials* 2023, 13, 1013, doi:10.3390/nano13061013 137

Zewen Lin, Zhenxu Lin, Yanqing Guo, Haixia Wu, Jie Song and Yi Zhang et al.
Effect of a-SiC$_x$N$_y$:H Encapsulation on the Stability and Photoluminescence Property of CsPbBr$_3$ Quantum Dots
Reprinted from: *Nanomaterials* 2023, 13, 1228, doi:10.3390/nano13071228 149

Xiaolin Sun, Lu Li, Shanshan Shen and Fang Wang
TiO$_2$/SnO$_2$ Bilayer Electron Transport Layer for High Efficiency Perovskite Solar Cells
Reprinted from: *Nanomaterials* 2023, 13, 249, doi:10.3390/nano13020249 160

Fan Ye, Rui-Tuo Hong, Yi-Bin Qiu, Yi-Zhu Xie, Dong-Ping Zhang and Ping Fan et al.
Nanocrystalline ZnSnN$_2$ Prepared by Reactive Sputtering, Its Schottky Diodes and Heterojunction Solar Cells
Reprinted from: *Nanomaterials* 2022, 13, 178, doi:10.3390/nano13010178 170

Beitong Cheng, Yong Zhou, Ruomei Jiang, Xule Wang, Shuai Huang and Xingyong Huang et al.
Structural, Electronic and Optical Properties of Some New Trilayer Van de Waals Heterostructures
Reprinted from: *Nanomaterials* 2023, 13, 1574, doi:10.3390/nano13091574 188

Min Wang, Xiaoxiao Huang, Zhiqian Yu, Pei Zhang, Chunyang Zhai and Hucheng Song et al.
A Stable Rechargeable Aqueous Zn–Air Battery Enabled by Heterogeneous MoS$_2$ Cathode Catalysts
Reprinted from: *Nanomaterials* 2022, 12, 4069, doi:10.3390/nano12224069 200

Shuming Zhang, Xidi Sun, Xin Guo, Jing Zhang, Hao Li and Luyao Chen et al.
A Wide-Range-Response Piezoresistive–Capacitive Dual-Sensing Breathable Sensor with Spherical-Shell Network of MWCNTs for Motion Detection and Language Assistance
Reprinted from: *Nanomaterials* 2023, 13, 843, doi:10.3390/nano13050843 211

Shu Ying, Jiean Li, Jinrong Huang, Jia-Han Zhang, Jing Zhang and Yongchang Jiang et al.
A Flexible Piezocapacitive Pressure Sensor with Microsphere-Array Electrodes
Reprinted from: *Nanomaterials* 2023, 13, 1702, doi:10.3390/nano13111702 225

About the Editors

Kunji Chen

Kunji Chen is a Professor of the School of Electronic Science and Engineering, Nanjing University. He graduated from the Department of Physics of Nanjing University in 1963 and was a visiting scholar at the University of Chicago from 1981 to 1983. Since then, he has been to relevant universities and research institutes in the United States, Japan, and other countries over 10 times to carry out cooperative research work. From 2002 to 2014, he served as the vice chairman of the Jiangsu Physical Society, and from 1996 to 2017, he was elected and served as a member of the International Advisory Committee of the International Conference on Amorphous and Nanocrystalline Semiconductors.

Research Interests: nano-semiconductor materials and devices, and nano-optoelectronics. He has presided over and completed more than 20 major national research projects, 973 projects, and key and general projects of the National Natural Science Foundation of China. He has published more than 350 papers, one book, and 15 national invention patents. As the first author, he won the second prize of the 2003 National Natural Science Award, and won the first and second prizes of the provincial and ministerial Science and Technology Progress Awards (1988, 1995, 2002, and 2023).

Shunri Oda

Shunri Oda is a Professor in the Department of Physical Electronics and Quantum Nanoelectronics Research Center, Tokyo Institute of Technology. He obtained his Doctor of Engineering (PhD) degree from the Department of Physical Processing of Information, Tokyo Institute of Technology, in 1979. Title of Doctoral Thesis: Preparation of Low-Resistivity ZnS and Its Application to Display Devices. From 1982 to 1983, he served as a Visiting Researcher at the Massachusetts Institute of Technology USA. His current research interests include the fabrication of silicon quantum dots by pulsed plasma processes, single-electron tunneling devices based on nanocrystalline silicon, ballistic transport in silicon nanodevices, silicon-based quantum information devices, and NEMS hybrid devices. He was a Research Leader of the MEXT Center of Innovation program. He has authored more than 700 technical papers in international journals and conferences including 200 invited papers. He edited books (with D. Ferry), including *Silicon Nanoelectronics* (CRC Press, 2006) and *Nanoscale Silicon Devices* (CRC Press, 2015).

Prof. Oda is a fellow of the IEEE and Japan Society for Applied Physics and a member of the Electrochemical Society and Materials Research Society. He is a Distinguished Lecturer of the IEEE Electron Devices Society. He won the Tejima Award in 1985, as well as the Performance Award, MRS, and ISTEC in 1995, and the Fujino Award in 2002.

Linwei Yu

Dr./Prof. Linwei Yu received his B.S. degree in semiconductor physics in 2001 and his Ph.D. degree in microelectronics from Nanjing University, Nanjing, China, in 2007. From 2007 to 2009, he was a Postdoctoral Researcher with LPICM, Ecole Polytechnique, France, and joined CNRS as a Permanent Researcher (CR2) in 2009. Since 2013, he has been a Full Professor in the School of Electronics Science and Engineering, Nanjing University. His research interests include 1) growth mechanism and engineering of Si-based nanostructures, 2) high-performance silicon-nanowire-based stretchable electronics and sensors, and 3) radial junction optoelectronics, photovoltaics, and energy storage.

He has now published more than 200+ SCI papers in variouis journals, including *Nature Communications*, *Physical Review Letters*, *Nano Letters*, *ACS Nano*, *Advanced Materials*, *Advanced Functional Materials*, and *Nano Energy*, while holding 3 PCT patents and 30+ Chinese patents. He is serving as the Editorial Board Members for *Nanotechnology*, as an advisory standing committee member of the International Conference of Amorphous and Nanocrystalline Semiconductors (ICANS), and an active referees for 30+ international journals, including *Nature Electronics*, *Advanced Materials*, *Nano Letters*, and *ACS Nano*.

Preface

The 29th International Conference on Amorphous and Nanocrystalline Semiconductors (ICANS 29) served as a continuation of the biennial conference that has been held since 1965. ICANS 29 was held from 23 to 26 August at the campus of Nanjing University—a great venue for global academic researchers, industrial partners, and policy makers to come together and share their latest progress, breakthroughs, and new ideas on a wide range of topics in the fields of amorphous and nanocrystalline thin films and other nanostructured materials, as well as device applications.

It was the first time that this prestigious event was held in China, and it provided the perfect opportunity for young Chinese researchers and students to participate more actively in academic exchange as a part of the conference and become familiarized with the latest developments in the fields in which they work. And despite a one-year delay due to the COVID-19 pandemic, ICANS 29 still attracted more than 300 paper submissions from 11 countries, including both on-site and virtual oral and poster presentations, which made it truly both a global and hybrid conference.

For this Special Issue, twenty-one papers presented at the ICANS 29 were selected by the journal *Nanomaterials* for publication following a rigorous peer-review process. The scope of this Special Issue is to include recent developments and research activities in the field of amorphous and nanocrystalline semiconductors. This encompasses topics such as Si-based, oxide, perovskite, 2D thin films and nanostructures; device applications for TFTs, solar cells, and LEDs; as well as memory devices and emerging flexible electronics and neuromorphic applications. The papers collected here only reflect a portion of the topics presented in this conference. We hope that this Special Issue will stimulate fruitful discussions and cooperation between experts in academia and industry who work in the field of amorphous and nanocrystalline semiconductors. It was also a wonderful gift to have hosted the ICANS29 in Nanjing, China.

We would like to thank all of the authors and the *Nanomaterials* Editorial Office for their great contributions and excellent work.

ICANS 29 Conference Commettees
Local Organizing Committee:
Honorary Chair: Kunji Chen, Nanjing University, Nanjing, China.
Chair: Yi Shi, Nanjing University, Nanjing, China.
Vice-Chair: Jun Xu, Nanjing University, Nanjing, China.
Secretary General: Linwei Yu, Nanjing University, Nanjing, China.
Xiaodong Pi, Zhejiang University, Hangzhou, China;
Xuegong Yu, Zhejiang University, Hangzhou, China;
Xinran Wang, Nanjing University, Nanjing, China;
Qing Wan, Nanjing University, Nanjing, China;
Junwei Luo, Institute of Semiconductor, CAS, Beijing, China;
Zhitang Song, Shanghai Inst. of Microsystem and Information Technology, Shanghai, China;
Hairen Tan, Nanjing University, Nanjing, China;
Yuzheng Guo, Wuhan University, Wuhan, China;
Yang Yang, Nanjing Tech University, Nanjing, China;
Rui Huang, Hanshan Normal University, Chaozhou, China.
Program Committee:
Chair: Linwei Yu, Nanjing University, Nanjing, China.

Co-chair: John Robertson, University of Cambridge, UK.
Vice-chair: Yuzheng Guo, Wuhan University, Wuhan, China.
Vice-chair: Xuegong Yu, Zhejiang University, Hangzhou, China.
José Alvarez, GeePs, Centrale-Supelec, France;
Richard Curry, University of Manchester, Manchester, UK;
Elvira Fortunato, Universidade NOVA de Lisboa, Portugal;
Sang-Hee Ko Park, Korea Advanced Institute of Science and Technology, Seoul, Korea;
Sung Kyu Park, Chung-Ang University, Seoul, Korea;
Ling Li, Institute of Microelectronics, CAS, Beijing, China;
Klaus Lips, Helmholtz-Zentrum Berlin, Germany;
Mario Moreno, INAOE, Puebla, Mexico;
Kosuke Nagashio, The University of Tokyo, Tokyo, Japan;
Guozhen Shen, Institute of Semiconductor, CAS, Beijing, China;
Jiaming Sun, Nankai University, Tianjin, China;
Jianpu Wang, Nanjing Tech University, Nanjing, China.

International Advisory Committee:
Chair: John Robertson, University of Cambridge, UK.
Pere Roca i Cabarrocas, Ecole Polytechnique, France;
Sergei Baranovski, University of Marburg, Germany;
Hiroyuki Fujiwara, Gifu University, Japan;
Hideo Hosono, Tokyo Institute of Technology, Japan;
Jin Jang, Kyung Hee University, Korea;
Safa Kasap, University of Saskatchewan, Canada;
Jan Kocka, Academy of Sciences of Czech Republic;
Rodrigo Martins, UNINOVA, Portugal;
Ruud Schropp, University of the Western Cape, South Africa;
Linwei Yu, Nanjing University, China;
Sigurd Wagner, Princeton University, USA;
Richard Curry, University of Manchester, UK;
Yuzheng Guo, Wuhan University, China;
Klaus Lips, Helmholtz-Zentrum Berlin, Germany.

Secretarial Group:
Secretary General: Linwei Yu, Nanjing University, Nanjing, China.
Members: Zongguang Liu, Nanjing University, Nanjing, China;
Hucheng Song, Nanjing University, Nanjing, China;
Junzhuang Wang, Nanjing University, Nanjing, China.

Kunji Chen, Shunri Oda, and Linwei Yu
Editors

Editorial

Editorial for the Special Issue "Amorphous and Nanocrystalline Semiconductors: Selected Papers from ICANS 29"

Kunji Chen [1,*], Shunri Oda [2] and Linwei Yu [3]

1. School of Electronic Science and Engineering, National Laboratory of Solid State Microstructures, Nanjing University, Nanjing 210093, China
2. Quantum Nanoelectronics Research Center, Tokyo Institute of Technology, 2-12-1 Ookayama, Meguro-ku, Tokyo 152-8550, Japan; soda@pe.titech.ac.jp
3. National Laboratory of Solid State Microstructures, School of Electronics Science and Engineering, Collaborative Innovation Center of Advanced Microstructures, Nanjing University, Nanjing 210093, China; yulinwei@nju.edu.cn
* Correspondence: kjchen@nju.edu.cn

Citation: Chen, K.; Oda, S.; Yu, L. Editorial for the Special Issue "Amorphous and Nanocrystalline Semiconductors: Selected Papers from ICANS 29". *Nanomaterials* 2023, *13*, 2594. https://doi.org/10.3390/nano13182594

Received: 25 August 2023
Accepted: 15 September 2023
Published: 20 September 2023

Copyright: © 2023 by the authors. Licensee MDPI, Basel, Switzerland. This article is an open access article distributed under the terms and conditions of the Creative Commons Attribution (CC BY) license (https:// creativecommons.org/licenses/by/ 4.0/).

The 29th International Conference on Amorphous and Nanocrystalline Semiconductors served as a continuation of the biennial conference that has been held since 1965. ICANS 29 was held from 23 to 26 August at the campus of Nanjing University—a great venue for global academic researchers, industrial partners, and policy-makers to come together and share their latest progress, breakthroughs, and new ideas on a wide range of topics in the fields of amorphous and nanocrystalline thin films and other nanostructured materials, as well as device applications.

It was the first time that this prestigious event was held in China, and it provided the perfect opportunity for young Chinese researchers and students to more actively participate in academic exchange as a part of the conference and get to know the latest developments in the fields in which they work. And despite a one-year delay due to the COVID-19 pandemic, ICANS 29 still attracted more than 300 paper submissions from 11 countries, including on-site or online oral and poster presentations, which made it truly both a global and hybrid conference.

For this Special Issue on "ICANS29", twenty-one papers presented at the 29th ICANS have been selected by the Journal of *Nanomaterials* for publication, following a rigorous peer-review process. The scope of this Special Issue is to provide recent developments and research activities in the field of amorphous and nanocrystalline semiconductors. This includes topics such as Si-based, oxide, perovskite, 2D thin films and nanostructures, device applications for TFTs, solar cells, and LEDs, as well as memory devices and emerging flexible electronics and neuromorphic applications. For example, new fabrication technologies of amorphous and nanocrystalline thin films, electronic and optical characteristics, and device applications are presented and discussed in [1–21], including theoretical work in an ab initio study [9], controllable growth and formation of Si nanowires [1], nanocrystals [2], and quantum dots [3,4]. There are also several papers that cover emerging memory devices. These include phase change memory [5–9] based on amorphous chalcogenide thin films and memristors based on transition metal oxides acting as an artificial synapse [10–12]; the improvement of electro-luminescence (EL) efficiency Er-doped oxide thin films [13–15]; 2D semiconductor thin films and perovskite for solar cells and aqueous Zn-air battery [16–19]; and flexible electronic materials for integrated strain sensors [20,21]. We anticipate that this Special Issue should interest a broad audience in these related fields.

In terms of the content of this Special Issue, the first paper, from Yu's group [1], details the process of growing a uniform ordered ultrathin Si nanowire (SiNW) array by a nano-stripe-confined approach to produce highly uniform indium catalyst droplets via a relatively new in-plane solid–liquid–solid growth model. The diameters of the ultrathin SiNWs can be scaled down to only 28 ± 4 nm, which opens up a reliable route to batch-fabricate and integrate ultrathin SiNW channels for various high-performance field effect transistors (FET), sensors, and display applications. Xu's group [2] describes using Si nanocrystals (NCs)/SiC multilayer structures to form uniform SiNCs by plasma-enhanced chemical vapor deposition (PECVD) via the interface-confinement growth approach. The results show that the main EL near 750 nm can be obtained from 4 nm sized NCs. Moreover, it was found that the external quantum efficiency (EQE) of SiNCs/SiC multilayer light-emitting diode (LED) can be improved by phosphor doping. Pi's group [3] used nonthermal plasma to synthesize Si quantum dots (QDs) hyper-doped with Er at the concentration of ~1% to obtain near-infrared (NIR) light emission at a wavelength of ~830 and ~1540 nm. Furthermore, an ultrahigh η_{ET} (~93%) was obtained owing to the effective energy transfer from SiQDs to Er^{3+}. The results suggest that Er-hyper-doped SiQDs have great potential for the fabrication of high-performance near-infrared (NIR) light-emitting devices. Additionally, Miyazaki's group [4] demonstrated the high-density formation of SiQDs with Ge-core on ultrathin SiO_2 with the control of highly selective chemical vapor deposition for NIRLED applications. The results show that light emission is attributed to radiative recombination between quantized states in the Ge-core with a deep potential well for holes caused by electron/hole simultaneous injection from the gate and substrate, respectively.

The study of amorphous chalcogenide phase change characteristics and phase change random access memory (PCRAM) devices is one of the most active areas of research featured in this conference. Song's group [5] investigated the microstructural characterization and electrical properties of the Ta-doped Sb_3Te_1 films. The results show that $Ta_{1.45}Sb_3Te_1$ has enhanced thermal stability, reduced grain size, and increased the switching speed of PCRAM devices. Wu's group [6] reported the effect of carbon doping on Sb_2Te_3. It was found that the face-centered cubic (FCC) phase of C-doped Sb_2Te_3 appeared at 200 °C and began to transform into the hexagonal (HEX) phase at 25 °C. Based on the first principle density functional theory calculation, it was found that the formation energy of the FCC–Sb_2Te_3 structure decreases gradually with an increase in C-doping concentration. In addition, Zheng's group [7] studied the relationship between electron transport and microstructure in mature $Ge_2Sb_2Te_5$ (GST) alloy. The results indicated that the first resistance dropping in GST films is related to the increase in carrier concentration. However, the second drop is related to the increase in carrier mobility. In order to suppress the crosstalk and provide a high on-current to melt the incorporated phase change materials in the PCRAM array and 3D stacking chips, the authors [8] investigated the influence of Si concentration on the electrical properties of the Si-Te ovonic threshold selector. The results showed that the threshold voltage and leakage current remain basically unchanged with the electrode size scaling down. In addition, Kolobov's group [9] studied the effect of doping in typical chalcogenide glass As_2S_3 with transition metals (TM), such as Mo, W, and V, by using first-principal scaling simulations. The results indicated a strong effect of TM deposits on electronic structures of the a-As_2S_3 as well as an appearance of a magnetic response, which suggests that chalcogenide glasses doped with TMs may become a technologically important material. Besides the PCRAM devices, the memristive devices based on a metal–insolate–metal (MIM) structure with a transition metal oxide dielectric layer are yet another kind of emerging memory device discussed in this Special Issue. Ma's group [10] reported that the controllable memory window of the HfO_2/TiO_x memristive device could be obtained by tuning the thickness ratio of sublayers. As an artificial synapse based on HfO_2/TiO_x memristor, stable, controllable biological functions have been observed, which provides a hardware basis for their integration into the next generation of brain-inspired chips. In addition, they [11] described the a-SiNx:H-film-based resistive switching memory and used the transient current measurements to discover the Si dangling bonds nanopathway in a-SiNx:H resistive switching memory. Moreover, they [12] also employed the capacitance–voltage (C-V) measurements to investigate how to control the carrier injection efficiency in 3D nanocrystalline Si floating gate memory.

Sun's group [13,14] studied the luminescence characteristics of metal oxides doped by rare-earth elements. Firstly, the amorphous Al_2O_3-Y_2O_3:Er nanolaminate films prepared by atomic layer deposition with 1530 nm EL were obtained. It was found that the introduction of Y_2O_3 into Al_2O_3 reduced the electric field for Er excitation, and the EL performance was significantly enhanced. Additionally, a YBO_3 phosphor co-doped by Bi^{3+} and Gd^{3+} was prepared via high-temperature solid-state synthesis. The strong photon emissions were found under both ultraviolet and visible radiation, which could serve as a potential application for skin treatment. On the other hand, an intriguing phenomenon was discovered by Huang's group [15]. The PL properties and stability of $CsPbBr_3$ QDs films can be enhanced by an a-SiC_xN_y:H encapsulation layer prepared using a very-high-frequency PECVD technique. This method not only reduced the impact of air, light, and water on the QDs but also effectively passivated their surface defect states, leading to improved PL efficiency.

There are also two papers on solar cells from Wang's group [16], which focus on how to employ TiO_2/SnO_2 bilayer electron transport layer (ETL) to construct high-performance perovskite solar cells (PSCs). Such a bilayer structure ETL can not only produce a larger grain size of PSCs but also provide a high current density and reduced hysteresis. Their other paper [17] demonstrated high-performance $Ag/ITO/CuO_2/ZnSnN_2/Au$ heterojunction solar cells. The crucial technique employed was a nanocrystallization process, which can greatly reduce the electron density of the $ZnSnN_2$ sublayer. Song's group [18] designed 2D van der Waal (vdW) heterostructures, such as $BP/BP/MoS_2$, $BlueP/BlueP/MoS_2$, $BP/graphene/Mos_2$ and $BlueP/graphene/MoS_2$, etc., trilayer structures, and stimulated using the first-principles calculation to discover new optoelectronic properties. It is suggested that sophisticated 2D trilayer vdW heterostructures can provide further optimized optoelectronic devices. For the study of 2D materials, Song's group [19] reported a stable rechargeable aqueous ZN-air battery by using a heterogeneous 2D MoS_2 cathode catalyst, which demonstrated a capacity of $330 mAhg^{-1}$ and a durability of 500 cycles (~180 h) at $0.5\ mAcm^{-2}$. The hydrophilic and heterogeneous MoS_2 catalysts were prepared through a simple hydrothermal synthesis method.

Finally, this Special Issue features a study of the flexible electronic materials for sensor device applications. Shi and Pan's group [20,21] presented a scalable porous piezoresistive/capacitive dual-mode sensor with a porous structure in polydimethylsiloxane (PDMS) and with multi-walled carbon nanotubes (MLCNTs) embedded on its internal surface to form a 3D spherical-shell-structured conductive flexible network. This kind of polymer flexible sensor can be assembled into a wearable sensor that has a good ability to detect human motion and can be used for simple gesture and sign language recognition. Moreover, they also proposed a novel approach that combines self-assembled technology to prepare a high-performance flexible capacitive pressure sensor with a microsphere-array gold electrode and a nanofiber nonwoven dielectric material. The results of COMSOL simulations and the experiments showed that the flexible capacitive pressure sensor exhibits excellent performance in pressure measurements and has significant potential for electronic skin applications.

The papers collected here only reflect a portion of the topics presented in this conference. We hope that this Special Issue will stimulate fruitful discussions and cooperation between experts in academia and industry who work in the field of amorphous and nanocrystalline semiconductors. It has also been a wonderful gift to have hosted the ICANS29 in Nanjing, China.

We would like to thank all of the authors and the *Nanomaterials* Editorial Office for their great contributors and excellent work.

Conflicts of Interest: The authors declare no conflict of interest.

References

1. Cheng, Y.; Gan, X.; Liu, Z.; Wang, J.; Xu, J.; Chen, K.; Yu, L. Nanostripe-Confined Catalyst Formation for Uniform Growth of Ultrathin Silicon nanowires. *Nanomaterials* **2023**, *13*, 121. [CrossRef]
2. Sun, T.; Li, D.; Chen, J.; Wang, Y.; Han, J.; Zhu, T.; Li, W.; Xu, J.; Chen, K. Enhanced Electroluminescence from a Silicon Nanocrystal/Silicon Carbide Multilayer Light-Emitting Diode. *Nanomaterials* **2023**, *13*, 1109. [CrossRef] [PubMed]

3. Wang, K.; He, Q.; Yang, D.; Pi, X. Highly Efficient Energy Transfer from Silicon to Erbium in Erbium-Hyperdoped Silicon Quantum Dots. *Nanomaterials* **2023**, *13*, 277. [CrossRef]
4. Makihara, K.; Yamamoto, Y.; Imai, Y.; Taoka, N.; Schubert, M.A.; Tillack, B.; Miyazaki, S. Room Temperature Light Emission from Superatom-like Ge–Core/Si–Shell Quantum Dots. *Nanomaterials* **2023**, *13*, 1475. [CrossRef] [PubMed]
5. Shao, M.; Qiao, Y.; Xue, Y.; Song, S.; Song, Z.; Li, X. Advantages of Ta-Doped Sb_3Te_1 Materials for Phase Change Memory Applications. *Nanomaterials* **2023**, *13*, 633. [CrossRef] [PubMed]
6. Zhang, J.; Rong, N.; Xu, P.; Xiao, Y.; Lu, A.; Song, W.; Song, S.; Song, Z.; Liang, Y.; Wu, L. The Effect of Carbon Doping on the Crystal Structure and Electrical Properties of Sb_2Te_3. *Nanomaterials* **2023**, *13*, 671. [CrossRef]
7. Liu, C.; Zheng, Y.; Xin, T.; Zheng, Y.; Wang, R.; Cheng, Y. The Relationship between Electron Transport and Microstructure in $Ge_2Sb_2Te_5$ Alloy. *Nanomaterials* **2023**, *13*, 582. [CrossRef]
8. Wu, R.; Sun, Y.; Zhang, S.; Zhao, Z.; Song, Z. Great Potential of Si-Te Ovonic Threshold Selector in Electrical Performance and Scalability. *Nanomaterials* **2023**, *13*, 1114. [CrossRef]
9. Kuznetsov, V.G.; Gavrikov, A.A.; Krbal, M.; Trepakov, V.A.; Kolobov, A.V. Amorphous As_2S_3 Doped with Transition Metals: An Ab Initio Study of Electronic Structure and Magnetic Properties. *Nanomaterials* **2023**, *13*, 896. [CrossRef]
10. Yang, Y.; Zhu, X.; Ma, Z.; Hu, H.; Chen, T.; Li, W.; Xu, J.; Xu, L.; Chen, K. Artificial HfO_2/TiO_x Synapses with Controllable Memory Window and High Uniformity for Brain-Inspired Computing. *Nanomaterials* **2023**, *13*, 605. [CrossRef]
11. Chen, T.; Leng, K.; Ma, Z.; Jiang, X.; Chen, K.; Li, W.; Xu, J.; Xu, L. Tracing the Si Dangling Bond Nanopathway Evolution in a-SiNx:H Resistive Switching Memory by the Transient Current. *Nanomaterials* **2023**, *13*, 85. [CrossRef] [PubMed]
12. Hu, H.; Ma, Z.; Yu, X.; Chen, T.; Zhou, C.; Li, W.; Chen, K.; Xu, J.; Xu, L. Controlling the Carrier Injection Efficiency in 3D Nanocrystalline Silicon Floating Gate Memory by Novel Design of Control Layer. *Nanomaterials* **2023**, *13*, 962. [CrossRef] [PubMed]
13. Yang, Y.; Pei, H.; Ye, Z.; Sun, J. Enhancement of the Electroluminescence from Amorphous Er-Doped Al_2O_3 Nanolaminate Films by Y_2O_3 Cladding Layers Using Atomic Layer Deposition. *Nanomaterials* **2023**, *13*, 849. [CrossRef] [PubMed]
14. Yu, Z.; Yang, Y.; Sun, J. Narrow UVB-Emitted YBO3 Phosphor Activated by Bi^{3+} and Gd^{3+} Co-Doping. *Nanomaterials* **2023**, *13*, 1013. [CrossRef]
15. Lin, Z.; Lin, Z.; Guo, Y.; Wu, H.; Song, J.; Zhang, Y.; Zhang, W.; Li, H.; Hou, D.; Huang, R. Effect of a-SiC_xN_y:H Encapsulation on the Stability and Photoluminescence Property of $CsPbBr_3$ Quantum Dots. *Nanomaterials* **2023**, *13*, 1228. [CrossRef]
16. Sun, X.; Li, L.; Shen, S.; Wang, F. TiO_2/SnO_2 Bilayer Electron Transport Layer for High Efficiency Perovskite Solar Cells. *Nanomaterials* **2023**, *13*, 249. [CrossRef]
17. Ye, F.; Hong, R.-T.; Qiu, Y.-B.; Xie, Y.-Z.; Zhang, D.-P.; Fan, P.; Cai, X.-M. Nanocrystalline $ZnSnN_2$ Prepared by Reactive Sputtering, Its Schottky Diodes and Heterojunction Solar Cells. *Nanomaterials* **2023**, *13*, 178. [CrossRef]
18. Cheng, B.; Zhou, Y.; Jiang, R.; Wang, X.; Huang, S.; Huang, X.; Zhang, W.; Dai, Q.; Zhou, L.; Lu, P.; et al. Structural, Electronic and Optical Properties of Some New Trilayer Van de Waals Heterostructures. *Nanomaterials* **2023**, *13*, 1574. [CrossRef]
19. Wang, M.; Huang, X.; Yu, Z.; Zhang, P.; Zhai, C.; Song, H.; Xu, J.; Chen, K. A Stable Rechargeable Aqueous Zn–Air Battery Enabled by Heterogeneous MoS_2 Cathode Catalysts. *Nanomaterials* **2022**, *12*, 4069. [CrossRef]
20. Zhang, S.; Sun, X.; Guo, X.; Zhang, J.; Li, H.; Chen, L.; Wu, J.; Shi, Y.; Pan, L. A Wide-Range-Response Piezoresistive–Capacitive Dual-Sensing Breathable Sensor with Spherical-Shell Network of MWCNTs for Motion Detection and Language Assistance. *Nanomaterials* **2023**, *13*, 843. [CrossRef]
21. Ying, S.; Li, J.; Huang, J.; Zhang, J.-H.; Zhang, J.; Jiang, S.; Sun, X.; Pan, L.; Shi, Y. A Flexible Piezocapacitive Pressure Sensor with Microsphere-Array Electrodes. *Nanomaterials* **2023**, *13*, 1702. [CrossRef] [PubMed]

Disclaimer/Publisher's Note: The statements, opinions and data contained in all publications are solely those of the individual author(s) and contributor(s) and not of MDPI and/or the editor(s). MDPI and/or the editor(s) disclaim responsibility for any injury to people or property resulting from any ideas, methods, instructions or products referred to in the content.

Article

Nanostripe-Confined Catalyst Formation for Uniform Growth of Ultrathin Silicon Nanowires

Yinzi Cheng, Xin Gan, Zongguang Liu *, Junzhuan Wang, Jun Xu, Kunji Chen and Linwei Yu *

School of Electronic Science and Engineering/National Laboratory of Solid-State Microstructures, Nanjing University, Nanjing 210093, China
* Correspondence: liuzongguang@nju.edu.cn (Z.L.); yulinwei@nju.edu.cn (L.Y.)

Abstract: Uniform growth of ultrathin silicon nanowire (SiNW) channels is the key to accomplishing reliable integration of various SiNW-based electronics, but remains a formidable challenge for catalytic synthesis, largely due to the lack of uniform size control of the leading metallic droplets. In this work, we explored a nanostripe-confined approach to produce highly uniform indium (In) catalyst droplets that enabled the uniform growth of an orderly SiNW array via an in-plane solid–liquid–solid (IPSLS) guided growth directed by simple step edges. It was found that the size dispersion of the In droplets could be reduced substantially from $D_{cat}^{pl} = 20 \pm 96$ nm on a planar surface to only $D_{cat}^{ns} = 88 \pm 13$ nm when the width of the In nanostripe was narrowed to $W_{str} = 100$ nm, which could be qualitatively explained in a confined diffusion and nucleation model. The improved droplet uniformity was then translated into a more uniform growth of ultrathin SiNWs, with diameter of only $D_{nw} = 28 \pm 4$ nm, which has not been reported for single-edge guided IPSLS growth. These results lay a solid basis for the construction of advanced SiNW-derived field-effect transistors, sensors and display applications.

Keywords: silicon nanowires; confined catalyst formation; in-plane solid–liquid–solid growth

Citation: Cheng, Y.; Gan, X.; Liu, Z.; Wang, J.; Xu, J.; Chen, K.; Yu, L. Nanostripe-Confined Catalyst Formation for Uniform Growth of Ultrathin Silicon Nanowires. *Nanomaterials* **2023**, *13*, 121. https://doi.org/10.3390/nano13010121

Academic Editor: Adriano Sacco

Received: 17 November 2022
Revised: 19 December 2022
Accepted: 21 December 2022
Published: 26 December 2022

Copyright: © 2022 by the authors. Licensee MDPI, Basel, Switzerland. This article is an open access article distributed under the terms and conditions of the Creative Commons Attribution (CC BY) license (https://creativecommons.org/licenses/by/4.0/).

1. Introduction

Silicon nanowires (SiNWs) are one of the promising 1D channel materials for the construction of high-performance field-effect transistors (FETs) [1–3], display logics [4–6] and sensors [7–10], due to their large surface-to-volume ratio and excellent electrostatic modulation capability in a fin or gate-all-around gating configuration. Compared to the sophisticated top-down patterning and etching strategy, the bottom-up catalytic growth of SiNWs, led by metal catalyst droplets, for example via the famous vapor–liquid–solid (VLS) growth mechanism [11–15], offers a low-cost, high-yield and diverse fabrication strategy. Many advanced nanoelectronics have been successfully demonstrated based on VLS-grown SiNWs serving as semiconducting channels [16–19]. However, a major challenge for the catalytic growth of SiNWs, in view of scalable electronic applications on planar substrate, is how to integrate or grow them directly into pre-designed locations without the use of post-growth transferring and alignment, while a uniform diameter control of the as-grown SiNWs, determined usually by the leading catalyst droplets [20,21], is also highly desirable.

In order to gain better control of the catalytic growth of SiNWs, an in-plane solid–liquid–solid (IPSLS) growth strategy has been developed in our previous works [22–24], where a hydrogenated amorphous Si (a-Si) thin film is deposited upon a substrate surface to serve as the precursor layer for the indium (In) catalyst droplets to absorb and produce continuous polycrystalline SiNWs. In principle, the IPSLS growth of SiNWs is driven by the higher Gibbs energy in the disordered a-Si precursor layer, with respect to the crystalline phase of SiNWs, where the In droplets serve as a moving mediator to facilitate the phase conversion [22,23,25,26]. More importantly, the In catalyst droplets can be guided by pre-patterned edge lines to produce SiNW arrays along their moving courses [27–29], which is a key aspect that enables the scalable and precise integration of an orderly SiNW array.

Similarly to that in VLS growth, the diameter of the IPSLS SiNWs is basically determined by the size of the leading catalyst droplets [22,30,31]. However, for the formation of the In catalyst droplets by using an H_2 plasma treatment on a free planar surface or in relatively wide In stripes with a width >2 µm, as seen, for example, in Figure 1d(i,iii), the size dispersion of the In droplets is usually quite large (see, for instance, the SEM image in Figure 1d(iv)). Then, this random size variation in the catalyst droplets will be passed on to the as-grown SiNWs (Figure 1a), with larger diameter D_{nw} variation, which will pose a disadvantage to the device reliability [32], causing large fluctuations in the $I_{on/off}$ ratio [33], mobility [34] and subthreshold swing (SS) [35,36], as diagrammed in Figure 1b. Therefore, it is of paramount importance to seek a new approach to greatly improve the diameter uniformity of the IPSLS SiNWs, which is indispensable for achieving a reliable electronic integration (Figure 1c). Actually, for the VLS growth of SiNWs, many catalyst formation control technologies have been explored to control the initial size of the catalyst metal droplets, such as via temperature control [37] and EBL pattern [38,39], which have proven rather efficient to produce highly uniform vertical VLS-grown SiNWs. Following these insights, the key to control the uniformity of IPSLS SiNWs is to accomplish, first and foremost, a uniform size control of the leading catalyst droplets, which unfortunately remains an unexplored topic for the IPSLS growth of SiNWs. In this work, a new nanostripe-confined catalyst formation approach is proposed and tested to obtain (Figure 1e), first, a uniform formation of In catalyst droplets, and then, use them to produce more uniform and ultrathin SiNWs with diameter of $D_{nw} = 28 \pm 4$ nm, as a key basis for the future scalable and reliable integration of SiNW-based electronics.

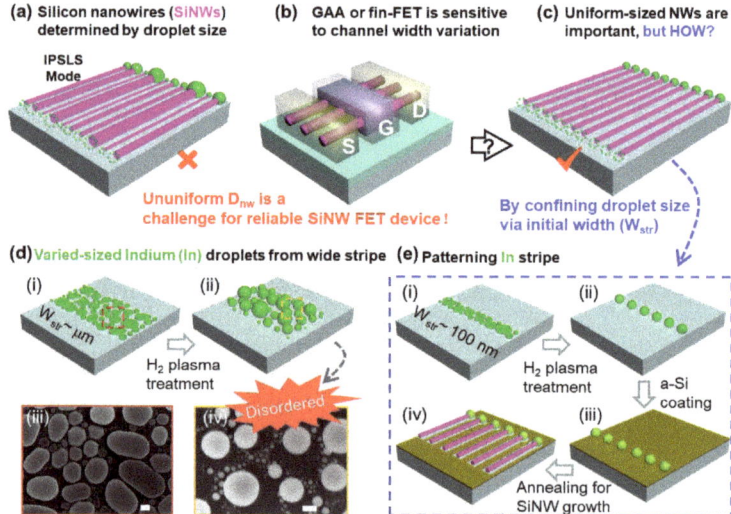

Figure 1. (**a**) Schematic illustration showing that the diameters of the SiNWs grown via IPSLS mode are determined by the size of the leading droplets. In view of serving as FET channels, in fin-gate or gate-all-around configurations, as depicted in (**b**), a uniform array of SiNW channels, for instance, those in (**c**), is highly desirable. (**d**) Panel (**i,ii**) diagram of the formation of random In catalyst droplets within the wide In stripes or planar surface, as observed in SEM images of the In grains before (**iii**) and after (**iv**) H_2 plasma treatment. In comparison, (**e**) depicts the formation of rather uniform In droplets by using a narrow In stripe with a width down to 100 nm, with the key fabrication steps depicted in (**i,ii**) catalyst formation by H_2 plasma treatment, (**iii**) amorphous silicon (a-Si) precursor coating, and (**iv**) annealing to activate the SiNW to grow by consuming the a-Si layer. Scale bars in panels (**iii,iv**) in (**d**) are all for 200 nm.

2. Materials and Methods
2.1. Preparation of Catalyst Nanostripes

A silicon wafer coated with 500 nm thick SiO_2 was first cleaned using acetone, alcohol and deionized water. The guiding edges, single-sided step edges, with etching depth of ~150 nm were prepared on the SiO_2 surface by using standard photolithography and RIE etching procedures, as depicted in Figure 2a. Then, polymethyl methacrylate (PMMA) photoresist with thickness of 200 nm was spin-coated on the substrate at 3000 r/min. After that, a series of empty stripe regions with width (W_{str}) ranging from 100 nm to 500 nm were patterned by electron beam lithography (EBL) in scanning electron microscopy with a dose value of 240. Finally, 16 nm In stripes were deposited via thermal evaporation, and the In stripes were obtained via standard lift-off procedure.

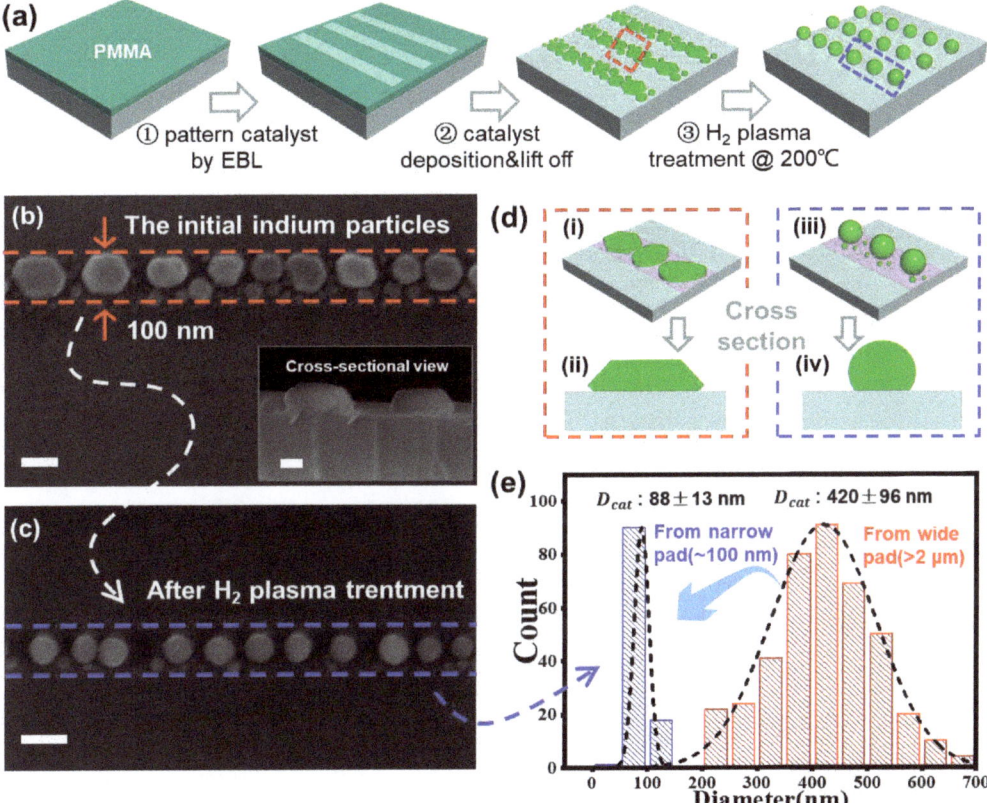

Figure 2. (a) Schematic illustration of the formation of uniform In droplets from narrow In stripes prepared by using EBL. (b) The SEM image of the initial discrete In catalyst grains evaporated on the nanostripe region with a width of 100 nm, while the inset shows the cross-sectional view of these pancake-shaped catalyst grains. (c) The formation of a chain of uniform In droplets after H_2 plasma treatment. (d) Tilted and cross-sectional view diagrams of the In grains evaporated (i,ii) and after H_2 plasma treatment (iii,iv) within the nanostripe regions. (e) Statistics of the droplet diameters formed within a strong nanostripe confinement (light blue, for narrow W_{str} = 100 nm) or on a much wider In stripe (light red, with W_{str} > 2 μm). Scale bars in (b,c) stand for 100 nm.

2.2. Growth of Ultrathin SiNW Array

The samples were loaded into a PECVD system and treated by H_2 plasma under 200 °C, with H_2 flow of 14 SCCM and pressure of 140 Pa. An H_2 plasma treatment was performed to produce H^+ radicals and reduce the thin In_2O_3 oxide layer formed on the surface of the indium grains, so as to release them to deform and merge into discrete and spherical droplets. Then, the 10 nm thick a-Si:H precursor layer was deposited at 100 °C, with a gas flow rate and chamber pressure of 5 SCCM and 20 Pa, respectively. After that, the substrate temperature was raised to ~350 °C and kept in vacuum for 1 h. The extra a-Si supply on the vertical sidewall surfaces attracted the In droplets to move along the edge lines to produce SiNW arrays by converting the a-Si layer into crystalline SiNWs. Finally, the remnant a-Si layer was selectively etched off by using H_2 plasma at ~120 °C for 5 min, with typical gas flow rate and chamber pressure of 15 SCCM and 140 Pa, respectively.

3. Results and Discussion

3.1. Formation of Uniform In Droplets from Narrow In Stripes

After H_2 plasma treatment, the surface oxide layers of the In catalysts were removed, and the initial faceted and irregular In grains (see the SEM characterization in Figure 2b, for example) were transformed into discrete spherical droplets (Figure 2c). Actually, for the In stripe of W_{str} = 100 nm, the initial In grains were of flat pancake shapes, as witnessed in the cross-sectional SEM view provided in the inset of Figure 2b and illustrated in Figure 2d(i,ii), with many random small grains resting among the larger ones. After H_2 plasma treatment at a temperature higher than the In melting point of 156 °C, the In droplets were allowed to transform into energetically more favorable spherical shapes, as observed in Figure 2c and depicted in Figure 2d(iii,iv), while the closely neighbored In grains could also agglomerate into bigger ones. Interestingly, compared to the size distribution of the In droplets formed on the planar surface (or within much wider stripes W_{str} > 2 µm, the red columns in Figure 2e), the size dispersion of the In droplets within narrow stripes (blue, for W_{str} = 100 nm) could be substantially reduced from $D_{In}^{2\mu m}$ = 20 ± 96 nm to D_{In}^{100nm} = 88 ± 13 nm, which is highly beneficial for achieving uniform size control of the In catalyst droplets and thus the as-grown SiNWs.

3.2. In Droplet Formation on Nanostripes of Different Widths

In order to understand how the nanostripe confinement helped to improve the uniformity of the catalyst formation, a series of In stripes with different widths of 200 nm, 300 nm and 500 nm were prepared by using EBL, with a constant In thickness of 16 nm. Indeed, as seen in Figure 3a–c and the corresponding statistics in Figure 3d–f, there was a clear trend where the size dispersion of the relatively large In droplets, highlighted by different colors, could be gradually improved from D_{In}^{500nm} = 218 ± 31 nm to D_{In}^{300nm} = 151 ± 27 nm and D_{In}^{200nm} = 172 ± 16 nm, with the decrease in the In stripe width. Meanwhile, there was also a high-density population of much smaller droplets, with typical diameter of ~50 nm, among the larger ones, which all had a similar size distribution for different In stripe widths. In addition, for the same stripe width of 300 nm (Figure 3b), if the In thickness was increased from 16 nm to 32 nm, the In grains were found to merge with their neighbors, as seen in Figure 3g, while the resulting In droplets after H_2 plasma treatment became not only larger in diameter but also far more uniform than those observed in Figure 3b, with $D_{In\_32nm}^{300nm}$ = 270 ± 15 nm (Figure 3h).

These catalyst droplet formation phenomena could be explained based on the following considerations: at the very beginning of the evaporation, the In atoms, evaporated by thermal evaporation, will land on the sample surface and diffuse on the empty surface until they run into each other to form more stable nuclei (as depicted schematically by the left panel of Figure 3i). In the next step, more adatoms continue to arrive and get trapped by the nearest nuclei (the right panel of Figure 3i) if they fall within the corresponding surface collection zone, which is roughly measured by the typical diffusion length of the In atoms λ_{In}. On a planar surface, the area of the collection zone can be approximated by $S_{ad\_planar} \sim d_s^2 \sim 4\lambda_{In}^2$, where d_s is the separation between the initial nucleation sites or grains $d_s \sim 2\lambda_{In}$. According to the average large grain-to-grain separations, extracted from the droplets on the planar surface, as seen, for example, in Figure S1, the average separation is estimated from the density by $d_s = n_{In}^{-1/2} \sim 244$ nm, or a $\lambda_{In} \sim 122$ nm.

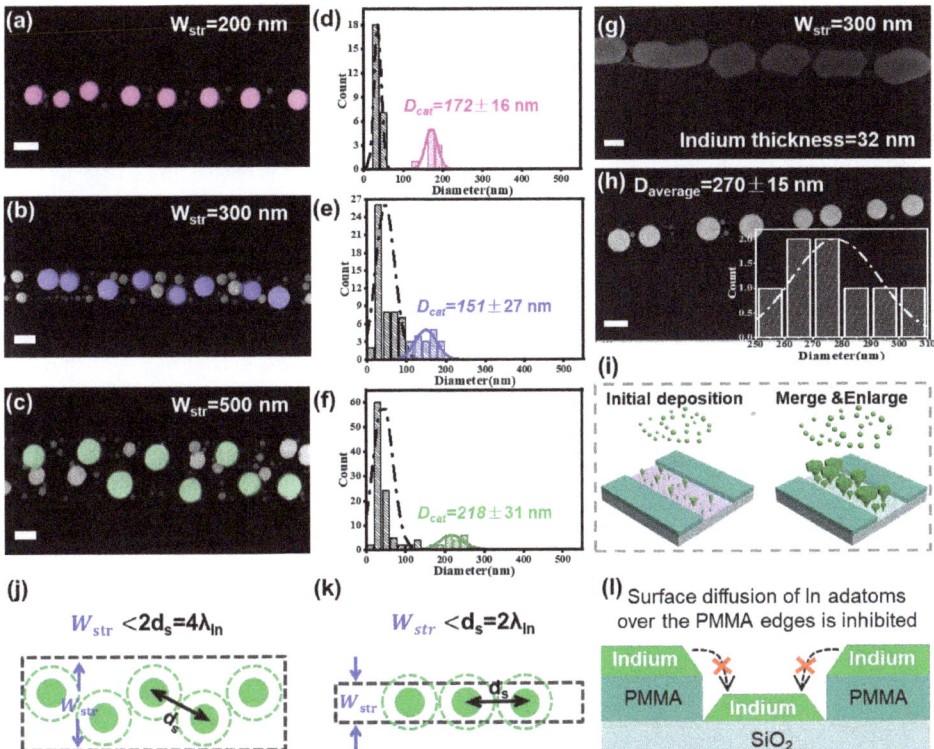

Figure 3. (**a**–**c**) SEM images of the In droplets formed on the nanostripes with the same In thickness of 16 nm but different W_{str} = 200 nm, 300 nm and 500 nm, with corresponding diameter statistics displayed in (**d**–**f**). In comparison, the In catalyst formations within a stripe width of 300 nm, with a thicker In thickness of 32 nm, are presented in (**g**,**h**) for the situations of initial grains and the droplets formed after H$_2$ plasma treatment. (**i**) depicts schematically the (left) initial In atom nucleation and (right) subsequent merging stages during the thermal evaporation. (**j**,**k**) illustrate the formation of discrete In nucleation sites and their collection zones, delineated by the dashed circles, within nanostripe widths of $W_{str} \sim 3\lambda_{In}$ and $W_{str} \sim 2\lambda_{In}$, respectively. (**l**) illustrates that the surface diffusion of In adatoms over the PMMA pattern edges is very inefficient or inhibited. Scale bars in (**a**–**c**,**g**,**h**) are all for 200 nm.

Moreover, for the nanostripe-confined catalyst formation, the surface diffusion of the In adatoms on the PMMA resistor polymer surface is considered to be very inefficient, and thus, there is likely little flux contribution coming from the PMMA surface to the stripe regions, as schematically depicted in Figure 3l. So, for the initial nucleation formation within a nanostripe region with $W_{str} < d_s$ or $2\lambda_{In}$, the effective area of the adatom collection zone is reduced and becomes $S_{ad\_str} = d_s W_{str} < d_s^2$.

In this scenario, the volume of the final catalyst droplets formed within a nanostripe or on a planar surface can be written as $D_{In\_str}^3 \sim S_{ad} t_{In} = d_s W_{str} t_{In}$ and $D_{In\_planar}^3 \sim S_{ad} t_{In} = d_s^2 t_{In}$, respectively. Considering the random variation of the grain-to-grain separation δd_s, that is $d_s = \overline{d_s} + \delta d_s$, as the major source of size fluctuations, its influence on the final diameter dispersions/variations of the In droplets within a nanostripe or on a planar surface can be derived as,

$$\delta D_{In\_str} \sim \delta d_s \frac{W_{str}}{3D_{In}^2/t_{In}} \quad (1)$$

$$\delta D_{In\_planar} \sim \delta d_s \frac{2d_s}{3D_{In}^2/t_{In}} \quad (2)$$

Obviously, for a narrow In stripe confinement, $W_{str} \ll d_s$, the random diameter fluctuation passed to the catalyst droplets can be greatly suppressed within the nanostripe, as $\delta D_{In\_str} \ll \delta D_{In\_planar}$. Additionally, it can be seen from both Equations (1) and (2) that, with a larger In catalyst droplet, the diameter fluctuation should also decrease, because of the denominator term of $3D_{In}^2/t_{In}$.

Meanwhile, the formation of the tiny grains among the large ones should happen at a later stage, that is, after the formation of the large grains. This can be clearly seen from the cross-sectional SEM view in Figure S2, where the small grains only exist in the spare regions among the large grains, with a much lower height than their larger neighbors. The emergence of such small grains seems to indicate that, with the increase in more In coverage on the sample surface, the irradiation heating (from the evaporation crucible below) on the sample is gradually decreased, probably due to the enhanced reflection of In layer coating deposited on the sample surface. This thus leads to a temperature decrease on the sample surface that quickly reduces the diffusion distance of the adatoms on the surface, as $\lambda_{In} \sim \lambda_0 e^{-E_d/kT}$ is highly temperature-dependent, where E_d is the diffusion barrier height on the SiO$_2$ substrate surface. So, confining the catalyst formation within a narrow stripe region, with $W_{str} \ll 2\,d_s$ at least during the initial nucleation stage, provides indeed a convenient and efficient way to suppress the random size fluctuation of the catalyst droplets (seen in Figure 3j,k). It is also predicted that maintaining suitable heating on the substrate holder, though not available for the current experimental setup in this work, could help to further improve the uniformity of the In catalyst droplets.

3.3. Growth of Ultrathin SiNWs Led by the Uniform In Droplets

Since the SiNW diameter was basically determined by the leading In catalyst droplets via IPSLS strategy, with $D_{nw} \sim D_{In}$ [22,30,31] (for instance, that in Figure 4e), ultrathin SiNW arrays, as seen in Figure 4a–c, could now be grown by using the pre-patterned In nanostripes with W_{str} = 70 nm; see Figure 4d for more information on the starting growth location (and Figure S3 for the initial catalyst formation prior to annealing growth). Remarkably, rather thin and uniform SiNWs with an average diameter of D_{nw} = 28 ± 4 nm, according to the statistics in Figure 4f, could be grown directly by using only single-step guided IPSLS growth, which is far more uniform and thinner, compared to the SiNWs grown by using catalysts on much wider micro stripes (W_{str} > 2 μm), as seen also with the corresponding droplets in Figure S4 with an average diameter of D_{nw} = 83 ± 22 nm. The diameter fluctuation along the SiNWs was caused mostly by the roughness of the guiding edge. In addition, the catalyst droplets will remain at the end of SiNWs after growth by using the IPSLS strategy (Figure 4e) and can be easily removed by using dilute hydrochloric acid before the fabrication of high-performance devices. These results indicate a new narrow-stripe-confined catalyst formation strategy to obtain uniform indium catalyst droplets, as a key to substantially suppress the SiNW-to-SiNW diameter variation. As a matter of fact, the construction of high-performance SiNW-based logics and sensors [31,40] all demand rather thin, uniform and orderly SiNW arrays as 1D channels to achieve stronger electrostatic modulation control [41] or higher field-effect sensitivity [42].

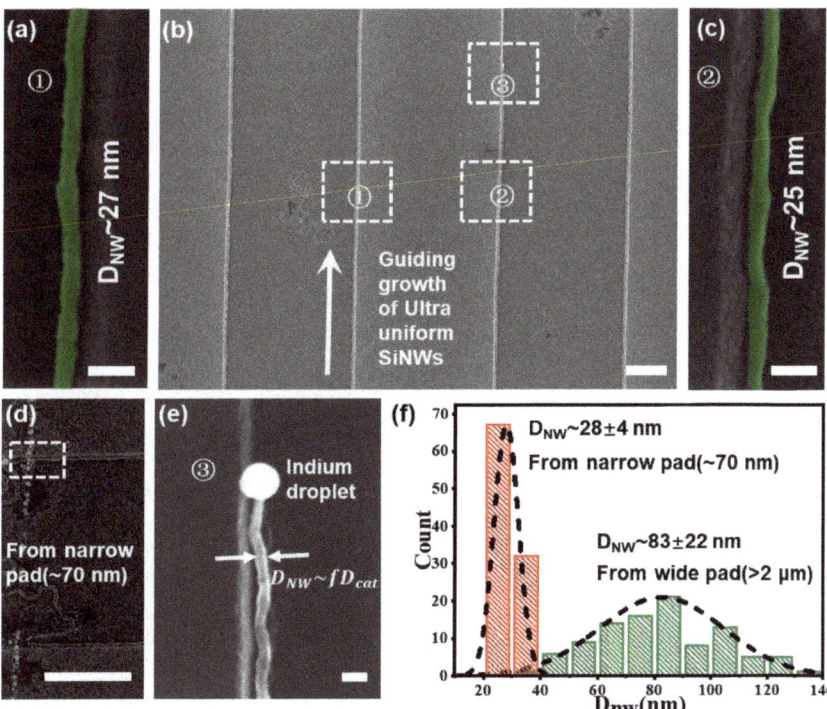

Figure 4. (a–c) SEM images of the IPSLS growth of uniform ultrathin SiNWs led by the uniform In droplets formed within nanostripes of W_{str} ~ 70 nm, with diameters of 27 nm and 25 nm in (a) and (c), respectively. Close view of the (d) starting and (e) ending of a specific SiNW guided along the step edge. (f) The diameter statistics of the SiNWs grown from narrow In stripe (red) and wide stripe (green). Scale bars in (a,c,e) are for 100 nm, (b,d) stand for 1 μm.

4. Conclusions

In summary, we have established a nanostripe-confined approach for rather uniform In catalyst formation and demonstrated an orderly growth of ultrathin SiNW arrays, with a narrow size dispersion of only $D_{nw} = 28 \pm 4$ nm, enabled by the largely improved catalyst droplet diameter uniformity in pre-patterned nanostripe regions, which can help to largely suppress the random diffusion and nucleation of the In catalyst adatoms particularly during the early grain formation stage within the tightly confined nanostripe. This study opens up a reliable route to batch fabricate and integrate ultrathin SiNW channels for various high-performance electronics and sensor applications.

Supplementary Materials: The following supporting information can be downloaded at: https://www.mdpi.com/article/10.3390/nano13010121/s1, Figure S1: SEM image of the catalyst droplet dispersion on planar surface after H_2 plasma treatment. The average separation can be estimated from the density by $d_s = n_{In}^{-1/2} \sim 244$ nm; Figure S2: (a) SEM image of In grains deposited on planar surface. (b) Cross-sectional view of the deposited In grains. The small grains only exist in the spare regions among the large grains, with a much lower height than their larger neighbors; Figure S3: (a) SEM image of In droplets formed on 70 nm stripe after H_2 plasma treatment. (b) Diameter statistics of the as-received In droplets, with D_{cat} of 55 nm ± 11 nm; Figure S4: Guided growth of SiNW arrays via the IPSLS mode by using catalysts formed within wide stripe (>2 μm), which shows the large diameter variation of formed catalysts (a) and the as-grown SiNW (b).

Author Contributions: Conceptualization, L.Y. and Z.L.; methodology, L.Y., Z.L., Y.C. and X.G.; validation, L.Y., Z.L., Y.C., J.W., J.X. and K.C.; formal analysis, L.Y., Z.L. and Y.C.; investigation, L.Y., Z.L., Y.C., X.G., J.W. and J.X.; resources, L.Y. and K.C.; data curation, L.Y., Z.L. and Y.C.; writing—original draft preparation, L.Y., Z.L., Y.C. and X.G.; writing—review and editing, L.Y., Z.L. and J.W.; visualization, L.Y., Z.L., Y.C. and J.X.; supervision, L.Y., Z.L., J.W., J.X. and K.C.; project administration, L.Y., J.X. and K.C.; funding acquisition, L.Y., Z.L., J.W. and J.X. All authors have read and agreed to the published version of the manuscript.

Funding: This research was funded by the National Key Research Program of China under grant Nos. 92164201 and 61921005, National Natural Science Foundation of China under Nos. 11874198, 62104100 and 61974064.

Data Availability Statement: Data are contained within the article and Supplementary Materials.

Conflicts of Interest: The authors declare no conflict of interest.

Abbreviations

SiNW: Silicon nanowire; NW: Nanowire; GAA: Gate-All-Around; cat: Catalyst; str: Stripe; VLS: Vapor-Liquid-Solid; FETs: Field effect transistors; IPSLS: In-plane solid-liquid-solid; SS: Subthreshold swing; EBL: Electron beam lithography; PMMA: Polymethyl methacrylate; RIE: Reactive ion etching; a-Si:H: Hydrogenated amorphous silicon; a-Si: Amorphous silicon; In: Indium; W_{str}: Width of stripe; D_{nw}: Diameter of the nanowire; D_{In}: Diameter of the indium; λ_{In}: The diffusion length of indium; d_s: The separation between the nucleation.

References

1. Cui, Y.; Zhong, Z.; Wang, D.; Wang, W.U.; Lieber, C. High performance silicon nanowire field effect transistors. *Nano Lett.* **2003**, *3*, 149–152. [CrossRef]
2. Goldberger, J.; Hochbaum, A.I.; Fan, R.; Yang, P. Silicon Vertically Integrated Nanowire Field Effect Transistors. *Nano Lett.* **2006**, *6*, 973–977. [CrossRef]
3. Song, X.; Zhang, T.; Wu, L.; Hu, R.; Qian, W.; Liu, Z.; Wang, J.; Shi, Y.; Xu, J.; Chen, K.; et al. Highly Stretchable High-Performance Silicon Nanowire Field Effect Transistors Integrated on Elastomer Substrates. *Adv. Sci.* **2022**, *9*, e2105623. [CrossRef] [PubMed]
4. McAlpine, M.C.; Friedman, R.S.; Jin, S.; Lin, K.-h.; Wang, W.U.; Lieber, C.M. High-performance nanowire electronics and photonics on glass and plastic substrates. *Nano Lett.* **2003**, *3*, 1531–1535. [CrossRef]
5. Yan, H.; Choe, H.S.; Nam, S.; Hu, Y.; Das, S.; Klemic, J.F.; Ellenbogen, J.C.; Lieber, C.M. Programmable nanowire circuits for nanoprocessors. *Nature* **2011**, *470*, 240–244. [CrossRef]
6. Lee, M.; Jeon, Y.; Moon, T.; Kim, S. Top-Down Fabrication of Fully CMOS-Compatible Silicon Nanowire Arrays and Their Integration into CMOS Inverters on Plastic. *ACS Nano* **2011**, *5*, 2629–2636. [CrossRef]
7. Hsu, J.F.; Huang, B.R.; Huang, C.S.; Chen, H.L. Silicon nanowires as pH sensor. *Jpn. J. Appl. Phys.* **2005**, *44*, 2626–2629. [CrossRef]
8. Lu, Y.; Peng, S.; Luo, D.; Lal, A. Low-concentration mechanical biosensor based on a photonic crystal nanowire array. *Nat. Commun.* **2011**, *2*, 578. [CrossRef]
9. Tian, B.; Cohen-Karni, T.; Qing, Q.; Duan, X.; Xie, P.; Lieber, C.M. Three-dimensional, flexible nanoscale field-effect transistors as localized bioprobes. *Science* **2010**, *329*, 830–834. [CrossRef]
10. Huang, S.; Zhang, B.; Lin, Y.; Lee, C.-S.; Zhang, X.J. Compact biomimetic hair sensors based on single silicon nanowires for ultrafast and highly-sensitive airflow detection. *Nano Lett.* **2021**, *21*, 4684–4691. [CrossRef]
11. Wagner, R.S.; Ellis, W.C. Vapor-Liquid-Solid Mechanism of Single Crystal Growth. *Appl. Phys. Lett.* **1964**, *4*, 89–90. [CrossRef]
12. Wu, Y.; Xiang, J.; Yang, C.; Lu, W.; Lieber, C.M. Single-crystal metallic nanowires and metal/semiconductor nanowire heterostructures. *Nature* **2004**, *430*, 61–65. [CrossRef] [PubMed]
13. Zhong, Z.; Qian, F.; Wang, D.; Lieber, C.M. Synthesis of p-Type Gallium Nitride Nanowires for Electronic and Photonic Nanodevices. *Nano Lett.* **2003**, *3*, 343–346. [CrossRef]
14. Huang, M.H.; Mao, S.; Feick, H.; Yan, H.; Wu, Y.; Kind, H.; Weber, E.; Russo, R.; Yang, P. Room-temperature ultraviolet nanowire nanolasers. *Science* **2001**, *292*, 1897–1899. [CrossRef] [PubMed]
15. Khan, A.; Huang, K.; Hu, M.; Yu, X.; Yang, D. Wetting Behavior of Metal-Catalyzed Chemical Vapor Deposition-Grown One-Dimensional Cubic-SiC Nanostructures. *Langmuir* **2018**, *34*, 5214–5224. [CrossRef] [PubMed]
16. Mongillo, M.; Spathis, P.; Katsaros, G.; Gentile, P.; De Franceschi, S. Multifunctional devices and logic gates with undoped silicon nanowires. *Nano Lett.* **2012**, *12*, 3074–3079. [CrossRef]

17. Zhao, Y.; You, S.S.; Zhang, A.; Lee, J.-H.; Huang, J.; Lieber, C.M. Scalable ultrasmall three-dimensional nanowire transistor probes for intracellular recording. *Nat. Nanotechnol.* **2019**, *14*, 783–790. [CrossRef]
18. Paska, Y.; Stelzner, T.; Christiansen, S.; Haick, H.J.A. Enhanced sensing of nonpolar volatile organic compounds by silicon nanowire field effect transistors. *ACS Nano* **2011**, *5*, 5620–5626. [CrossRef]
19. Qing, Q.; Jiang, Z.; Xu, L.; Gao, R.; Mai, L.; Lieber, C.M. Free-standing kinked nanowire transistor probes for targeted intracellular recording in three dimensions. *Nat. Nanotechnol.* **2014**, *9*, 142–147. [CrossRef]
20. Cui, Y.; Lauhon, L.J.; Gudiksen, M.S.; Wang, J.; Lieber, C.M. Diameter-controlled synthesis of single-crystal silicon nanowires. *Appl. Phys. Lett.* **2001**, *78*, 2214–2216. [CrossRef]
21. Schmid, H.; Björk, M.T.; Knoch, J.; Riel, H.; Riess, W.; Rice, P.; Topuria, T. Patterned epitaxial vapor-liquid-solid growth of silicon nanowires on Si(111) using silane. *J. Appl. Phys.* **2008**, *103*, 024304. [CrossRef]
22. Yu, L.; Alet, P.J.; Picardi, G.; Roca i Cabarrocas, P. An in-plane solid-liquid-solid growth mode for self-avoiding lateral silicon nanowires. *Phys. Rev. Lett.* **2009**, *102*, 125501. [CrossRef] [PubMed]
23. Yu, L.; Roca i Cabarrocas, P. Initial nucleation and growth of in-plane solid-liquid-solid silicon nanowires catalyzed by indium. *Phys. Rev. B* **2009**, *80*, 085313. [CrossRef]
24. Yu, L.; Roca i Cabarrocas, P. Growth mechanism and dynamics of in-plane solid-liquid-solid silicon nanowires. *Phys. Rev. B* **2010**, *81*, 085323. [CrossRef]
25. Xue, Z.; Xu, M.; Zhao, Y.; Wang, J.; Jiang, X.; Yu, L.; Wang, J.; Xu, J.; Shi, Y.; Chen, K.; et al. Engineering island-chain silicon nanowires via a droplet mediated Plateau-Rayleigh transformation. *Nat. Commun.* **2016**, *7*, 12836. [CrossRef]
26. Zhao, Y.; Ma, H.; Dong, T.; Wang, J.; Yu, L.; Xu, J.; Shi, Y.; Chen, K.; Roca, I.C.P. Nanodroplet Hydrodynamic Transformation of Uniform Amorphous Bilayer into Highly Modulated Ge/Si Island-Chains. *Nano Lett.* **2018**, *18*, 6931–6940. [CrossRef]
27. Yu, L.; Oudwan, M.; Moustapha, O.; Fortuna, F.; Roca i Cabarrocas, P. Guided growth of in-plane silicon nanowires. *Appl. Phys. Lett.* **2009**, *95*, 113106. [CrossRef]
28. Xu, M.; Xue, Z.; Yu, L.; Qian, S.; Fan, Z.; Wang, J.; Xu, J.; Shi, Y.; Chen, K.; Roca i Cabarrocas, P. Operating principles of in-plane silicon nanowires at simple step-edges. *Nanoscale* **2015**, *7*, 5197–5202. [CrossRef]
29. Liu, Z.; Yan, J.; Ma, H.; Hu, T.; Wang, J.; Shi, Y.; Xu, J.; Chen, K.; Yu, L. Ab Initio Design, Shaping, and Assembly of Free-Standing Silicon Nanoprobes. *Nano Lett.* **2021**, *21*, 2773–2779. [CrossRef]
30. Hu, R.; Xu, S.; Wang, J.; Shi, Y.; Xu, J.; Chen, K.; Yu, L. Unprecedented Uniform 3D Growth Integration of 10-Layer Stacked Si Nanowires on Tightly Confined Sidewall Grooves. *Nano Lett.* **2020**, *20*, 7489–7497. [CrossRef]
31. Sun, Y.; Dong, T.; Yu, L.; Xu, J.; Chen, K. Planar Growth, Integration, and Applications of Semiconducting Nanowires. *Adv. Mater.* **2020**, *32*, e1903945. [CrossRef] [PubMed]
32. Zafar, S.; D'Emic, C.; Jagtiani, A.; Kratschmer, E.; Miao, X.; Zhu, Y.; Mo, R.; Sosa, N.; Hamann, H.; Shahidi, G.; et al. Silicon Nanowire Field Effect Transistor Sensors with Minimal Sensor-to-Sensor Variations and Enhanced Sensing Characteristics. *ACS Nano* **2018**, *12*, 6577–6587. [CrossRef] [PubMed]
33. Lee, B.H.; Kang, M.H.; Ahn, D.C.; Park, J.Y.; Bang, T.; Jeon, S.B.; Hur, J.; Lee, D.; Choi, Y.K. Vertically Integrated Multiple Nanowire Field Effect Transistor. *Nano Lett.* **2015**, *15*, 8056–8061. [CrossRef] [PubMed]
34. Nguyen, B.M.; Taur, Y.; Picraux, S.T.; Dayeh, S.A. Diameter-independent hole mobility in Ge/Si core/shell nanowire field effect transistors. *Nano Lett.* **2014**, *14*, 585–591. [CrossRef] [PubMed]
35. Yang, B.; Buddharaju, K.; Teo, S.; Singh, N.; Lo, G.; Kwong, D.J. Vertical silicon-nanowire formation and gate-all-around MOSFET. *IEEE Electron Device Lett.* **2008**, *29*, 791–794. [CrossRef]
36. Yoon, J.-S.; Rim, T.; Kim, J.; Kim, K.; Baek, C.-K.; Jeong, Y.-H. Statistical variability study of random dopant fluctuation on gate-all-around inversion-mode silicon nanowire field-effect transistors. *Appl. Phys. Lett.* **2015**, *106*, 103507. [CrossRef]
37. Akhtar, S.; Usami, K.; Tsuchiya, Y.; Mizuta, H.; Oda, S. Vapor–Liquid–Solid Growth of Small- and Uniform-Diameter Silicon Nanowires at Low Temperature from Si2H6. *Appl. Phys. Exp.* **2008**, *1*, 014003. [CrossRef]
38. Mårtensson, T.; Borgström, M.; Seifert, W.; Ohlsson, B.J.; Samuelson, L. Fabrication of individually seeded nanowire arrays by vapour–liquid–solid growth. *Nanotechnology* **2003**, *14*, 1255–1258. [CrossRef]
39. Wang, C.; Murphy, P.F.; Yao, N.; McIlwrath, K.; Chou, S.Y. Growth of straight silicon nanowires on amorphous substrates with uniform diameter, length, orientation, and location using nanopatterned host-mediated catalyst. *Nano Lett.* **2011**, *11*, 5247–5251. [CrossRef]
40. Doucey, M.A.; Carrara, S. Nanowire Sensors in Cancer. *Trends Biotechnol.* **2019**, *37*, 86–99. [CrossRef]
41. Lee, S.-Y.; Chen, H.-W.; Shen, C.-H.; Kuo, P.-Y.; Chung, C.-C.; Huang, Y.-E.; Chen, H.-Y.; Chao, T.-S. Experimental demonstration of stacked gate-all-around poly-Si nanowires negative capacitance FETs with internal gate featuring seed layer and free of post-metal annealing process. *IEEE Electron Device Lett.* **2019**, *40*, 1708–1711. [CrossRef]
42. Zhang, H.; Kikuchi, N.; Ohshima, N.; Kajisa, T.; Sakata, T.; Izumi, T.; Sone, H. Interfaces, Design and fabrication of silicon nanowire-based biosensors with integration of critical factors: Toward ultrasensitive specific detection of biomolecules. *ACS Appl. Mater. Interfaces* **2020**, *12*, 51808–51819. [CrossRef] [PubMed]

Disclaimer/Publisher's Note: The statements, opinions and data contained in all publications are solely those of the individual author(s) and contributor(s) and not of MDPI and/or the editor(s). MDPI and/or the editor(s) disclaim responsibility for any injury to people or property resulting from any ideas, methods, instructions or products referred to in the content.

Article

Enhanced Electroluminescence from a Silicon Nanocrystal/Silicon Carbide Multilayer Light-Emitting Diode

Teng Sun [1], Dongke Li [1,2], Jiaming Chen [1], Yuhao Wang [1], Junnan Han [1], Ting Zhu [1], Wei Li [1], Jun Xu [1,*] and Kunji Chen [1]

[1] School of Electrical Science and Engineering, Collaborative Innovation Centre of Advanced Microstructures, Jiangsu Provincial Key Laboratory of Advanced Photonic and Electrical Materials, Nanjing University, Nanjing 210000, China
[2] ZJU-Hangzhou Global Scientific and Technological Innovation Centre, School of Materials Science and Engineering, Zhejiang University, Hangzhou 311200, China
* Correspondence: junxu@nju.edu.cn

Abstract: Developing high-performance Si-based light-emitting devices is the key step to realizing all-Si-based optical telecommunication. Usually, silica (SiO_2) as the host matrix is used to passivate silicon nanocrystals, and a strong quantum confinement effect can be observed due to the large band offset between Si and SiO_2 (~8.9 eV). Here, for further development of device properties, we fabricate Si nanocrystals (NCs)/SiC multilayers and study the changes in photoelectric properties of the LEDs induced by P dopants. PL peaks centered at 500 nm, 650 nm and 800 nm can be detected, which are attributed to surface states between SiC and Si NCs, amorphous SiC and Si NCs, respectively. PL intensities are first enhanced and then decreased after introducing P dopants. It is believed that the enhancement is due to passivation of the Si dangling bonds at the surface of Si NCs, while the suppression is ascribed to enhanced Auger recombination and new defects induced by excessive P dopants. Un-doped and P-doped LEDs based on Si NCs/SiC multilayers are fabricated and the performance is enhanced greatly after doping. As fitted, emission peaks near 500 nm and 750 nm can be detected. The current density-voltage properties indicate that the carrier transport process is dominated by FN tunneling mechanisms, while the linear relationship between the integrated EL intensity and injection current illustrates that the EL mechanism is attributed to recombination of electron–hole pairs at Si NCs induced by bipolar injection. After doping, the integrated EL intensities are enhanced by about an order of magnitude, indicating that EQE is greatly improved.

Keywords: Si nanocrystals; SiC; phosphorous; LED

1. Introduction

In order to meet the growing demand of data-carrying capacity, optical telecommunication has attracted much attention [1–3]. In this regard, developing efficient Si-based light-emitting devices (LED) is the key issue to be addressed. Silicon nanocrystals (Si NCs) are believed to be the most promising option to realize this due to their novel physical properties [4,5]. Photoluminescence (PL) spectral shift and enhanced quantum efficiency were observed in Si NCs with various dot sizes, while the stable quantum yield of Si NCs with polymer can reach 60–70% [5–8]. Recently, Si NCs-based LEDs with different emitting wavelengths have been achieved [9–12]. Further research is still necessary to explore the potential of Si NCs-based LEDs.

Doping intentionally in a semiconductor is a key method for enhancement of electrical and optical properties. Aside from the great improvement in conductivity, phosphorous (P) and boron (B) dopants can change the electronic structures of Si NCs, thus affecting the optical properties [13–17]. More interestingly, P dopants will introduce a deep level in 2 nm sized Si NCs and emit near-infrared light near 1200 nm [18–20]. In addition, carrier tunneling between Si NCs is dependent on the barrier of the host matrix in stacked structures [21]. Silicon carbide (SiC) as a host matrix is a promising candidate due to a narrower and modulative bandgap [22–24].

In our previous work, we investigated electron spin resonance of size-varied Si NCs/SiC multilayers induced by P dopants and carrier transport behaviors of various P-doped Si NCs/SiC multilayers, respectively [16,24]. Electroluminescence (EL) of Si NCs/SiC multilayers and Si NCs embedded in a SiC matrix have also been discussed in detail [25–27]. In the present work, 4 nm sized Si NCs/SiC multilayers with various P-doping ratios are fabricated. PL intensities are enhanced first and then decreased after gradually introducing P dopants. A similar tendency happens to Hall mobility. As fitted, PL peaks at 500 nm, 650 nm and 800 nm are observed, which are ascribed to amorphous SiC (650 nm), Si NCs (800 nm) and the surface states between SiC and Si NCs (500 nm), respectively. The enhanced PL intensities are caused by passivation of Si dangling bonds at the surface of Si NCs by P dopants. The quenched PL intensities are ascribed to enhanced carriers-induced Auger recombination and the emerging defects caused by excessive P dopants. We also fabricate un-doped and P-doped LEDs with the structure of aluminum (Al)/4 nm sized Si NCs/SiC multilayers/indium-tin oxide (ITO). The EL intensities of both the un-doped and P-doped LEDs are enhanced after gradually increasing the applied current. The fitted EL spectra have emission peaks near 500 nm and 750 nm, which are similar to the PL spectra. It is found that the radiative recombination in Si NCs accounts for the majority (above 90%) of emissions, manifesting that it is easier for electrons and holes being recombined radiatively in Si NCs. Only part of the radiative recombination induced by quantum-confined Si NCs occurs in the surface states between Si NCs and SiC during the tunneling process. After doping, not only are the electrical properties of the devices greatly improved but the EL intensities at the same applied current are also enhanced by about an order of magnitude, indicating an order of magnitude increase in external quantum efficiency (EQE).

2. Experiment

a-Si/SiC multilayers are deposited on p-Si ((1 0 0), 0.01 $\Omega \cdot$cm) wafer and quartz substrates in plasma-enhanced chemical vapor deposition (PECVD) system. The radio frequency, RF power and chamber temperature are kept at 13.56 MHz, 5 W and 250 °C, respectively. As for a-Si/SiC multilayers, the gas mixtures of phosphine (PH_3) gas (5%, H_2 dilution) and silane (SiH_4) gas are introduced into the chamber to deposit the P-doped a-Si sublayers. Various doping levels are achieved by changing nominal gas ratio between silane and phosphine ($[PH_3]/[SiH_4]$). The gas flow of SiH_4 during the Si sublayer deposition process is fixed at 10 sccm, while the flow of PH_3 is varied from 0, 2, 5 and 10 sccm. Gas mixtures of CH_4 (50 sccm) and SiH_4 (5 sccm) are introduced to deposit SiC sublayers. The as-deposited samples are annealed at 450 °C for 1 h under nitrogen (N_2) ambient for dehydrogenation and then thermally annealed at 1000 °C for 1 h under nitrogen (N_2) ambient to form Si NCs/SiC multilayers and annealed-SiC layers. Then, Al electrode is deposited at the back of p-Si substrate through magnetron sputtering under room temperature and the samples are annealed at 420 °C for 0.5 h under N_2 ambient to reduce the dangling bonds and heat defects in the films and also to promote ohmic contact between the film and the electrodes. At last, ITO electrode is deposited also by magnetron sputtering under a shadow mask and the active area was about 0.03 cm^2. The temperature of the ITO deposition process is kept at 200 °C.

Images of Si NCs/SiC multilayers are measured by transmission electron microscopy (TEM, TECNAI G^{2}F20 FEI, Hillsboro, TX, USA), while the TEM samples are fabricated by focused ion beam scanning electron microscopy (FIB-SEM) system. Hall mobility of Si NCs with various doping ratios is detected by van der Pauw (VDP) geometry (LakeShore 8400 series, Lorain, OH, USA) at room temperature. Steady-state PL spectra equipped with 325 nm continuous He-Cd laser by Edinburgh FLS 980 spectrophotometer at room temperature. Fourier transform infrared (FTIR, Thermo Scientific Nicolet iS20, Waltham, MA, USA) spectra are measured to analyze chemical bond composition of samples. Absorption spectra are measured at room temperature by Shimadzu UV-3600 spectrophotometer (Shimadzu Co., Kyoto, Japan). Finally, the EL measurements of the Si NCs/SiC multilayers were carried out at room temperature with Edinburgh FLS 980 spectrophotometer by applying positive bias to the Al back electrode, and ITO is used as the negative electrode. The PL and EL spectra are all corrected by the fundamental correction file of the instrument because the detected spectra are wide. The original PL of Si NCs/SiC multilayers with various doping ratios, EL spectra of un-doped LED and EL spectra of 1% P-doped LED are shown in Figure S1a–c, respectively. In addition, the photometer used in this work can only cover the visible range well; the EL emissions in near-infrared (NIR) range require further research.

3. Results and Discussion

The cross-sectional TEM images of Si NCs/SiC multilayers are shown in Figure 1a,b. Periodic structures and the abrupt interface between Si and SiC sublayers can be obviously observed in Figure 1a. On top of the Si NCs/SiC multilayers is platinum (Pt), which is used to protect the TEM samples from being etched by gallium (Ga) ions during the focused ion beam process. As measured, the thicknesses of the Si and SiC sublayers are ~4.0 nm and ~2.0 nm, respectively. The high-resolution TEM image of a single Si NC in the Si sublayer is also shown in the inset of Figure 1a, manifesting that the ~4 nm sized Si NCs are formed in Si sublayers after a high-temperature annealing process (1000 °C, N_2 ambient). The lattice distance is ~0.314 nm, which is corresponding to the lattice distances of Si (1 1 1). Si NCs with well constrained dot sizes are formed in Si sublayers. Meanwhile, Si NCs are close to each other and separated by amorphous structure according to Figure 1b, indicating good size-controllability of Si NCs in the stacked structures [25].

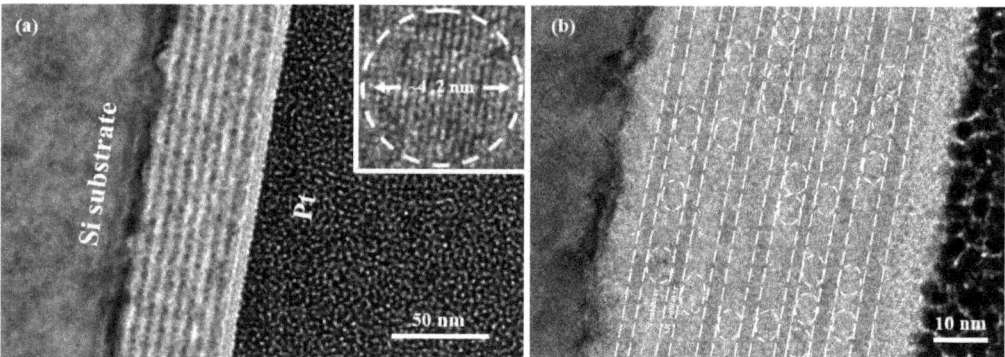

Figure 1. Cross-sectional TEM images of Si NCs/SiC multilayers at 50 nm scale (**a**) and 10 nm scale (**b**). Inset is the high-resolution TEM images of a single Si NC.

Figure 2a shows PL intensities tendency of Si NCs/SiC multilayers with various P-doping ratios excited by the 325 nm laser. It is found that PL intensities are first enhanced and then decreased after gradually introducing P dopants into our samples. The maximum PL intensities are achieved when the P-doping ratio is 1%. Meanwhile, Hall mobility, as measured, increases first and then decreases with increasing P-doping ratios, and a maximum Hall mobility of 1.74 $cm^2/V \cdot s$ is achieved at 2.5% P-doped samples according to Figure 2b. As noted, there exist considerable defects, such as Si dangling bonds at the surface of Si NCs, which will trap free carriers and suppress recombination of electrons and holes. Interestingly, P dopants can passivate the dangling bonds at the surface of Si NCs, leading to luminescence being improved. Excessive P dopants, however, will introduce amounts of free carriers and do damage to the lattice, which will enhance Auger recombination and introduce new defects, respectively [18,28,29]. Similar doping properties are also found in Si NCs/SiC systems, and P dopants will introduce further defects/Si vacancies in an ultra-small Si NCs/SiC system [16]. The enhanced PL intensity is, thus, attributed to passivation of Si dangling bonds by P dopants. Although the highest mobility is achieved when the P-doping ratio is 2.5%, P dopants will provide considerable free carriers, leading to stronger Auger recombinations, which will quench the PL of samples [30,31]. When doping ratio continues to increase, Hall mobility is decreased, manifesting that enhanced impurity scattering and lattice damage may occur, further suppressing luminescence. After fitting the PL spectrum of the 1% P-doped samples, PL peaks at ~500 nm (~2.5 eV), ~650 nm (~1.9 eV) and ~800 nm (1.6 eV) can fit the spectrum well. Inspired by the effective mass approximation (EMA) model, the theoretical bandgap of 4 nm sized Si NCs is ~1.7 eV(~750 nm) and similar EL spectra originated from 4 nm sized Si NCs were also observed in our previous work [27,32–35]. We attribute the PL peak near 800 nm to the 4 nm sized Si NCs in the Si NCs/SiC system.

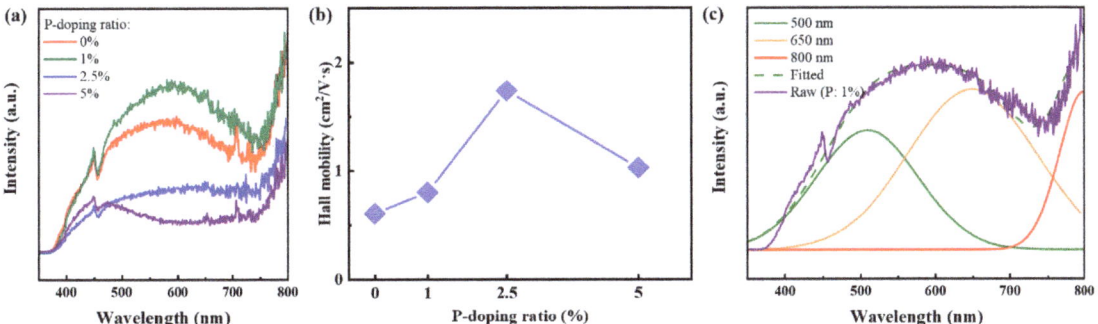

Figure 2. (a) PL spectra of Si NCs/SiC multilayers with various P-doping ratios; (b) Hall mobility of Si NCs/SiC multilayers with various P-doping ratios; (c) fitted PL spectra of 1% P-doped Si NCs/SiC multilayers.

Here, we also deposit SiC single layers. The annealing process of the amorphous SiC single layers is in consistence with that of Si NCs/SiC multilayers. We first measure the PL spectra of un-annealed SiC (R = 10). Pronounced PL peaks near 650 nm can be detected according to Figure 3. As reported, 1200 °C or 1250 °C is believed to be the crystallization temperature for SiC, and there is no crystallization after 100 h at 1000 °C in dry air [36,37]. Our previous work suggests that there is also no crystallization for SiC in Si NCs/SiC multilayers through the same annealing process, as measured by X-ray diffraction [24]. Chen et al. have reported similar results in amorphous SiC films, and the emissions can be enhanced via nitrogen doping [38]. Based on our existing knowledge, we attribute the PL peak near 650 nm to amorphous SiC. Annealed SiC with various R ([CH$_4$]/[SiH$_4$]), where the gas flow of SiH$_4$ is fixed at 5 sccm, have also been fabricated. In Figure S2a, PL peaks around 500 nm can be detected in all the annealed SiC single layers with various R. The maximum PL intensities of SiC single layers are reached when R is 12. Figure S2b exhibits the absorption spectra of annealed SiC single layers with various R. It is found that the absorption spectra change slightly with R. Guided by Tauc's plot, we use the relationship of $(\alpha h \nu)^2 \propto (h\nu - E_g)$ to simulate the absorption edge of the annealed SiC single layers. As provided in the inset of Figure S2b, the estimated optical bandgap of SiC also changes slightly with R and is ~2.5 eV (~500 nm) [39]. In order to study the chemical composition of the annealed SiC with various R, the FTIR spectra of the un-annealed and annealed SiC single layers with various R are measured in Figure S2c,d, respectively. Comparing such spectra before and after annealing, we can find that the function group bond of Si–H near 2090 cm^{-1} disappears after thermally annealing regardless of R. It demonstrates that the thermal annealing process excludes hydrogen atoms completely. Signals of the functional group bond of Si–C near the wavenumber of 790 cm^{-1} can be detected in the samples with R = 10, 12 and 14. The highest intensities of the functional group bond of Si–C are reached when R is 12. Meanwhile, all the samples show the functional group bond of Si–O near the wavenumber of 1110 cm^{-1} and have similar intensities of signals. As for samples with R = 12 and 14, the functional group bond of C=O near the wavenumber of 2400 cm^{-1} can be detected, indicating that such samples may be C-rich SiC layers [40–42]. The PL emissions near 500 nm may be related to the functional group bond of Si–C. In addition, similar PL peaks near 500 nm have been found in annealed amorphous SiC alloys in our previous work [43]. It is also found that pronounced peaks near 500 nm can be observed in SiC$_x$, SiC$_x$O$_y$ nanoclusters and silicon-rich silicon oxide enriched with C [44–46]. They attribute such PL or EL peaks to the surface states of SiC nanoclusters. As for the annealed SiC single layers, Si NCs may be formed in the SiC dielectric, although the annealing temperature (1000 °C) is not enough to form SiC particles. Si NCs embedded in a SiC matrix have been widely investigated in our previous work [25–27]. Consequently, we attribute the PL peaks near 500 nm in the Si NCs/SiC multilayers to the surface states between amorphous SiC and Si NCs.

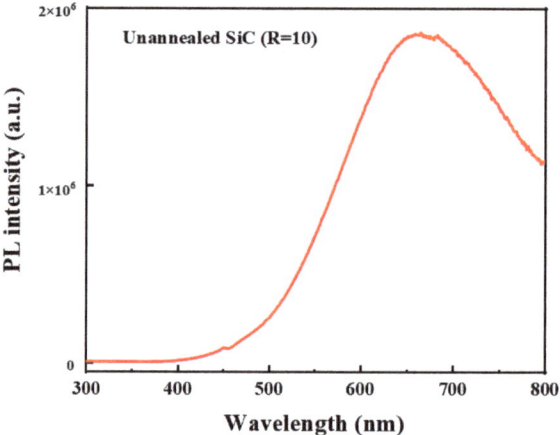

Figure 3. PL spectrum of the un-annealed SiC (R = 10).

Figure 4a shows the schematic representation of the Si NCs/SiC multilayer LED prepared in this work, the structure of which is ITO/Si NCs/SiC multilayers/p-Si/Al. At the back of the p-Si substrate is an Al electrode, which is connected to the positive terminal of a power meter. On top of the p-Si substrate are Si NCs/SiC multilayers and then the circle ITO electrode, which is connected to the negative terminal of the power meter. The area of the active circular regions upon the samples is 0.03 cm^2. Further, 1% P-doped Si NCs/SiC multilayers-based LEDs are selected due to the tendency of PL intensities and P doping behaviors. Figure 4b displays injection current density as a function of applied voltage (J-V) curves of Si NCs/SiC multilayers before and after doping. The current density increases slowly at first and then increases dramatically when the applied voltage is high enough in both un-doped and doped LEDs. It is also found that the electrical properties of the Si NCs/SiC multilayer LED are greatly improved by P dopants, i.e., on-set voltage (improved to ~5.7 V) and conductivity. In order to better understand the vertical carrier transport mechanisms of our devices, we estimate the quantity J/V^2 as a function of $1/V$ in Figure 4c. According to the inset of Figure 4c, we can see that a straight line can fit the data at high applied voltage well, indicating that carrier transport is dominated by a Fowler–Nordheim (FN) tunneling mechanism when applied voltage is high [27,47].

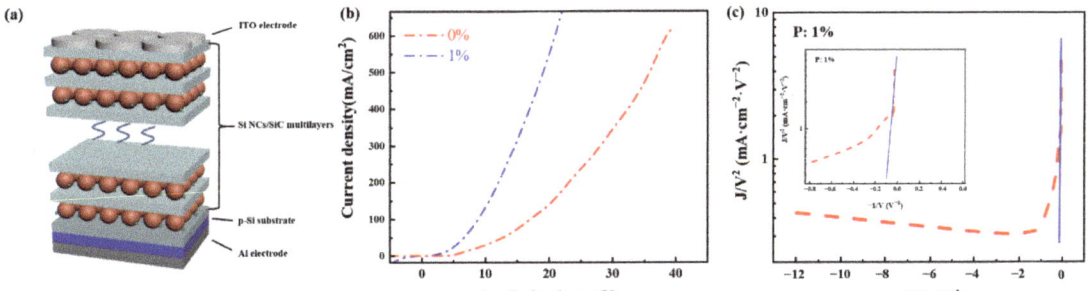

Figure 4. (a) Schematic structure of the Si NCs/SiC multilayer LED fabricated in this work; (b) current density of 0% and 1% P-doped LEDs as a function of applied voltage; (c) FN tunneling plot of 1% P-doped data; inset is the zoom of the region fitting the FN mechanism.

Furthermore, EL spectra of 0% and 1% P-doped LEDs with various applied currents are measured, as shown in Figure 5a,b, respectively. EL intensities are gradually enhanced with increasing applied current both in un-doped and P-doped LED. To better understand the EL emissions of the LEDs, we select the EL spectrum of 1% P-doped LED with 13 mA applied current to fit. It is found that EL peaks near 500 nm and 750 nm can fit the spectrum well according to Figure 5c. As for EL peaks near 500 nm, we attribute it to the surface states between SiC and Si NCs, which has been discussed above. Interestingly, such EL peaks near 500 nm cannot be observed in 8 nm sized Si NCs/SiC multilayer LED devices regardless of doping. Figure S3 shows the EL spectrum of 1% P-doped 8 nm sized Si NCs/SiC multilayers-based LED at the applied current of 10 mA. The EL emissions, accordingly, from the surface states between SiC and Si NCs (500 nm) may be attributed to quantum-confined ultra-small Si NCs [48,49]. Further, the main EL peaks near 750 nm account for the majority of the EL spectrum. Similar results were observed in Si NCs/SiC multilayers crystallized by a KrF pulsed excimer laser in our previous work [35]. In addition, the peaks near 750 nm are consistent with the bandgap of 4 nm sized Si NCs, which has been estimated before [27,50]. The peak near 750 nm, consequently, is ascribed to the 4 nm sized Si NCs. As shown in the inset of Figure 5c, the bright light, which can be seen by naked eyes when our LED works, indicates that our P-doped LEDs have better performance than in our previous work. Figure 5d shows the main luminescence center (above 90%) emissions of Si NCs (750 nm). The proportion of the integrated EL peak near 750 nm is decreased slightly, while that of the integrated EL peak near 500 nm is increased slightly. Electrons and holes can be injected by FN tunneling into 4 nm sized Si NCs and then radiative recombination occurs, leading to EL peaks near 750 nm being emitted from the LED. With applied voltage increasing, more electrons can tunnel through SiC, making radiative recombination of the surface states between SiC and Si NCs easier to occur.

Figure 6 exhibits the energy diagram of the LED. The energy of ITO, Al, p-Si and the bandgap of Si NCs are taken from the previous work [27,35,51,52]. The energy bands of Al and ITO are −4.1 eV and −4.7 eV, respectively. The energy of conduction band and valence band of p-Si substrate are −4.1 eV and −5.2 eV, respectively. As measured before and extracted by PL and EL spectra, the bandgap of 4 nm sized Si NCs is 1.7 eV. According to the following relation (1) and (2):

$$E_g = 3\Delta CB + 1.1 eV \qquad (1)$$

$$\Delta VB = 2\Delta CB \qquad (2)$$

where E_g, ΔCB and ΔVB are the bandgap of Si NCs, the energy shift of conduction band and the energy shift of valence band, respectively, the estimated energy shifts of conduction band and valence band are 0.2 eV and 0.4 eV, respectively [10,53]. The energy of conduction band and valence band of Si NCs, thus, are −3.9 eV and −5.6 eV, respectively. The barrier height between Si and SiC dielectric matrix has been estimated in our previous work: V_{0e} (0.4 eV) and V_{0h} (0.8 eV) for electron and hole, respectively [27]. The energy of conduction band and valence band of SiC are −3.5 eV and −6.4 eV, respectively. Figure 6 also displays the schematic of EL recombination mechanisms in our samples. With applied voltage, large amounts of electrons and holes are injected into Si NCs by FN tunneling. Most of the electrons and holes are recombined radiatively and emit the EL peak near 750 nm (~1.7 eV). During the tunneling process, some of the electrons are recombined radiatively in SiC dielectric and emit an EL peak near 500 nm (~2.5 eV).

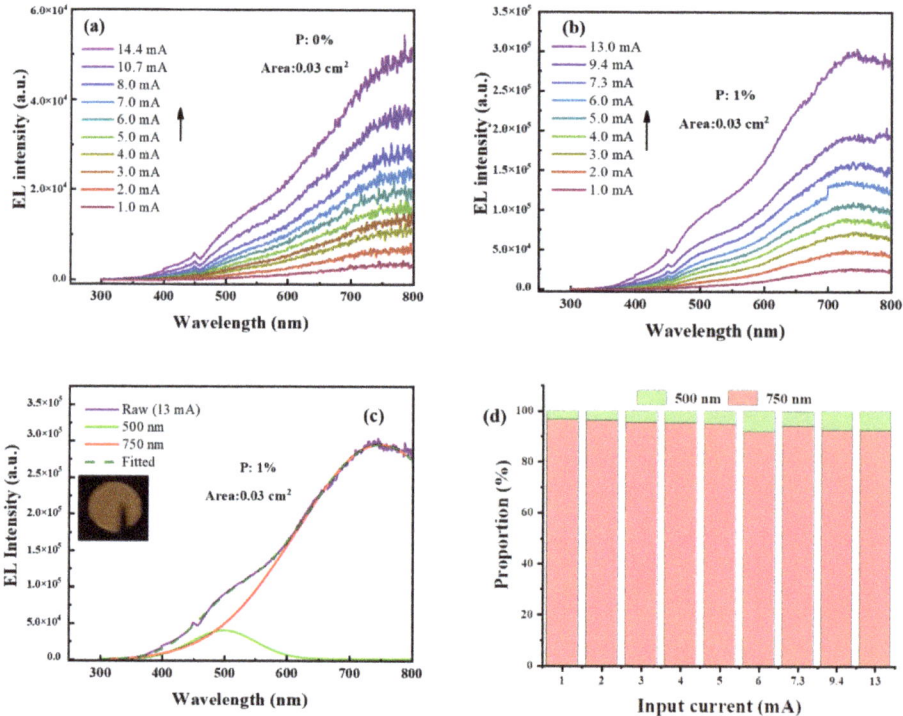

Figure 5. EL spectra of 0% (**a**) and 1% (**b**) P-doped Si NCs/SiC multilayer LEDs with various applied currents; (**c**) fitted spectra of 1% P-doped LED with 13 mA applied current. Inset is the photograph of the EL emissions from the LED; (**d**) proportion of different EL peaks with various applied currents.

As noted, FN tunneling will occur with a voltage as low as 5.7 V (the on-set voltage for 1% P-doped LED), which is consistent with the turn-on voltage when EL signals are observed. It should be noted that injection current will increase with increasing applied voltage accordingly. A linear relationship between integrated EL intensity and applied current can be observed both in the un-doped and P-doped LEDs according to Figure 7a. It indicates that bipolar injection of electrons and holes rather than impact ionization occurs in our devices regardless of doping [27,54]. Thus, it is believed that the EL emissions originated from recombination of bipolar injected electrons and holes through FN tunneling, as shown in Figure 6. By comparing the integrated EL intensities under the same applied current between 0% and 1% P-doped devices in Figure 7b, the EL intensities in 1% P-doped devices are ~eight times those in 0% P-doped devices. Meanwhile, the EQE of the P-doped devices is improved by ~eight times 0% P-doped devices. It is illustrated that P dopants can not only greatly improve LED performance, including conductive properties and on-set voltage, but also EL emissions due to the doping effect of P dopants, namely providing considerable free carriers and improving surface structures of Si NCs.

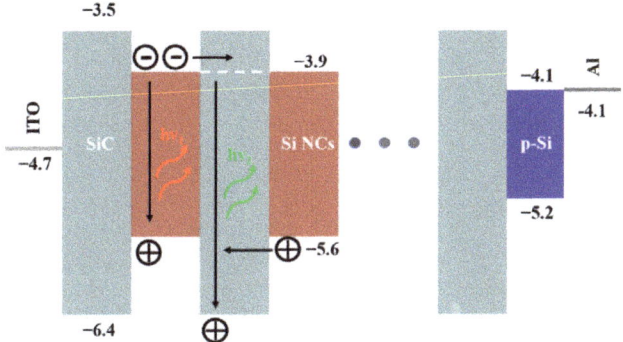

Figure 6. Schematic of EL recombination mechanisms and energy diagrams of the Si NCs/SiC multilayer LED.

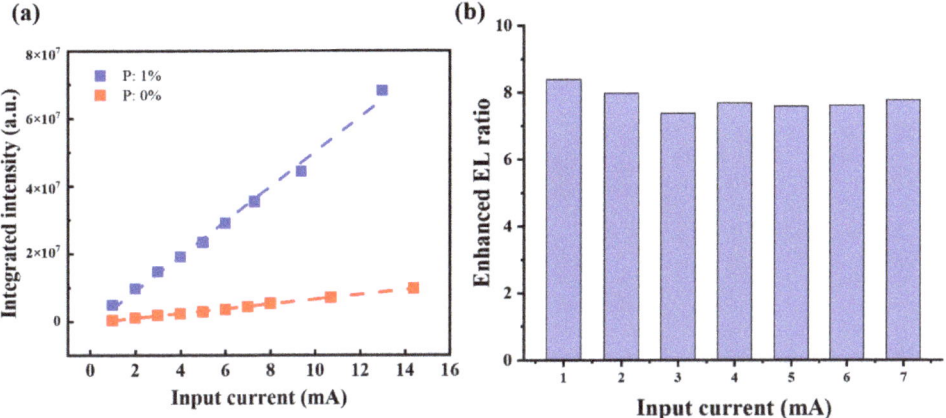

Figure 7. (a) Integrated intensity of EL spectra of 0% and 1% P-doped LEDs with various applied currents; (b) enhanced EL ratio between 0% and 1% P-doped LEDs with various applied currents.

4. Conclusions

In conclusion, 4 nm sized Si NCs/SiC multilayers are fabricated in this work. PL peaks originated from 4 nm sized Si NCs (800 nm), amorphous SiC (620 nm) and surface states between Si NCs and SiC (500 nm) can be observed. PL intensities can be enhanced after P doping due to passivation of Si dangling bonds by P dopants and quenched when P dopants are excessive. Bright LEDs with the structure of ITO/4 nm sized Si NCs/SiC multilayers/p-Si substrate/Al are finally fabricated. The EL band can be fitted by 500 nm peak and 750 nm peak, which are attributed to surface states between Si NCs and SiC and Si NCs, respectively. It is indicated that the carrier transport process is dominated by an FN tunneling mechanism. EL mainly comes from recombination between electrons and holes in Si NCs due to bipolar injection. After doping, performance of devices is greatly improved. Aside from conductivity and on-set voltage, integrated EL intensities under the same applied current are enhanced by about an order of magnitude, indicating the EQE of Si NCs/SiC multilayer LED can be improved greatly by doping. It is demonstrated that P-doped Si NCs/SiC multilayers is a promising method to realize more efficient Si-NCs-based LED devices and Si-based optical telecommunication.

Supplementary Materials: The following supporting information can be downloaded at: https://www.mdpi.com/article/10.3390/nano13061109/s1, Figure S1: Original detected spectra of (a) PL spectra with various doping ratios, (b) EL spectra of un-doped Si NCs LED and (c) EL spectra of 1% P-doped Si NCs LED; Figure S2: (a) PL spectrum of unanealed SiC (R=10) excited by 325 nm laser; (b) Absorption spectra of anealed SiC with various R. Inset is the Tauc's plot against photon energy; FTIR spectra of un-anealed (c) and annealed (d) SiC with various R; Figure S3: EL spectrum of 1% P-doped 8 nm sized Si NCs/SiC multilayers at 10 mA applied current.

Author Contributions: Conceptualization, T.S. and J.X.; methodology, T.Z.; software, T.S. and J.H.; validation, W.L., T.S. and J.X.; formal analysis, T.S., J.C., D.L. and Y.W.; investigation, T.S.; resources, T.S. and J.X.; data curation, T.S.; writing—original draft preparation, T.S. and J.X.; writing—review and editing, T.S. and J.X.; visualization, T.S.; supervision, K.C.; project administration, J.X.; funding acquisition, J.X. All authors have read and agreed to the published version of the manuscript.

Funding: This research was funded by National Key R&D program of China (2018YFB2200101), NSFC (61921005 and 62004078) and NSF of Jiangsu Province (BK20201073). And the APC was funded by National Key R&D program of China (2018YFB2200101).

Data Availability Statement: Data will be made available on request.

Conflicts of Interest: The authors declare no conflict of interest.

References

1. Whitworth, G.L.; Dalmases, M.; Taghipour, N.; Konstantatos, G. Solution-Processed PbS Quantum Dot Infrared Laser with Room-Temperature Tunable Emission in the Optical Telecommunications Window. *Nat. Photonics* **2021**, *15*, 738–742. [CrossRef]
2. Tatsuura, S.; Furuki, M.; Sato, Y.; Iwasa, I.; Tian, M.; Mitsu, H. Semiconductor Carbon Nanotubes as Ultrafast Switching Materials for Optical Telecommunications. *Adv. Mater.* **2003**, *15*, 534–537. [CrossRef]
3. Fernandez-Gonzalvo, X.; Chen, Y.-H.; Yin, C.; Rogge, S.; Longdell, J.J. Coherent Frequency Up-Conversion of Microwaves to the Optical Telecommunications Band in an Er:YSO Crystal. *Phys. Rev. A* **2015**, *92*, 062313. [CrossRef]
4. Leonardi, A.A.; Lo Faro, M.J.; Irrera, A. Biosensing Platforms Based on Silicon Nanostructures: A Critical Review. *Anal. Chim. Acta* **2021**, *1160*, 338393. [CrossRef]
5. Dohnalová, K.; Poddubny, A.N.; Prokofiev, A.A.; de Boer, W.D.; Umesh, C.P.; Paulusse, J.M.; Zuilhof, H.; Gregorkiewicz, T. Surface Brightens up Si Quantum Dots: Direct Bandgap-like Size-Tunable Emission. *Light Sci. Appl.* **2013**, *2*, e47. [CrossRef]
6. Marinins, A.; Zandi Shafagh, R.; van der Wijngaart, W.; Haraldsson, T.; Linnros, J.; Veinot, J.G.C.; Popov, S.; Sychugov, I. Light-Converting Polymer/Si Nanocrystal Composites with Stable 60–70% Quantum Efficiency and Their Glass Laminates. *ACS Appl. Mater. Interfaces* **2017**, *9*, 30267–30272. [CrossRef]
7. de Boer, W.D.A.M.; Timmerman, D.; Dohnalová, K.; Yassievich, I.N.; Zhang, H.; Buma, W.J.; Gregorkiewicz, T. Red Spectral Shift and Enhanced Quantum Efficiency in Phonon-Free Photoluminescence from Silicon Nanocrystals. *Nat. Nanotechnol.* **2010**, *5*, 878–884. [CrossRef]
8. Sugimoto, H.; Zhou, H.; Takada, M.; Fushimi, J.; Fujii, M. Visible-Light Driven Photocatalytic Hydrogen Generation by Water-Soluble All-Inorganic Core–Shell Silicon Quantum Dots. *J. Mater. Chem. A* **2020**, *8*, 15789–15794. [CrossRef]
9. Ono, T.; Xu, Y.; Sakata, T.; Saitow, K. Designing Efficient Si Quantum Dots and LEDs by Quantifying Ligand Effects. *ACS Appl. Mater. Interfaces* **2022**, *14*, 1373–1388. [CrossRef] [PubMed]
10. Xin, Y.; Nishio, K.; Saitow, K. White-Blue Electroluminescence from a Si Quantum Dot Hybrid Light-Emitting Diode. *Appl. Phys. Lett.* **2015**, *106*, 201102. [CrossRef]
11. Maier-Flaig, F.; Rinck, J.; Stephan, M.; Bocksrocker, T.; Bruns, M.; Kübel, C.; Powell, A.K.; Ozin, G.A.; Lemmer, U. Multicolor Silicon Light-Emitting Diodes (SiLEDs). *Nano Lett.* **2013**, *13*, 475–480. [CrossRef]
12. Sarkar, A.; Bar, R.; Singh, S.; Chowdhury, R.K.; Bhattacharya, S.; Das, A.K.; Ray, S.K. Size-Tunable Electroluminescence Characteristics of Quantum Confined Si Nanocrystals Embedded in Si-Rich Oxide Matrix. *Appl. Phys. Lett.* **2020**, *116*, 231105. [CrossRef]
13. Hao, X.J.; Cho, E.-C.; Scardera, G.; Shen, Y.S.; Bellet-Amalric, E.; Bellet, D.; Conibeer, G.; Green, M.A. Phosphorus-Doped Silicon Quantum Dots for All-Silicon Quantum Dot Tandem Solar Cells. *Sol. Energy Mater. Sol. Cells* **2009**, *93*, 1524–1530. [CrossRef]
14. Hao, X.J.; Cho, E.-C.; Flynn, C.; Shen, Y.S.; Park, S.C.; Conibeer, G.; Green, M.A. Synthesis and Characterization of Boron-Doped Si Quantum Dots for All-Si Quantum Dot Tandem Solar Cells. *Sol. Energy Mater. Sol. Cells* **2009**, *93*, 273–279. [CrossRef]
15. Fujii, M.; Mimura, A.; Hayashi, S.; Yamamoto, Y.; Murakami, K. Hyperfine Structure of the Electron Spin Resonance of Phosphorus-Doped Si Nanocrystals. *Phys. Rev. Lett.* **2002**, *89*, 206805. [CrossRef] [PubMed]
16. Sun, T.; Li, D.; Chen, J.; Han, J.; Zhu, T.; Li, W.; Xu, J.; Chen, K. Electron Spin Resonance in P-Doped Si Nanocrystals/SiC Stacked Structures with Various Dot Sizes. *Appl. Surf. Sci.* **2023**, *613*, 155983. [CrossRef]
17. Oliva-Chatelain, L.; Ticich, M.; Barron, R. Doping Silicon Nanocrystals and Quantum Dots. *Nanoscale* **2016**, *8*, 1733–1745. [CrossRef] [PubMed]

18. Lu, P.; Mu, W.; Xu, J.; Zhang, X.; Zhang, W.; Li, W.; Xu, L.; Chen, K. Phosphorus Doping in Si Nanocrystals/SiO$_2$ Multilayers and Light Emission with Wavelength Compatible for Optical Telecommunication. *Sci. Rep.* **2016**, *6*, 22888. [CrossRef]
19. Li, D.; Chen, J.; Sun, T.; Zhang, Y.; Xu, J.; Li, W.; Chen, K. Enhanced Subband Light Emission from Si Quantum Dots/SiO$_2$ Multilayers via Phosphorus and Boron Co-Doping. *Opt. Express* **2022**, *30*, 12308. [CrossRef]
20. Jiang, Y.; Li, D.; Xu, J.; Li, W.; Chen, K. Size-Dependent Phosphorus Doping Effect in Nanocrystalline-Si-Based Multilayers. *Appl. Surf. Sci.* **2018**, *461*, 66–71. [CrossRef]
21. Kocevski, V.; Eriksson, O.; Rusz, J. Band Alignment Switching and the Interaction between Neighboring Silicon Nanocrystals Embedded in a SiC Matrix. *Phys. Rev. B* **2015**, *91*, 165429. [CrossRef]
22. Molla, M.Z.; Zhigunov, D.; Noda, S.; Samukawa, S. Structural Optimization and Quantum Size Effect of Si-Nanocrystals in SiC Interlayer Fabricated with Bio-Template. *Mater. Res. Express* **2019**, *6*, 065059. [CrossRef]
23. Das, D.; Sain, B. Electrical Transport Phenomena Prevailing in Undoped Nc-Si/a-SiN$_x$:H Thin Films Prepared by Inductively Coupled Plasma Chemical Vapor Deposition. *J. Appl. Phys.* **2013**, *114*, 073708. [CrossRef]
24. Sun, T.; Li, D.; Chen, J.; Han, J.; Li, W.; Xu, J.; Chen, K. Temperature-Dependent Carrier Transport Behaviors in Phosphorus-Doped Silicon Nanocrystals/Silicon Carbide Multilayers. *Vacuum* **2023**, *207*, 111657. [CrossRef]
25. Rui, Y.; Li, S.; Xu, J.; Cao, Y.; Li, W.; Chen, K. Comparative Study of Electroluminescence from Annealed Amorphous SiC Single Layer and Amorphous Si/SiC Multilayers. *J. Non-Cryst. Solids* **2012**, *358*, 2114–2117. [CrossRef]
26. Rui, Y.; Li, S.; Cao, Y.; Xu, J.; Li, W.; Chen, K. Structural and Electroluminescent Properties of Si Quantum Dots/SiC Multilayers. *Appl. Surf. Sci.* **2013**, *269*, 37–40. [CrossRef]
27. Rui, Y.; Li, S.; Xu, J.; Song, C.; Jiang, X.; Li, W.; Chen, K.; Wang, Q.; Zuo, Y. Size-Dependent Electroluminescence from Si Quantum Dots Embedded in Amorphous SiC Matrix. *J. Appl. Phys.* **2011**, *110*, 064322. [CrossRef]
28. Fujii, M.; Mimura, A.; Hayashi, S.; Yamamoto, K.; Urakawa, C.; Ohta, H. Improvement in Photoluminescence Efficiency of SiO$_2$ Films Containing Si Nanocrystals by P Doping: An Electron Spin Resonance Study. *J. Appl. Phys.* **2000**, *87*, 1855–1857. [CrossRef]
29. Li, D.; Jiang, Y.; Zhang, P.; Shan, D.; Xu, J.; Li, W.; Chen, K. The Phosphorus and Boron Co-Doping Behaviors at Nanoscale in Si Nanocrystals/SiO$_2$ Multilayers. *Appl. Phys. Lett.* **2017**, *110*, 233105. [CrossRef]
30. Joo, B.S.; Jang, S.; Gu, M.; Jung, N.; Han, M. Effect of Auger Recombination Induced by Donor and Acceptor States on Luminescence Properties of Silicon Quantum Dots/SiO$_2$ Multilayers. *J. Alloys Compd.* **2019**, *801*, 568–572. [CrossRef]
31. Fujii, M.; Hayashi, S.; Yamamoto, K. Photoluminescence from B-Doped Si Nanocrystals. *J. Appl. Phys.* **1998**, *83*, 7953–7957. [CrossRef]
32. Kayanuma, Y. Quantum-Size Effects of Interacting Electrons and Holes in Semiconductor Microcrystals with Spherical Shape. *Phys. Rev. B* **1988**, *38*, 9797–9805. [CrossRef]
33. Brus, L.E. Electron-Electron and Electron-hole Interactions in Small Semiconductor Crystallites: The Size Dependence of the Lowest Excited Electronic State. *J. Chem. Phys.* **1984**, *80*, 4403–4409. [CrossRef]
34. Jiang, C.-W.; Green, M.A. Silicon Quantum Dot Superlattices: Modeling of Energy Bands, Densities of States, and Mobilities for Silicon Tandem Solar Cell Applications. *J. Appl. Phys.* **2006**, *99*, 114902. [CrossRef]
35. Xu, X.; Cao, Y.Q.; Lu, P.; Xu, J.; Li, W.; Chen, K.J. Electroluminescence Devices Based on Si Quantum Dots/SiC Multilayers Embedded in PN Junction. *IEEE Photonics J.* **2014**, *6*, 1–7. [CrossRef]
36. Hay, R.S.; Fair, G.E.; Bouffioux, R.; Urban, E.; Morrow, J.; Hart, A.; Wilson, M. Hi-Nicalon™-S SiC Fiber Oxidation and Scale Crystallization Kinetics. *J. Am. Ceram. Soc.* **2011**, *94*, 3983–3991. [CrossRef]
37. Takeda, M.; Urano, A.; Sakamoto, J.; Imai, Y. Microstructure and Oxidation Behavior of Silicon Carbide Fibers Derived from Polycarbosilane. *J. Am. Ceram. Soc.* **2004**, *83*, 1171–1176. [CrossRef]
38. Chen, G.; Chen, S.; Lin, Z.; Huang, R.; Guo, Y. Enhanced Red Emission from Amorphous Silicon Carbide Films via Nitrogen Doping. *Micromachines* **2022**, *13*, 2043. [CrossRef]
39. Zanatta, A.R. Revisiting the Optical Bandgap of Semiconductors and the Proposal of a Unified Methodology to Its Determination. *Sci. Rep.* **2019**, *9*, 11225. [CrossRef]
40. Ni, Z.; Zhou, S.; Zhao, S.; Peng, W.; Yang, D.; Pi, X. Silicon Nanocrystals: Unfading Silicon Materials for Optoelectronics. *Mater. Sci. Eng. R Rep.* **2019**, *138*, 85–117. [CrossRef]
41. Swain, B.P.; Dusane, R.O. Multiphase Structure of Hydrogen Diluted A-SiC:H Deposited by HWCVD. *Mater. Chem. Phys.* **2006**, *99*, 240–246. [CrossRef]
42. Fuad, A.; Kultsum, U.; Taufiq, A.; Hartatiek. Low-Temperature Synthesis of α-SiC (6H-SiC) Nanoparticles with Magnesium Catalyst. *Mater. Today Proc.* **2019**, *17*, 1451–1457. [CrossRef]
43. Xu, J.; Mei, J.; Rui, Y.; Chen, D.; Cen, Z.; Li, W.; Ma, Z.; Xu, L.; Huang, X.; Chen, K. UV and Blue Light Emission from SiC Nanoclusters in Annealed Amorphous SiC Alloys. *J. Non-Cryst. Solids* **2006**, *352*, 1398–1401. [CrossRef]
44. Kořínek, M.; Schnabel, M.; Canino, M.; Kozák, M.; Trojánek, F.; Salava, J.; Löper, P.; Janz, S.; Summonte, C.; Malý, P. Influence of Boron Doping and Hydrogen Passivation on Recombination of Photoexcited Charge Carriers in Silicon Nanocrystal/SiC Multilayers. *J. Appl. Phys.* **2013**, *114*, 073101. [CrossRef]
45. Gallis, S.; Nikas, V.; Suhag, H.; Huang, M.; Kaloyeros, A.E. White Light Emission from Amorphous Silicon Oxycarbide (a-SiC$_x$O$_y$) Thin Films: Role of Composition and Postdeposition Annealing. *Appl. Phys. Lett.* **2010**, *97*, 081905. [CrossRef]
46. Wang, C.; Zhou, J.; Song, M.; Chen, X.; Zheng, Y.; Xia, W. Modification of Plasma-Generated SiC Nanoparticles by Heat Treatment under Air Atmosphere. *J. Alloys Compd.* **2022**, *900*, 163507. [CrossRef]

47. Lin, G.-R.; Lin, C.-J.; Kuo, H.-C. Improving Carrier Transport and Light Emission in a Silicon-Nanocrystal Based MOS Light-Emitting Diode on Silicon Nanopillar Array. *Appl. Phys. Lett.* **2007**, *91*, 093122. [CrossRef]
48. Takagahara, T.; Takeda, K. Theory of the Quantum Confinement Effect on Excitons in Quantum Dots of Indirect-Gap Materials. *Phys. Rev. B* **1992**, *46*, 15578–15581. [CrossRef] [PubMed]
49. Vinciguerra, V.; Franzò, G.; Priolo, F.; Iacona, F.; Spinella, C. Quantum Confinement and Recombination Dynamics in Silicon Nanocrystals Embedded in Si/SiO$_2$ Superlattices. *J. Appl. Phys.* **2000**, *87*, 8165–8173. [CrossRef]
50. Cao, Y.; Xu, J.; Ge, Z.; Zhai, Y.; Li, W.; Jiang, X.; Chen, K. Enhanced Broadband Spectral Response and Energy Conversion Efficiency for Hetero-Junction Solar Cells with Graded-Sized Si Quantum Dots/SiC Multilayers. *J. Mater. Chem. C* **2015**, *3*, 12061–12067. [CrossRef]
51. Xu, X.; Zhu, T.; Xiao, K.; Zhu, Y.; Chen, J.; Li, D.; Xu, L.; Xu, J.; Chen, K. High-Efficiency Air-Processed Si-Based Perovskite Light-Emitting Devices via PMMA-TBAPF$_6$ Co-Doping. *Adv. Opt. Mater.* **2022**, *10*, 2102848. [CrossRef]
52. Xu, X.; Xiao, K.; Hou, G.; Zhu, Y.; Zhu, T.; Xu, L.; Xu, J.; Chen, K. Air-Processed Stable near-Infrared Si-Based Perovskite Light-Emitting Devices with Efficiency Exceeding 7.5%. *J. Mater. Chem. C* **2022**, *10*, 1276–1281. [CrossRef]
53. van Buuren, T.; Dinh, L.N.; Chase, L.L.; Siekhaus, W.J.; Terminello, L.J. Changes in the Electronic Properties of Si Nanocrystals as a Function of Particle Size. *Phys. Rev. Lett.* **1998**, *80*, 3803–3806. [CrossRef]
54. Warga, J.; Li, R.; Basu, S.N.; Dal Negro, L. Electroluminescence from Silicon-Rich Nitride/Silicon Superlattice Structures. *Appl. Phys. Lett.* **2008**, *93*, 151116. [CrossRef]

Disclaimer/Publisher's Note: The statements, opinions and data contained in all publications are solely those of the individual author(s) and contributor(s) and not of MDPI and/or the editor(s). MDPI and/or the editor(s) disclaim responsibility for any injury to people or property resulting from any ideas, methods, instructions or products referred to in the content.

Highly Efficient Energy Transfer from Silicon to Erbium in Erbium-Hyperdoped Silicon Quantum Dots

Kun Wang [1], Qiang He [1], Deren Yang [1,2] and Xiaodong Pi [1,2,*]

[1] State Key Laboratory of Silicon and Advanced Semiconductor Materials & School of Materials Science and Engineering, Zhejiang University, Hangzhou 310027, China
[2] Institute of Advanced Semiconductors & Zhejiang Provincial Key Laboratory of Power Semiconductor Materials and Devices, Hangzhou Innovation Center, Zhejiang University, Hangzhou 311215, China
* Correspondence: xdpi@zju.edu.cn

Abstract: Erbium-doped silicon (Er-doped Si) materials hold great potential for advancing Si photonic devices. For Er-doped Si, the efficiency of energy transfer (η_{ET}) from Si to Er^{3+} is crucial. In order to achieve high η_{ET}, we used nonthermal plasma to synthesize Si quantum dots (QDs) hyperdoped with Er at the concentration of ~1% (i.e., ~5 × 10^{20} cm^{-3}). The QD surface was subsequently modified by hydrosilylation using 1-dodecene. The Er-hyperdoped Si QDs emitted near-infrared (NIR) light at wavelengths of ~830 and ~1540 nm. An ultrahigh η_{ET} (~93%) was obtained owing to the effective energy transfer from Si QDs to Er^{3+}, which led to the weakening of the NIR emission at ~830 nm and the enhancement of the NIR emission at ~1540 nm. The coupling constant (γ) between Si QDs and Er^{3+} was comparable to or greater than 1.8×10^{-12} $cm^3 \cdot s^{-1}$. The temperature-dependent photoluminescence and excitation rate of Er-hyperdoped Si QDs indicate that strong coupling between Si QDs and Er^{3+} allows Er^{3+} to be efficiently excited.

Keywords: nonthermal plasma; Er-hyperdoped Si QDs; efficiency of energy transfer; coupling constant; strong coupling

1. Introduction

Silicon (Si) photonics has been intensively investigated due to its excellent compatibility with Si-based complementary metal oxide semiconductor (CMOS) technologies [1,2]. While monolithic Si-compatible solutions for various devices such as detectors, waveguides, and modulators have been available for years, the development of Si photonics has been impeded by the lack of monolithic energy-efficient and cost-effective light sources [3,4]. Owing to a spectroscopically sharp and environment-stable radiative transition at the wavelength (λ) of 1.54 μm that is highly desired for Si photonics [5], Er-doped Si is an ideal material for the fabrication of light sources in Si photonics. However, the significant thermal quenching of Er-doped bulk Si restricts or even forbids its use in optoelectronic devices [6,7]. Because Er-doped Si QDs can effectively emit at a wavelength of 1.54 μm at room temperature with mild thermal quenching, they have attracted enormous attention [8]. The photoluminescence (PL) of 1.54 μm originates from the transfer from the exciton energy of Si QDs to Er^{3+} and the following inter-4f transition of Er^{3+}. Hence, the η_{ET} plays a significant role in the practical applications of Er-doped Si QDs. The η_{ET} can be described by [9]

$$\eta_{ET} = 1 - \frac{\tau_{Er-Si\ QDs}}{\tau_{Si\ QDs}} \qquad (1)$$

where $\tau_{Er-Si\ QDs}$ and $\tau_{Si\ QDs}$ are the PL lifetime of Er-doped and undoped Si QDs, respectively. H. Rinnert et al. reported Er-doped Si QDs embedded in SiO/SiO_2 multilayers, whose η_{ET} was estimated as ~50% [10]. Timoshenko et al. also reported similar η_{ET} (~50%) for Er-doped Si QDs embedded in SiO/SiO_2 multilayers [11]. Despite this, earlier studies demonstrated that Er^{3+} was only found in close proximity to Si QDs inside a dielectric matrix, resulting in a system with a low degree of coupling [12]. The weakly coupled system hindered exciton-to-Er^{3+} energy transfer in Er-doped Si QDs because of the variable distance between Si QDs and Er^{3+}. Notably, the exciton-to-Er^{3+} energy transfer should be most effective if Er^{3+} is present in the Si QD in order to facilitate the overlap of electron wavefunctions [13]. This motivates the development of Si QDs into which Er^{3+} is incorporated as a dopant.

Here, nonthermal plasma was used to fabricate Er-hyperdoped Si QDs at an atomic concentration of 1% (5×10^{20} cm^{-3}), surpassing its solubility. Er-hyperdoped Si QDs were found to emit NIR light at 830 and 1540 nm, with the majority of the Er^{3+} located in the subsurface area. As a result of the effective transfer of energy from Si QDs to Er^{3+}, the Er-hyperdoped Si QDs exhibited an ultrahigh η_{ET} of ~93%, emitting at 1540 nm. Meanwhile, the efficient energy transfer from Si QDs to Er^{3+} greatly reduced the 830 nm NIR emission. Additionally, a high effective excitation cross section of Er^{3+} (σ_{Er}) was obtained, with a value of 1.5×10^{-17} cm^2. Furthermore, the coupling constant (γ) between Si QD and Er^{3+} was greater than or equal to 1.8×10^{-12} $cm^3 \cdot s^{-1}$. The temperature-dependent photoluminescence and excitation rate of Er-hyperdoped Si QDs indicate that strong coupling between Si QDs and Er^{3+} allows Er^{3+} to be efficiently excited.

2. Materials and Methods

2.1. Materials

Er(tmhd)$_3$ (tris(2,2,6,6-tetramethyl-3,5-heptanedionate)) (99.999%) was purchased from Nanjing Aimouyuan Scientific equipment Co., Ltd. (Nanjing, China). SiH$_4$/Ar (20%/80% in volume) was obtained from Linde Electronic & Specialty Gases Co., Ltd. (Suzhou, China). Mesitylene (97%) and 1-dodecene (97%) were purchased from Aladdin (Shanghai, China). Methanol ($\geq$98.5%), hydrofluoric (HF) acid ($\geq$40%), and toluene ($\geq$99.5%) were obtained from J&K Scientific (Beijing, China). Reference solutions of Er (100 µg·mL^{-1} in nitric acid) and Si (1000 µg·mL^{-1} in nitric acid) were purchased from Qingdao Qingyao Biological Engineering Co., Ltd. (Qingdao, China) and Sigma-Aldrich Trading Co., Ltd. (Shanghai, China), respectively.

2.2. Synthesis of Er-Hyperdoped Si QDs

A mechanical pump was used to pump the pressure within the plasma chamber to 8×10^{-2} mBar, and a heating strip was used to increase the temperature of the pipeline and the erbium precursor Er(tmhd)$_3$ to 200 °C. The plasma chamber was filled with 4.8 sccm of a 20% SiH$_4$/Ar mixture and 500 sccm of Ar-loaded Er(tmhd)$_3$ (Figure 1a). The pressure of the plasma chamber was adjusted to ~3 mBar. Making use of a matching network and a power source operating at 13.56 MHz, the plasma was generated. The actual output power was stabilized at 70 W. Methanol was used to disperse the powder of Er-hyperhoped Si QDs for surface hydrosilylation (Figure 1b). After that, HF acid was used to strip the surface oxide off. An amount of 15 mL of mesitylene and 5 mL of 1-dodecene were combined, and then the precipitate was added into the mixture after centrifugation. The solution was heated at 180 °C for 3 h in an Ar environment. In order to remove the hydrosilylated Er-hyperdoped Si QDs from the solution, rotary evaporation was used. Finally, the toluene was used to disperse the prepared hydrosilylated Er-hyperdoped Si QDs.

2.3. Characterization

An FEI Tecnai G2 F20 S-TWIN (FEI, Hillsboro, OR, USA) was used to capture the TEM images, operating at an acceleration voltage of 200 kV. High-angle annular darkfield scanning transmission electron microscopy (HAADF-STEM) was performed and element map images were obtained using the FEI Titan G2 80-200 (FEI, Hillsboro, OR, USA) operated at an acceleration voltage of 200 kV. X-ray photoelectron spectroscopy (XPS) was used to examine the oxidation states in each sample (Kratos Shimadzu, AXIS Supra, Manchester, UK). Er concentration was calculated using ICP-MS (iCAP6300, Thermo, Waltham, MA,

USA). The Shimadzu-7000 (Shimadzu, Kyoto, Japan) was used for the X-ray diffraction (XRD) tests. Raman spectra were obtained to analyze the strain (Alpha300R, WITec, Ulm, Germany). Using FLS1000 PL equipment (Edinburgh Instruments Ltd., Edinburgh, UK), we obtained the PL and PLE spectra. A 405 nm laser was used to excite the transient PL, and an NIR photomultiplier (PMT928, Hamamatsu, Kyoto, Japan) was used to detect it. The effect of temperature on PL was investigated using a cryostat (OptistatDN2, Oxford Instruments, Abingdon, UK). Electron paramagnetic resonance (EPR) spectroscopy on a Bruker ESRA-300 (Bruker, Berlin-Adlershof, Germany) was used to analyze the Si dangling bond. Details of spectral analysis can be found in [14].

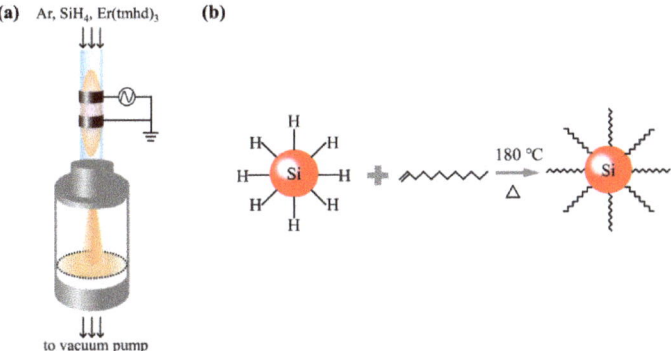

Figure 1. (a) Diagram depicting the nonthermal plasma. (b) Schematic of the surface hydrosilylation process.

3. Results and Discussion

Figure 1 shows the process for the fabrication of Er-hyperdoped Si QDs. Data in prior publications have shown that 1% Er is a comparatively optimal concentration [15,16]. Nonthermal plasma, representative of far from thermal equilibrium, was used to produce Er-hyperdoped Si QDs with a 1% Er concentration (5×10^{20} cm^{-3}) [17]. The QD surface was subsequently modified by hydrosilylation using 1-dodecene (Figure 1b) [18]. To put this in perspective, the Er solubility in crystalline Si is only ~10^{18} cm^{-3} [19], and therefore the Er concentration in this work was increased by two orders of magnitude. In the following, we discuss the 1% Er-hyperdoped Si QDs considered in this work.

The transmission electron microscopy (TEM) images of undoped and Er-hyperdoped Si QD are presented in Figure 2a,d. All of them show a sphere-like morphology, demonstrating that the morphological alteration induced by the hyperdoping of Er in Si QDs was negligible. The inset of Figure 2d shows a high-resolution TEM (HR-TEM) image of Er-hyperdoped Si QDs. The clear lattice fringe exhibits the excellent crystallinity in the Er-hyperdoped Si QDs. Moreover, good crystallinity was also observed in undoped Si QDs, as seen in the inset of Figure 2a. Figure 2a,d show that the Si (111) lattice spacing (d) increased from 0.314 nm for undoped Si QDs to 0.328 nm for Er-hyperdoped Si QDs. This may have resulted from the presence of sites of interstitial tetrahedral Er [20]. As shown in Figure 2b,e, the average diameters of both undoped and Er-hyperdoped Si QDs were 4.1 ± 0.4 nm. In addition, Figure 2f shows an HAADF-STEM image of an Er-hyperdoped Si QD, in which the brightest point is an Er atom. The Er atom was not present in undoped Si QDs (Figure 2c). The element maps indicate that that Er was closely associated to the Si element (Figure 2g–i), proving the Er atoms were incorporated into the Si QDs. Furthermore, the XRD patterns are presented in Figure S1. The pure diamond phases of Si QDs were detected in all samples, proving their high crystallinity (Figure S1). We thus draw the conclusion that hyperdoping did not alter the intrinsic diamond structure of Si QDs.

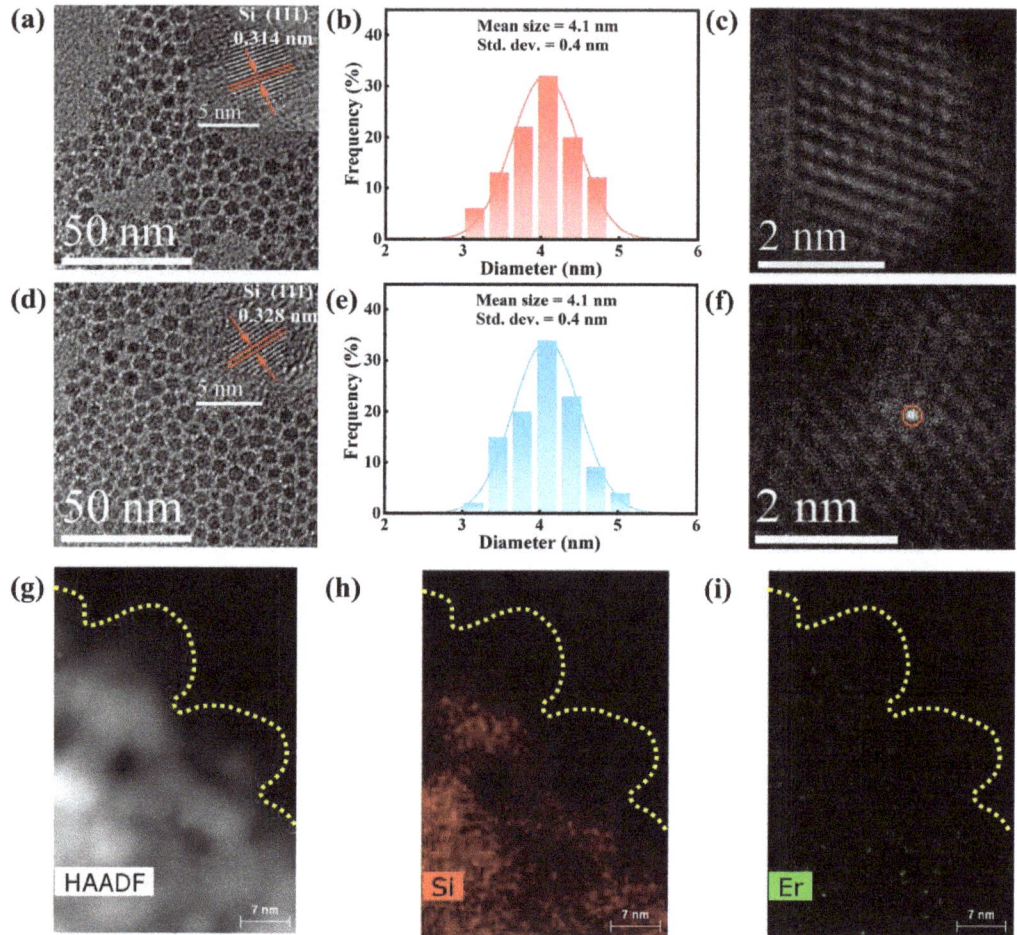

Figure 2. TEM images, size distribution, and HAADF-STEM images of (**a**–**c**) undoped and (**d**–**f**) Er-hyperdoped Si QDs. (**g**) HAADF-STEM image of Er-hyperdoped Si QDs and corresponding element maps for (**h**) Si and (**i**) Er.

In undoped Si QDs, Si-Si bonds were related to the Raman signal at ~506 cm^{-1} (Figure 3a). The phonon confinement effect is responsible for the redshift of the Raman signal for Si-Si bonds in bulk Si from its original position (~520 cm^{-1}) [21]. The Si-Si bond Raman peak was found to have blueshifted from ~506 to 515 cm^{-1} after Er hyperdoping. This blueshift was brought on by the compressive strain caused by the presence of sites of tetrahedral interstitial Er [21]. It is reasonable to make an approximation of the compressive strain (ε) using [22]

$$\varepsilon = \left(\frac{\Delta \nu}{691.2}\right) \times 100\% \tag{2}$$

where $\Delta \nu$ denotes the shift of the Raman peak. We observed a 9 cm^{-1} shift in the Raman peak, which suggests ε = ~1.3%. On the other hand, Raman peak broadening of the Si-Si bonds suggests that these bonds were deformed when Er^{3+} was doped [22]. This provides more evidence that Er was present in the Si QDs. After 15 days of exposure in ambient air, Si 2p XPS spectra were obtained and are presented in Figure 3b. An in-depth analysis of the XPS data was conducted using the method given in [9]. The XPS data are consistent with the fact that Er hyperdoping generates a compressive strain in Si QDs [23], suggesting that Er greatly reduced oxidation in the Si QDs, as shown in Figure 3b. The surface SiO_x thickness of Si QDs was reduced from 0.8 nm to 0.6 nm after Er hyperdoping, allowing for a corresponding SiO_x stoichiometric shift from $SiO_{1.0}$ to $SiO_{0.7}$ (Table S1). XPS measurements of Er 4d are shown side by side in Figure 3c. Only the Er-hyperdoped Si QDs showed the characteristic binding energy peak corresponding to Er 4d. There have also been attempts to fit this Er 4d peak with mixed singlets. There is a 2 eV spin-orbit splitting in the Er 4d peak for Er-hyperdoped Si QDs [24], similar to that seen in Er_2O_3. As Er in Er_2O_3 is in the optically active valence of +3 [25], we may assume that the Er^{3+} was preserved throughout the hyperdoping process. The radial distribution of Er was evaluated after exposing Er-hyperdoped Si QDs to air at room temperature for varied durations of time. Figure 3d shows the Er concentration after the surface oxide had been etched away. Er concentrations clearly increased at the beginning of the oxidation process and decreased over time. Figure 3d shows that the maximum Er concentration of ~2.22% was attained after etching away the oxide produced over the period of 8 days. As a result, we can deduce that subsurface regions of the Si QDs may have contained the majority of the Er.

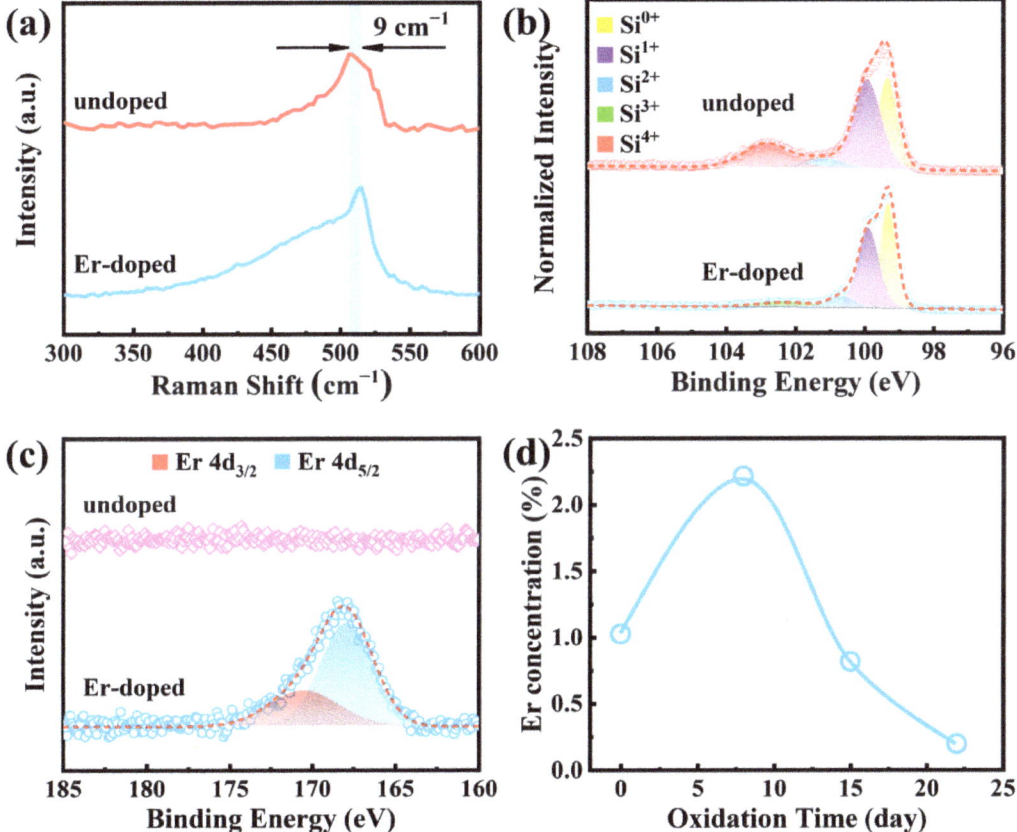

Figure 3. (a) Raman spectra, (b) Si 2p XPS spectra, and (c) Er 4d XPS spectra of Si QDs with pristine and Er hyperdoping. (d) Er concentration as a function of oxidation time.

As can be seen in Figure 4a, there was only one PL peak at 830 nm for undoped Si QDs. This peak was caused by the band gap transitions that occur in Si QDs [26,27]. Furthermore,

Figure 2b,e show that undoped and Er-hyperdoped Si QDs had almost the same size distribution, resulting in invariable emission at 830 nm for undoped and Er-hyperdoped Si QDs. In addition, this result is commensurate with the quantum confinement effect [28,29]. On the other hand, following Er hyperdoping, a distinct PL peak appeared at 1540 nm. The 1540 nm emission is ascribable to the $^4I_{13/2} \to {}^4I_{15/2}$ transition of Er^{3+} [30]. In Figure 4a, a significant reduction in the intensity of the 830 nm emission after hyperdoping can be seen. Energy transfer to the Er^{3+} ions, previously seen in the literature, may account for this decrease [10]. In terms of the PL excitation (PLE) spectra, the peak for the 830 nm emission occurred at 405 nm, as shown in Figure 4b. This is because the exciton occupancy probability at 1.49 eV (λ = 830 nm) drops precipitously after exceeding 3.06 eV (i.e., λ < 405 nm) [31]. The PLE spectrum shows a broad peak for the 1540 nm emission, superimposed over the resonance peaks at 408 nm ($^2H_{9/2} \to {}^4I_{15/2}$) and 460 nm ($^4F_{5/2} \to {}^4I_{15/2}$). The synchronicity of the PLE spectrum for the 830 and 1540 nm emissions suggests that Er^{3+} was excited through an efficient energy transfer from Si QDs to Er^{3+}, which is the critical attribute responsible for the emission of Er^{3+}. The PL lifetime was shown to drop from 275.1 s to 19.5 s due to Er hyperdoping (Figure 4c and Note S1). Using the first-derivative EPR spectra in Figure 4d, we estimated that the area density of dangling bonds at the surface of undoped Si QDs was 1.3×10^{12} cm^{-2} (corresponding to 0.6 dangling bonds per QD) whereas that at the surface of Er-hyperdoped Si QDs was 6.9×10^{11} cm^{-2} (corresponding to 0.4 dangling bonds per QD) (details can be found in [14]). Hence, the decreased area density of dangling bonds excluded the contribution of surface dangling bonds to the decrease of the PL lifetime.

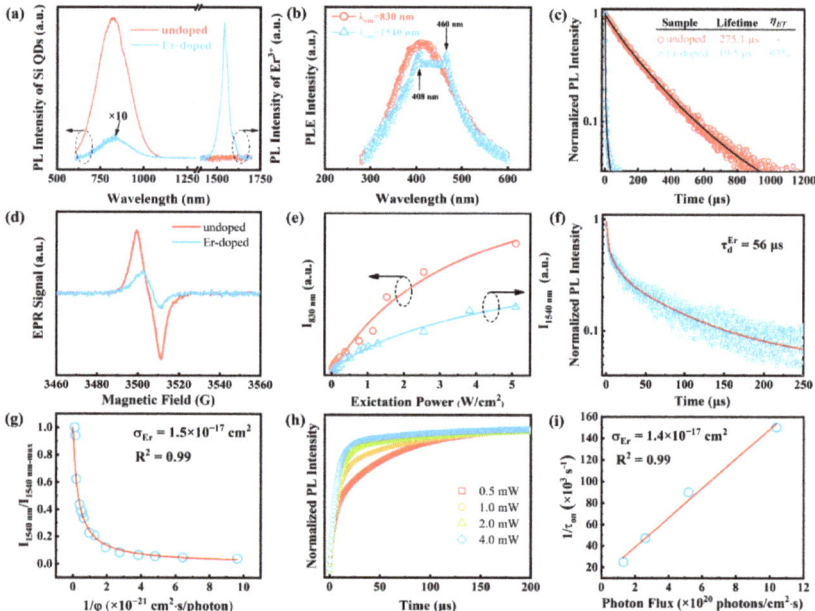

Figure 4. (a) NIR PL emission of Si QDs, pristine and with Er hyperdoping. (b) PLE spectra of Si QDs with Er hyperdoping. (c) The 830 nm PL decay curves of Si QDs, pristine and with Er hyperdoping. λ_{ex} = 405 nm. (d) First-derivative EPR spectra. (e) PL intensity detected at different excitation power, measured at 830 or 1540 nm. λ_{ex} = 405 nm. (f) PL decay monitored at 1540 nm (λ_{ex} = 405 nm) and the fitted curve. (g) The dependence of integrated PL intensity of the Er-related luminescence (1540 nm) on the reciprocal of photon flux. A line fitting the data is also shown. (h) Transient PL intensity at 1540 nm, normalized to the maximum value. (i) Photon-flux-dependent reciprocal of τ_{on}.

It was shown that the radiative recombination lifetime of Si QDs was not strongly influenced by compression because the compression hardly changed the HOMO–LUMO transition [32]. This means that the compressive strain hardly affected the PL lifetime. This suggest that the decrease of the PL lifetime is mainly induced by the transfer of energy from a Si QD to Er^{3+}, assuming that no other factors play a role. Using Equation (1), we obtain an ultrahigh η_{ET} of ~93%, which is substantially larger than the previously reported ~50% for Si QDs with neighboring Er^{3+} [10,11,33].

As seen in Figure 4e, the intensity of the PL at 830 nm ($I_{830\,nm}$) increased linearly with the increase of excitation power until the excitation power was so high that the PL tended to be saturated. For Si QDs, Auger recombination is responsible for this saturation [34]. Like $I_{830\,nm}$, the PL intensity at 1540 nm ($I_{1540\,nm}$) was proportional to the power of the 405 nm laser. Since this indicates that the $I_{1540\,nm}$ is constrained by the total amount of excitons transferred from Si QDs, this observation provides more evidence for the existence of excitonic energy transfer to Er^{3+} in Si QDs. We can calculate the effective excitation cross section of Er^{3+} (σ_{Er}) using [35]

$$\frac{I_{1540\,nm}}{I_{1540\,nm-max}} = \frac{1}{1+\frac{1}{\sigma_{Er}\tau_d^{Er}} \cdot \frac{1}{\varphi}} \quad (3)$$

where φ represents the photon flux and τ_d^{Er} is the total lifetime of Er^{3+} in level $^4I_{13/2}$. By fitting the data in Figure 4f,g, the value of σ_{Er} was obtained as 1.5×10^{-17} cm^2, an increase by four orders of magnitude compared with the value obtained for the direct excitation of Er^{3+} (8×10^{-21} cm^2) [35]. The following equation describes the rise time (τ_{on}) obtained for the 1540 nm light emission [36]

$$\frac{1}{\tau_{on}} = \sigma_{Er}\varphi + \frac{1}{\tau_d^{Er}} \quad (4)$$

Hence, we can calculate σ_{Er} by fitting Equation (4). We monitored the excitation-power-dependent rise for the 1540 nm light emission over time (Figure 4h). Figure 4i shows the fitting results, leading to a value of 1.4×10^{-17} cm^2 for σ_{Er}, which is quite similar to the result obtained using the data in Figure 4g. By assuming a strong coupling between a Si QD and Er^{3+}, we were able to derive a consistent description on the ultrahigh η_{ET} and σ_{Er}. The coupling constant (γ) between a Si QD and Er^{3+} was calculated using the formula in the literature [37] with the assumption of $\sigma_{Si\,QD} = 10^{-16}$ cm^2, $\varphi = 10^{20}$ cm$^{-2}\cdot$s^{-1}, $\tau_d^{Si\,QD} = 10^{-5}$ s, $\tau_d^{Er} = 5.6 \times 10^{-5}$ s and $A_0 < 10^{17}$ cm^{-3}, where $\sigma_{Si\,QD}$ is the absorption cross section of Si QD, $\tau_d^{Si\,QD}$ is the Si QD PL lifetime, and A_0 is the initial state of the Si QD. Finally, we obtained that the γ was comparable to or greater than the value of 1.8×10^{-12} cm$^3\cdot$s^{-1}, an increase by three orders of magnitude compared with the γ in Er-doped silicon-rich silica (3×10^{-15} cm$^3\cdot$s^{-1}) [36]. Therefore, the strong coupling between Si QDs and Er^{3+} allows Er^{3+} to be efficiently excited, resulting in the ultrahigh η_{ET} and σ_{Er}.

The dependence of the integrated PL intensity on temperature was studied to further verify the strong coupling. Figure 5a shows that $I_{830\,nm}$ decreased with the increase of temperature from 77 to 237 K while it increased with the increase of temperature from 237 K to room temperature. However, there was only a monotonic decrease in $I_{830\,nm}$ with the increase of temperature in undoped Si QDs (Figure S2). Thermal quenching accounts for the decrease of $I_{830\,nm}$ between 77 K and 237 K [38]. In contrast, $I_{830\,nm}$ increased as the temperature increased from 237 K to room temperature due to the phonon-mediated energy backtransfer from Er^{3+} [39]. Please note that the reverse trend was found for the $I_{1540\,nm}$. In the weak excitation regime, we have [40]

$$I \sim \sigma_{Er}\varphi N_{Er}\frac{\tau_d^{Er}}{\tau_{rad}^{Er}} \quad (5)$$

where N_{Er} is the excited Er^{3+} concentration and τ_{rad}^{Er} is the radiative lifetime of excited Er^{3+}. Furthermore, assuming that N_{Er} and τ_{rad}^{Er} are temperature-independent, σ_{Er} can be obtained by Equation (5). Thus, σ_{Er} is proportional to I/τ_d^{Er}. As shown in Figure 5b, the temperature dependence of $\sigma_{Er} \sim I/\tau_d^{Er}$ has the same tendency as $I_{1540\,nm}$. Therefore, the temperature-dependent σ_{Er} results in the trend of $I_{1540\,nm}$ with the increase of temperature. Moreover, the reverse temperature-dependent PL tendency between $I_{830\,nm}$ and $I_{1540\,nm}$ demonstrates the effective energy transfer of excitons between Si QDs and Er^{3+}, further supporting the strong coupling mechanism.

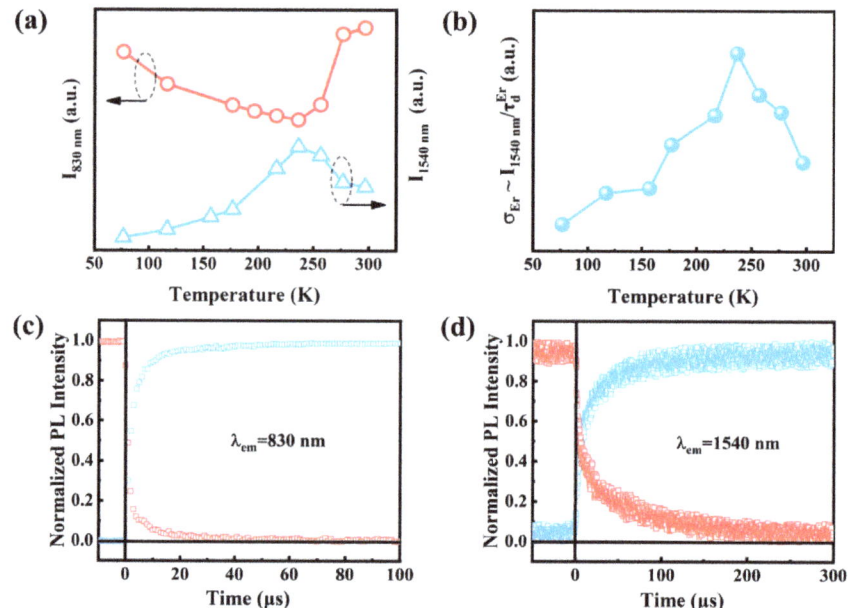

Figure 5. (a) Temperature dependence of the integrated PL intensity. (b) Temperature dependence of $\sigma_{Er} \sim I_{1540\,nm}/\tau_d^{Er}$. PL rise time and decay time for (c) Si QDs (830 nm) and (d) Er^{3+} (1540 nm) at a pump power of 2 mW.

Both the rise time (τ_{on}) and decay time (τ_d) of Si QDs (Figure 5c) and Er^{3+} (Figure 5d) were monitored in order to calculate the excitation rate. The excitation rate, denoted by $R = (1/\tau_{on} - 1/\tau_d)$, was calculated, showing that $R_{Si\,QD} = 7962$ s^{-1} for Si QDs and $R_{Er} = 7768$ s^{-1} for Er^{3+}. The observed R_{Er} is close to $R_{Si\,QD}$, which is also in agreement with the strong coupling mechanism [33].

4. Conclusions

The synthesis of Er-hyperdoped Si QDs was accomplished by employing nonthermal plasma. Both 830 and 1540 nm NIR light were emitted from Er-hyperdoped Si QDs under 405 nm excitation. PLE and PL decay measurements showed that energy transfer occurred between Si QDs and Er^{3+}. For Er-hyperdoped Si QDs, an ultrahigh η_{ET} and a high σ_{Er} were obtained. Moreover, the quantitative study of the coupling constant demonstrates the strong coupling between Si QDs and Er^{3+}. The strong coupling was also manifested by the temperature-dependent PL and excitation rate of Er-hyperdoped Si QDs. All these findings suggest that Er-hyperdoped Si QDs have great potential for the fabrication of high-performance NIR light-emitting devices.

Supplementary Materials: The following supporting information can be downloaded at: https://www.mdpi.com/article/10.3390/nano13020277/s1, Figure S1: XRD patterns of Si QDs, pristine and with Er hyperdoping; Figure S2: Temperature-dependent integrated emission intensities of undoped

Si QDs; Table S1: Atomic fraction of various charge states of Si; Note S1: The fitting of PL decay curves and the determination of PL lifetime. Reference [9] is cited in the supplementary materials.

Author Contributions: Conceptualization, K.W. and X.P.; Formal analysis, K.W., Q.H. and X.P.; Funding acquisition, D.Y. and X.P.; Investigation, K.W.; Methodology, K.W. and Q.H.; Project administration, D.Y. and X.P.; Supervision, D.Y. and X.P.; Visualization, K.W.; Writing—original draft, K.W.; Writing—review and editing, Q.H., D.Y. and X.P. All authors have read and agreed to the published version of the manuscript.

Funding: This research was funded by the National Key Research and Development Program of China (Grant no. 2018YFB2200101), the National Natural Science Foundation of China (Grant no. 91964107), the National Natural Science Foundation of China for Innovative Research Groups (grant no. 61721005), and Zhejiang University Education Foundation Global Partnership Fund. The authors also acknowledge the Instrumentation and Service Center for Molecular Sciences at Westlake University for technical assistance. The authors also acknowledge Dr. Xiaodong Zhu of Zhejiang University for her assistance during the XPS spectra analysis.

Institutional Review Board Statement: Not applicable.

Informed Consent Statement: Not applicable.

Data Availability Statement: The data that support the findings of this study are available from the corresponding author upon reasonable request.

Conflicts of Interest: The authors declare no conflict of interest.

References

1. Li, N.; Xin, M.; Su, Z.; Magden, E.S.; Singh, N.; Notaros, J.; Timurdogan, E.; Purnawirman, P.; Bradley, J.D.B.; Watts, M.R. A Silicon Photonic Data Link with a Monolithic Erbium-Doped Laser. *Sci. Rep.* **2020**, *10*, 1114. [CrossRef]
2. Fu, P.; Yang, D.C.; Jia, R.; Yi, Z.J.; Liu, Z.F.; Li, X.; Eglitis, R.I.; Su, Z.M. Metallic Subnanometer Porous Silicon: A Theoretical Prediction. *Phys. Rev. B* **2021**, *103*, 014117. [CrossRef]
3. Pavesi, L. Silicon-Based Light Sources for Silicon Integrated Circuits. *Adv. Opt. Technol.* **2008**, *2008*, 416926. [CrossRef]
4. Li, N.; Chen, G.; Ng, D.K.T.; Lim, L.W.; Xue, J.; Ho, C.P.; Fu, Y.H.; Lee, L.Y.T. Integrated Lasers on Silicon at Communication Wavelength: A Progress Review. *Adv. Opt. Mater.* **2022**, *10*, 2201008. [CrossRef]
5. Wojdak, M.; Klik, M.; Forcales, M.; Gusev, O.B.; Gregorkiewicz, T.; Pacifici, D.; Franzò, G.; Priolo, F.; Iacona, F. Sensitization of Er Luminescence by Si Nanoclusters. *Phys. Rev. B* **2004**, *69*, 233315. [CrossRef]
6. Schmidt, M.; Heitmann, J.; Scholz, R.; Zacharias, M. Bright Luminescence from Erbium Doped Nc-Si/SiO_2 Superlattices. *J. Non. Cryst. Solids* **2002**, *299–302*, 678–682. [CrossRef]
7. Wen, H.; He, J.; Hong, J.; Jin, S.; Xu, Z.; Zhu, H.; Liu, J.; Sha, G.; Yue, F.; Dan, Y. Efficient Er/O-Doped Silicon Light-Emitting Diodes at Communication Wavelength by Deep Cooling. *Adv. Opt. Mater.* **2020**, *8*, 2000720. [CrossRef]
8. Fujii, M.; Yoshida, M.; Hayashi, S.; Yamamoto, K. Photoluminescence from SiO_2 Films Containing Si Nanocrystals and Er: Effects of Nanocrystalline Size on the Photoluminescence Efficiency of Er^{3+}. *J. Appl. Phys.* **1998**, *84*, 4525–4531. [CrossRef]
9. Wang, K.; He, Q.; Yang, D.; Pi, X. Erbium-Hyperdoped Silicon Quantum Dots: A Platform of Ratiometric Near-Infrared Fluorescence. *Adv. Opt. Mater.* **2022**, *10*, 2201831. [CrossRef]
10. Rinnert, H.; Adeola, G.W.; Vergnat, M. Influence of the Silicon Nanocrystal Size on the 1.54 μm Luminescence of Er-Doped SiO/SiO_2 multilayers. *J. Appl. Phys.* **2009**, *105*, 2–5. [CrossRef]
11. Timoshenko, V.Y.; Lisachenko, M.G.; Shalygina, O.A.; Kamenev, B.V.; Zhigunov, D.M.; Teterukov, S.A.; Kashkarov, P.K.; Heitmann, J.; Schmidt, M.; Zacharias, M. Comparative Study of Photoluminescence of Undoped and Erbium-Doped Size-Controlled Nanocrystalline Si/SiO_2 Multilayered Structures. *J. Appl. Phys.* **2004**, *96*, 2254–2260. [CrossRef]
12. Izeddin, I.; Timmerman, D.; Gregorkiewicz, T.; Moskalenko, A.S.; Prokofiev, A.A.; Yassievich, I.N.; Fujii, M. Energy Transfer in Er-Doped SiO_2 Sensitized with Si Nanocrystals. *Phys. Rev. B* **2008**, *78*, 035327. [CrossRef]
13. Dexter, D.L. A Theory of Sensitized Luminescence in Solids. *J. Chem. Phys.* **1953**, *21*, 836–850. [CrossRef]
14. Liu, X.; Zhao, S.; Gu, W.; Zhang, Y.; Qiao, X.; Ni, Z.; Pi, X.; Yang, D. Light-Emitting Diodes Based on Colloidal Silicon Quantum Dots with Octyl and Phenylpropyl Ligands. *ACS Appl. Mater. Interfaces* **2018**, *10*, 5959–5966. [CrossRef]
15. Tessler, L.R.; Zanatta, A.R. Erbium Luminescence in A-Si:H. *J. Non. Cryst. Solids* **1998**, *227–230*, 399–402. [CrossRef]
16. Kenyon, A.J.; Chryssou, C.E.; Pitt, C.W.; Shimizu-Iwayama, T.; Hole, D.E.; Sharma, N.; Humphreys, C.J. Luminescence from Erbium-Doped Silicon Nanocrystals in Silica: Excitation Mechanisms. *J. Appl. Phys.* **2002**, *91*, 367–374. [CrossRef]
17. Mangolini, L.; Thimsen, E.; Kortshagen, U. High-Yield Plasma Synthesis of Luminescent Silicon Nanocrystals. *Nano Lett.* **2005**, *5*, 655–659. [CrossRef]
18. Dasog, M.; De Los Reyes, G.B.; Titova, L.V.; Hegmann, F.A.; Veinot, J.G.C. Size vs Surface: Tuning the Photoluminescence of Freestanding Silicon Nanocrystals across the Visible Spectrum via Surface Groups. *ACS Nano* **2014**, *8*, 9636–9648. [CrossRef]

19. Benton, J.L.; Michel, J.; Kimerling, L.C.; Jacobson, D.C.; Xie, Y.H.; Eaglesham, D.J.; Fitzgerald, E.A.; Poate, J.M. The Electrical and Defect Properties of Erbium-Implanted Silicon. *J. Appl. Phys.* **1991**, *70*, 2667–2671. [CrossRef]
20. Raffa, A.G.; Ballone, P. Equilibrium Structure of Erbium-Oxygen Complexes in Crystalline Silicon. *Phys. Rev. B* **2002**, *65*, 121309. [CrossRef]
21. Campbell, I.H.; Fauchet, P.M. The Effects of Microcrystal Size and Shape on the One Phonon Raman Spectra of Crystalline Semiconductors. *Solid State Commun.* **1986**, *58*, 739–741. [CrossRef]
22. Ni, Z.; Pi, X.; Zhou, S.; Nozaki, T.; Grandidier, B.; Yang, D. Size-Dependent Structures and Optical Absorption of Boron-Hyperdoped Silicon Nanocrystals. *Adv. Opt. Mater.* **2016**, *4*, 700–707. [CrossRef]
23. Zhou, S.; Pi, X.; Ni, Z.; Luan, Q.; Jiang, Y.; Jin, C.; Nozaki, T.; Yang, D. Boron- and Phosphorus-Hyperdoped Silicon Nanocrystals. *Part. Part. Syst. Charact.* **2015**, *32*, 213–221. [CrossRef]
24. Tewell, C.R.; King, S.H. Observation of Metastable Erbium Trihydride. *Appl. Surf. Sci.* **2006**, *253*, 2597–2602. [CrossRef]
25. Miritello, M.; Lo Savio, R.; Piro, A.M.; Franzò, G.; Priolo, F.; Iacona, F.; Bongiorno, C. Optical and Structural Properties of Er_2O_3 Films Grown by Magnetron Sputtering. *J. Appl. Phys.* **2006**, *100*, 013502. [CrossRef]
26. Hannah, D.C.; Yang, J.; Podsiadlo, P.; Chan, M.K.Y.; Demortière, A.; Gosztola, D.J.; Prakapenka, V.B.; Schatz, G.C.; Kortshagen, U.; Schaller, R.D. On the Origin of Photoluminescence in Silicon Nanocrystals: Pressure-Dependent Structural and Optical Studies. *Nano Lett.* **2012**, *12*, 4200–4205. [CrossRef] [PubMed]
27. Sangghaleh, F.; Sychugov, I.; Yang, Z.; Veinot, J.G.C.; Linnros, J. Near-Unity Internal Quantum Efficiency of Luminescent Silicon Nanocrystals with Ligand Passivation. *ACS Nano* **2015**, *9*, 7097–7104. [CrossRef] [PubMed]
28. Ni, Z.; Zhou, S.; Zhao, S.; Peng, W.; Yang, D.; Pi, X. Silicon Nanocrystals: Unfading Silicon Materials for Optoelectronics. *Mater. Sci. Eng. R Reports* **2019**, *138*, 85–117. [CrossRef]
29. Puzder, A.; Williamson, A.J.; Grossman, J.C.; Galli, G. Computational Studies of the Optical Emission of Silicon Nanocrystals. *J. Am. Chem. Soc.* **2003**, *125*, 2786–2791. [CrossRef]
30. Fujii, M.; Yoshida, M.; Kanzawa, Y.; Hayashi, S.; Yamamoto, K. 1.54 μm Photoluminescence of Er^{3+} Doped into SiO_2 Films Containing Si Nanocrystals: Evidence for Energy Transfer from Si Nanocrystals to Er^{3+}. *Appl. Phys. Lett.* **1997**, *71*, 1198–1200. [CrossRef]
31. Pradeep, J.A.; Agarwal, P. An Alternative Approach to Understand the Photoluminescence and the Photoluminescence Peak Shift with Excitation in Porous Silicon. *J. Appl. Phys.* **2008**, *104*, 123515. [CrossRef]
32. Weissker, H.C.; Ning, N.; Bechstedt, F.; Vach, H. Luminescence and Absorption in Germanium and Silicon Nanocrystals: The Influence of Compression, Surface Reconstruction, Optical Excitation, and Spin-Orbit Splitting. *Phys. Rev. B* **2011**, *83*, 4–9. [CrossRef]
33. Kik, P.G.; Brongersma, M.L.; Polman, A. Strong Exciton-Erbium Coupling in Si Nanocrystal-Doped SiO_2. *Appl. Phys. Lett.* **2000**, *76*, 2325–2327. [CrossRef]
34. Timmerman, D.; Izeddin, I.; Gregorkiewicz, T. Saturation of Luminescence from Si Nanocrystals Embedded in SiO_2. *Phys. Status Solidi Appl. Mater. Sci.* **2010**, *207*, 183–187. [CrossRef]
35. Gusev, O.B.; Bresler, M.S.; Pak, P.E.; Yassievich, I.N.; Forcales, M.; Vinh, N.Q.; Gregorkiewicz, T. Excitation Cross Section of Erbium in Semiconductor Matrices under Optical Pumping. *Phys. Rev. B* **2001**, *64*, 753021–753027. [CrossRef]
36. Pacifici, D.; Franzò, G.; Priolo, F.; Iacona, F.; Dal Negro, L. Modeling and Perspectives of the Si Nanocrystals–Er Interaction for Optical Amplification. *Phys. Rev. B* **2003**, *67*, 245301. [CrossRef]
37. Kenyon, A.J.; Wojdak, M.; Ahmad, I.; Loh, W.H.; Oton, C.J. Generalized Rate-Equation Analysis of Excitation Exchange between Silicon Nanoclusters and Erbium Ions. *Phys. Rev. B* **2008**, *77*, 035318. [CrossRef]
38. Ghosh, B.; Takeguchi, M.; Nakamura, J.; Nemoto, Y.; Hamaoka, T.; Chandra, S.; Shirahata, N. Origin of the Photoluminescence Quantum Yields Enhanced by Alkane-Termination of Freestanding Silicon Nanocrystals: Temperature-Dependence of Optical Properties. *Sci. Rep.* **2016**, *6*, 36951. [CrossRef]
39. Navarro-Urrios, D.; Pitanti, A.; Daldosso, N.; Gourbilleau, F.; Rizk, R.; Garrido, B.; Pavesi, L. Energy Transfer between Amorphous Si Nanoclusters and Er^{3+} Ions in a SiO_2 Matrix. *Phys. Rev. B* **2009**, *79*, 2–5. [CrossRef]
40. Hijazi, K.; Rizk, R.; Cardin, J.; Khomenkova, L.; Gourbilleau, F. Towards an Optimum Coupling between Er Ions and Si-Based Sensitizers for Integrated Active Photonics. *J. Appl. Phys.* **2009**, *106*, 024311. [CrossRef]

Disclaimer/Publisher's Note: The statements, opinions and data contained in all publications are solely those of the individual author(s) and contributor(s) and not of MDPI and/or the editor(s). MDPI and/or the editor(s) disclaim responsibility for any injury to people or property resulting from any ideas, methods, instructions or products referred to in the content.

Communication

Room Temperature Light Emission from Superatom-like Ge–Core/Si–Shell Quantum Dots

Katsunori Makihara [1,2,*], Yuji Yamamoto [2], Yuki Imai [1], Noriyuki Taoka [1], Markus Andreas Schubert [2], Bernd Tillack [2,3] and Seiichi Miyazaki [1]

1. Graduate School of Engineering, Nagoya University, Furo–cho, Chikusa–ku, Nagoya 464-8603, Japan
2. IHP–Leibniz-Institut für Innovative Mikroelektronik, Im Technologiepark 25, 15236 Frankfurt, Germany; yamamoto@ihp-microelectronics.com (Y.Y.)
3. Technische Universität Berlin, HFT4, Einsteinufer 25, 10587 Berlin, Germany
* Correspondence: makihara@nuee.nagoya-u.ac.jp; Tel.:+81-52-789-2727

Abstract: We have demonstrated the high–density formation of super–atom-like Si quantum dots with Ge–core on ultrathin SiO_2 with control of high–selective chemical–vapor deposition and applied them to an active layer of light–emitting diodes (LEDs). Through luminescence measurements, we have reported characteristics carrier confinement and recombination properties in the Ge–core, reflecting the type II energy band discontinuity between the Si–clad and Ge–core. Additionally, under forward bias conditions over a threshold bias for LEDs, electroluminescence becomes observable at room temperature in the near–infrared region and is attributed to radiative recombination between quantized states in the Ge–core with a deep potential well for holes caused by electron/hole simultaneous injection from the gate and substrate, respectively. The results will lead to the development of Si–based light–emitting devices that are highly compatible with Si–ultra–large–scale integration processing, which has been believed to have extreme difficulty in realizing silicon photonics.

Keywords: Si quantum dots; core/shell structure; CVD

1. Introduction

Monolithic integration of Si/Ge–based optoelectronic devices into Si–ULSI circuits is of great interest to further extend the functionality of complementary metal–oxide–semiconductor (CMOS) technologies according to a "more than Moore" approach [1–16]. Thus, light emission from Si/Ge–based quantum dots (QDs) has attracted much attention as an active element of optical applications because it has the advantages of photonic signal processing capabilities and combines them with electronic logic control and data storage. In recent years, intensive research has been dedicated to the growth and optoelectronic characterization of Si and/or Ge nanostructures using various fabrication techniques such as molecular beam epitaxy, chemical vapor deposition (CVD), and magnetron sputtering deposition [17–21]. However, the development of a growth technology for QDs with high areal density, uniform shape, and narrow size distribution for large–scale production is a crucial factor in realizing stable light emission from the QDs. In addition, to integrate optoelectronic devices into the CMOS technology, remarkable improvements in light emission efficiency and its stability for practical use are major technological challenges because of the difficulty in achieving a good balance between charge injection and confinement in the QDs. To satisfy both strong confinement and smooth injections of carriers, Ge–QDs and their self–aligned stack structures embedded in Si have so far been fabricated by controlling strain–induced self–assembling in the early stages of Ge heteroepitaxy on Si. So, we have focused on superatom-like Ge–core/Si-shell QDs (Si–QDs with Ge–core) formed on ultrathin SiO_2 because of their potential to realize a good balance between charge injection and confinement in the QDs, which can improve light emission efficiency and stability. Additionally, defect control at interfaces between Si and SiO_2 is established. In our previous

works, we have studied the highly selective growth of Ge on pre–grown Si–QDs and the subsequent coverage with a Si–cap that enables us to superatom–like Ge–core/Si–shell QDs and characterized their unique electron/hole storage properties, reflecting the type II energy band discontinuity between Si–QDs and Ge core [22–26]. More recently, we also demonstrated photoluminescence (PL) from the Si–QDs with Ge–core and discussed the origin of their PL properties, with PL having less dependence on temperature change. Therefore, we concluded that hole confinement in the Ge core plays an important role in radiative recombination in Si/Ge QDs. In addition, we also reported electroluminescence (EL) from the light–emitting diode structure with the 3–fold stacked Si–QDs with Ge core by application of continuous square–wave pulsed bias under cold light illumination, where EL signal was observed from the backside through the c–Si substrate caused by alternate electron/hole injection from the p–Si(100). In this work, we developed fabrication processes of the Si–QDs with Ge–core structure by means of a commercial reduced pressure (RP) CVD system [27–30] using SiH_4 and GeH_4 gases with H_2 carriers, which has been in practical use for the mass production of bipolar CMOS, for the fabrication of the Si–QDs with Ge–core on ultrathin SiO_2 layers, and evaluated their PL and EL characteristics without light illumination at room temperature.

2. Materials and Methods

The fabrication of Si–QDs with Ge–core was carried out using an RPCVD system. After a standard RCA cleaning step, ~7–nm–thick SiO_2 was grown on a p–Si(100) wafer by H_2/O_2 oxidation at 650°C. Furthermore, diluted HF dip is performed. At this point, residual SiO_2 thickness is ~2.0 nm. After that, the SiO_2/Si(100) wafer was loaded into the RPCVD reactor and baked at 850°C in RP–H_2 to control OH bond the SiO_2 surface. Subsequently, the wafer was cooled down to 650°C in RP–H_2. After temperature stabilization, Si–QDs were deposited using a H_2–SiH_4 mixture gas by controlling the early stages of Si nucleation. Directly after that the wafer was cooled down to 550 °C in the RP–H_2 environment and Ge was selectively grown on the pre–grown Si–QDs by H_2–GeH_4. Afterwards, the wafer was heated again up to 650 °C in a H_2 environment and Si cap was selectively deposited on the Ge on the Si–QDs by H_2–SiH_4 gas system. In order to maintain selectivity, lower SiH_4 partial pressure is used for the Si cap growth.

Areal density and average height of the Si–QD with Ge–core were evaluated by atomic force microscopy (AFM), where average dot heights were determined by log–normal functions. The dot height of the Si and Ge in the Si–QDs with a Ge–core was also characterized by cross–section transmission electron microscopy (TEM) and energy–dispersive X–ray spectroscopy (EDX) mapping. PL measurements were carried out by using a 976–nm light as an excitation source with an input power of 0.33 W/cm^2.

For the fabrication of the LED structure, ~200–nm–thick amorphous Si was deposited on the Si–QDs with Ge–core by electron beam evaporation at room temperature after the chemical oxidation of the dots surface. Then, phosphorus atom implantation at 45 keV was conducted. Subsequently, the sample was annealed at 300 °C. Then, the amorphous Si layer was etched by Cl_2 plasma to isolate each device. Finally, ring–patterned Al–top electrodes with an aperture of ~78.5 mm^2 and backside electrodes were fabricated by thermal evaporation. For the PL and EL measurements, thermoelectrically cooled InGaAs photodiodes were used as detectors in this work.

3. Results and Discussion

The formation of high–density Si–QDs with Ge–core on the ultrathin SiO_2 surface was confirmed by AFM and EDX mapping image measurements, as shown in Figures 1 and 2. The AFM topographic image taken after the RPCVD using a H_2–SiH_4 mixture gas at 650 °C confirms that hemispherical Si–QDs with an areal density as high as ~1.0×10^{11} cm^{-2} were self–assembled on the SiO_2 surface. Additionally, the AFM images taken after each deposition step confirm that the areal dot density remains unchanged after the Ge deposition and subsequent Si deposition using the high selectivity process condition. These results indicate that the Ge deposition and the Si–cap formation occur very selectively on each of the isolated dots, and no new dots nucleate on the SiO_2 layer during the Ge deposition and Si–cap formation. From dot height distributions evaluated from cross–sectional topographic images, we also confirm that the dot height is increased by ~1.6 nm and by ~1.0 nm after the Ge deposition and subsequent Si–cap formation, respectively. In our previous report on the formation of the Si–QDs with Ge–core using LPCVD [26], the full–width at half–maximum value of the size distribution after Si–cap formation is ~7 nm, whereas it is ~3 nm for RPCVD, indicating that dots of extremely uniform size can be formed. Highly selective deposition of Ge and subsequent Si on the dots was also verified from the cross–section TEM–EDX analysis, as indicated in Figure 2. After the H_2–GeH_4–RPCVD, an EDX mapping image shows that Ge was deposited on the pre–grown Si–QDs conformally, although there is no deposition on the SiO_2 surface in–between the pre–grown Si–QDs. After the subsequent H_2–SiH_4–RPCVD, Ge–core/Si–shell structure was clearly detected. In addition, the heights of the pre–grown Si–QD, Ge–core, and Si–cap are very consistent with the values estimated from the size distribution shown in Figure 1d. It is interesting to note that the EDX mapping image of the dots after the Si–cap deposition shows that Ge has a rather spherical shape in contrast to the as–deposited Ge selectively on pre–grown hemispherical–shaped Si–QDs. To get an insight into the change in the Ge–core shape, we also performed plane–view TEM–EDX analysis, as shown in Figure 3. After the Ge deposition, the color contrast of the Ge in the peripheral region of the dots is somewhat deeper than that in the center of the dots, which indicates that Ge was deposited conformally on the pre–grown Si–QDs because color contrast depends on the thickness. Contrary to this, after annealing at 650 °C and/or Si–cap deposition at 650 °C, it turned out to have a dark color in the central part. This result can be attributable to Ge reflow onto the pre–grown Si–QDs due to relaxation of surface energy or high structural strain at the Ge/Si interface. Consequently, a spherical–shaped Ge–core with an abrupt Ge/Si interface was formed. In fact, Raman scattering spectra for the dots indicate that compositional mixing hardly occurs at Si/Ge interfaces during the sample preparation, as verified by the relative intensity of the Si–Ge phonon mode with respect to the Si–Si and Ge–Ge phonon modes (not shown).

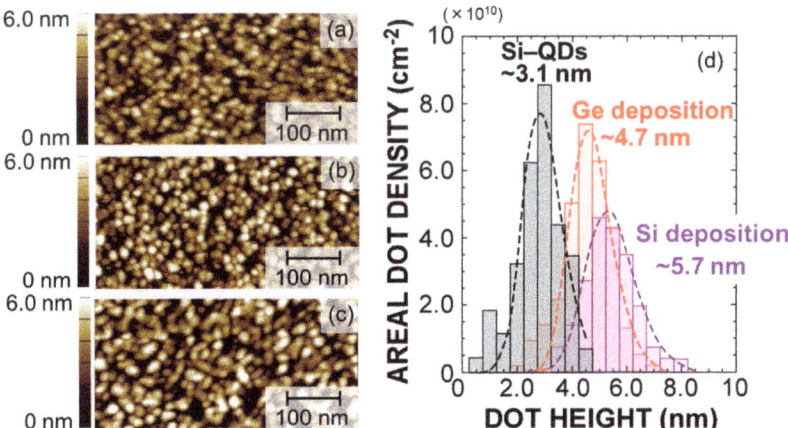

Figure 1. Typical AFM images of (**a**) pre–grown Si–QDs, (**b**) after Ge deposition, (**c**) after Si–cap formation, and (**d**) dot height distribution of the samples shown in (**a**–**c**). The corresponding curves denote log–normal functions well–fitted to the measured size distribution.

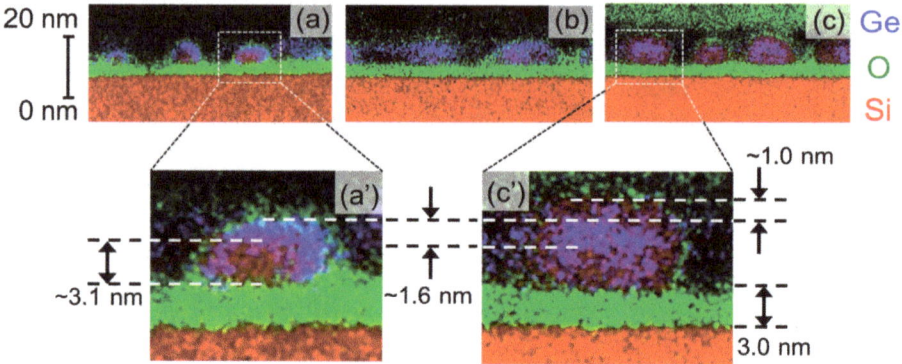

Figure 2. Cross–sectional EDX mapping images of (**a**) Ge deposition on the pre–grown Si–QDs, (**b**) after annealing at 650 °C, and (**c**) Si–cap formation at 650 °C. Images (**a′**) and (**c′**) are enlarged view corresponding to (**a**) and (**c**), respectively.

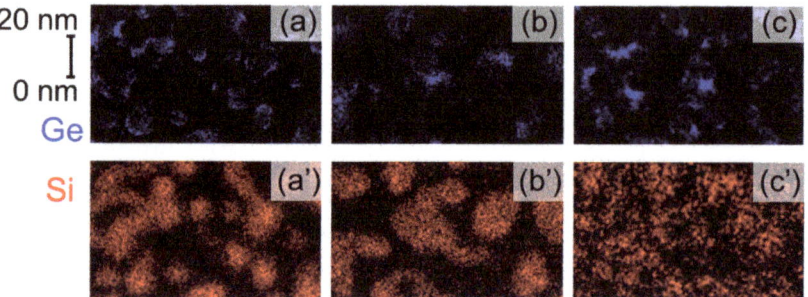

Figure 3. Room temperature PL spectra of the Si–QDs with Ge–core and their deconvoluted spectra evaluated from the spectral analysis using the Gaussian curve fitting method. In–plane EDX mapping images of (**a**) and (**a′**) Ge deposition on the pre–grown Si–QDs, (**b**) and (**b′**) after annealing at 650 °C, and (**c**) and (**c′**) Si–cap formation at 650 °C. Blue and red colors are corresponding to Ge and Si, respectively.

Under 976–nm light excitation at an input power of 0.33 W/cm^2 of the Si–QDs with Ge–core using a semiconductor laser, a stable PL signal consisting of four Gaussian components in the energy region from 0.66 to 0.83 eV was detected at room temperature, as shown in Figure 4. As verified from the dot size dependence of PL peak energy and the temperature dependence of PL properties discussed in [25], these components are attributable to radiative recombination through quantized states in QDs. Based on our previous discussion, Comp. 2 is attributable to the radiative recombination between the conduction and valence bands of the Ge–core through the first quantized states, as shown in Figure 5a. Therefore, providing that the selection rule in quantum mechanics is valid, the higher energy components, Comp. 3 and 4, can be explained by the radiative transition between the higher order quantized states in the conduction and balance bands of Ge core. Considering type II energy band diagram of the Ge–core/Si–shell structure, component 1 might be attributable to radiative recombination between the quantized state of electron in the Si clad and the quantized state of hole in the Ge core because the Si clad acts as a shallow potential well for electron in which electron wave function can penetrate the Ge core, as indicated in Figure 5b. It should be noted that the full width at half maximum (FWHM) values of these components are ~38 meV for all components, which can be explained by a size variation of the Ge–core. However, compared with our previous report, where PL signal in the range from 0.65 to 0.88eV was observed from the Si–QDs with Ge–core formed by a low–pressure CVD, the observed PL spectra in this work are narrower due to the formation of extremely uniform–sized dots.

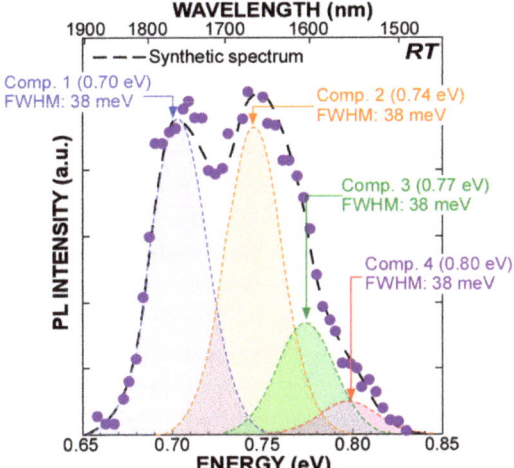

Figure 4. Room temperature PL spectra of the Si–QDs with Ge–core and their deconvoluted spectra evaluated from the spectral analysis using the Gaussian curve fitting method.

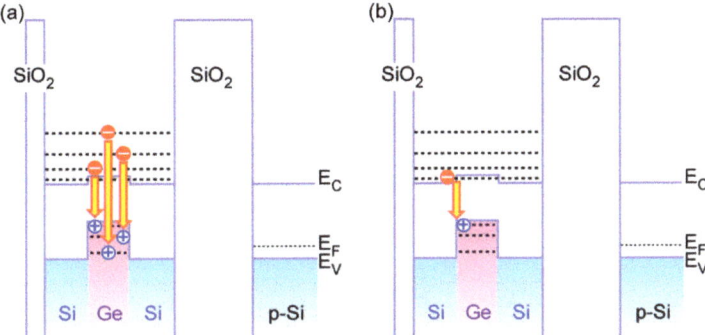

Figure 5. Energy band diagram of Si–QDs with Ge–core, illustrating radiative transition corresponding to (**a**) PL components 2–4, and (**b**) component 1 as shown in Figure 4.

The I–V characteristics of light–emitting diodes (LEDs) on p–Si(100), as schematically illustrated in the inset of Figure 6, show a clear rectification property reflecting the work function difference between the n–type amorphous Si electrode and p–Si(100) substrate (not shown). With the application of continuous square–wave pulsed bias in the negative half cycle with peak–to–peak amplitude over 1.0 V to the top electrodes, EL signals having similar components to PL signals became observable from the topside of the LED structure through the a–Si layer even at room temperature, as shown in Figure 5. Notice that the observed EL spectra consist of four Gaussian components. It should be noted that their peak energies and FWHM values are almost the same as those as the corresponding components evaluated from the PL signal shown in Figure 4 were detected at room temperature. This suggests that the recombination mechanism for the EL is the same as that of the PL. In addition, a significant change in each of the peak positions was confirmed with an increase in applied bias. No EL signals were detected in the positive half cycle bias condition. Therefore, the EL can be explained by radiative recombination of electrons and holes caused by electron injection from the P–doped a–Si and hole injection from the p–Si(100). With an increase in the bias amplitude, the EL intensity increased with almost no change in the peak energy of each component, and then higher emission energy components became dominant, as shown in Figure 7. It should be noted that, by applying a square–wave bias of −4 to 0 V, the EL signal peaked at ~0.8 eV, which is the same energy as that of component 4 evaluated from the PL spectrum is dominant, implying that electrons and holes were injected into higher–order quantized states. From the negative–bias amplitude dependence of the EL intensity, we have found that radiative recombinations between higher–order quantized states become a major factor for EL at −4 V as a result of which single–peak emission can be realized. Results obtained in this research will lead to the realization of Si–based optoelectronic devices by introducing a Ge–core into the Si–QDs (pseudo–super atom structure), which has the advantage of stimulating radiation due to a narrow emission wavelength spectrum in a low–voltage application.

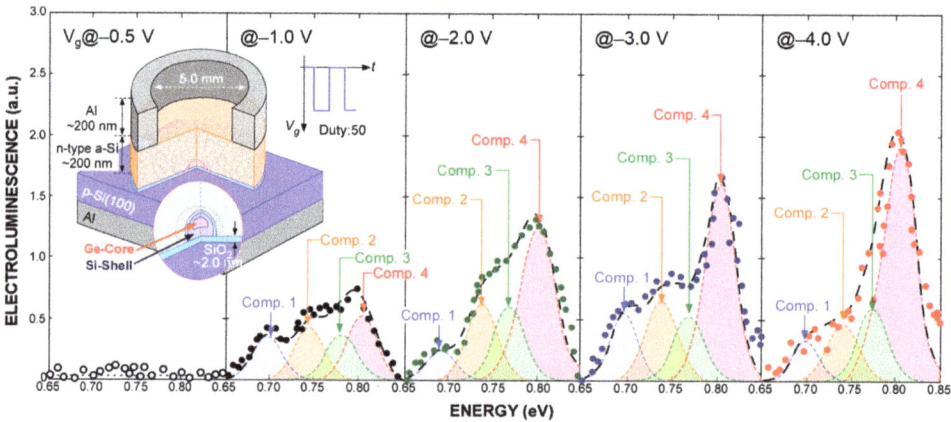

Figure 6. Room temperature EL spectra taken at different applied biases from LEDs having Si–QDs with Ge–core and their deconvoluted spectra evaluated from the spectral analysis using the Gaussian curve fitting method. A schematic illustration of the LED is also shown in the inset.

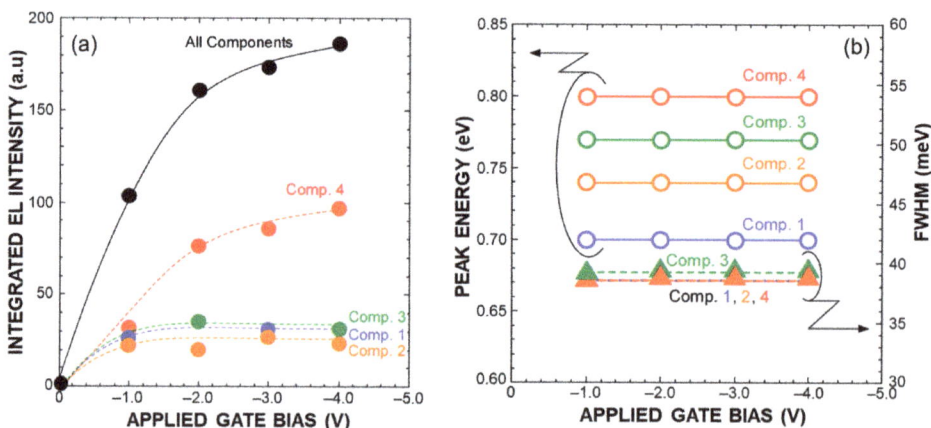

Figure 7. Integrated EL intensities (**a**) and EL peak energy and full width at half maximum of each EL component (**b**) evaluated from the spectral analysis using a Gaussian curve fitting method as a function of applied bias.

4. Conclusions

In summary, high–density Si–QDs with Ge–core show PL in the near–infrared region at room temperature, which indicates that the hole confinement in Ge–core plays an important role in radiative recombination into the Ge–core. We have also demonstrated stable EL in the near–infrared region from light–emitting devices having Si–QDs with a Ge–core caused by electron injection from the top electrodes simultaneously with hole injection from the substrate under continuous square–wave pulsed bias in negative half–cycle applications. From a technological point of view, it is quite important that such a stable EL caused by electron/hole simultaneous injection from the gate and substrate, respectively, at room temperature was realized by using Si–QDs with Ge core, which is compatible with Si–ULSI processing.

Author Contributions: K.M. and Y.Y. contributed equally to this work. Y.I., N.T. and M.A.S. conducted the sample characterizations. B.T. and S.M. contributed to the discussion. All authors have read and agreed to the published version of the manuscript.

Funding: This work was supported in part by the Grant–in–Aid for Scientific Research (A) 19H00762, 21H04559, and the Fund for the Promotion of Joint Intentional Research [Fostering Joint International Research (A)] 18KK0409 of MEXT Japan.

Data Availability Statement: Not applicable.

Acknowledgments: Authors thank the IHP cleanroom staff epitaxy process team for their excellent support.

Conflicts of Interest: The authors declare no conflict of interest.

References

1. Grom, G.F.; Lockwood, D.J.; McCaffrey, J.P.; Labbé, H.J.; Fauchet, P.M.; White, B., Jr.; Diener, J.; Kovalev, D.; Koch, F.; Tsybeskov, L. Ordering and self–organization in nanocrystalline silicon. *Nature* **2000**, *407*, 358–361. [CrossRef] [PubMed]
2. Zakharov, N.D.; Talalaev, V.G.; Werner, P.; Tonkikh, A.A.; Cirlin, G.E. Room–temperature light emission from a highly strained Si/Ge superlattice. *Appl. Phys. Lett.* **2003**, *83*, 3084. [CrossRef]
3. Jambois, O.; Rinnert, H.; Devaux, X.; Vergnat, M. Photoluminescence and electroluminescence of size–controlled silicon nanocrystallites embedded in SiO_2 thin films. *J. Appl. Phys.* **2005**, *98*, 046105. [CrossRef]
4. Sun, K.W.; Sue, S.H.; Liu, C.W. Visible photoluminescence from Ge quantum dots. *Phys. E Low-Dimens. Syst. Nanostruct.* **2005**, *28*, 525. [CrossRef]
5. Perez del Pino, A.; Gyorgy, E.; Marcus, I.C.; Roqueta, J.; Alonso, M.I. Effects of pulsed laser radiation on epitaxial self–assembled Ge quantum dots grown on Si substrates. *Nanotechnology* **2011**, *22*, 295304. [CrossRef]
6. Gassenq, A.; Guilloy, K.; Pauc, N.; Hartmann, J.-M.; Osvaldo Dias, G.; Rouchon, D.; Tardif, S.; Escalante, J.; Duchemin, I.; Niquet, Y.-M.; et al. Study of the light emission in Ge layers and strained membranes on Si substrates. *Thin Solid Films* **2016**, *613*, 64–67. [CrossRef]
7. Schlykow, V.; Zaumseil, P.; Schubert, M.A.; Skibitzki, O.; Yamamoto, Y.; Klesse, W.M.; Hou, Y.; Virgilio, M.; De Seta, M.; Di Gaspare, L.; et al. Photoluminescence from GeSn nano-heterostructures. *Nanotechnology* **2018**, *29*, 415702. [CrossRef]
8. Thai, Q.M.; Pauc, N.; Aubin, J.; Bertrand, M.; Chrétien, J.; Delaye, V.; Chelnokov, A.; Hartmann, J.-M.; Reboud, V.; Calvo, V. GeSn heterostructure micro-disk laser operating at 230 K. *Opt. Exp.* **2018**, *26*, 32500–32508. [CrossRef]
9. den Driesch, N.; Stange, D.; Rainko, D.; Breuer, U.; Capellini, G.; Hartmann, J.-M.; Sigg, H.; Mantl, S.; Grützmacher, D.; Buca, D. Epitaxy of Si-Ge-Sn-based heterostructures for CMOS-integratable light emitters. *Solid-State Electron.* **2019**, *155*, 139–143. [CrossRef]
10. Chrétien, J.; Pauc, N.; Pilon, F.A.; Bertrand, M.; Thai, Q.-M.; Casiez, L.; Bernier, N.; Dansas, H.; Gergaud, P.; Delamadeleine, E.; et al. GeSn Lasers Covering a Wide Wavelength Range Thanks to Uniaxial Tensile Strain. *ACS Photonics* **2019**, *6*, 2462–2469. [CrossRef]
11. Armand Pilon, F.T.; Lyasota, A.; Niquet, Y.-M.; Reboud, V.; Calvo, V.; Pauc, N.; Widiez, J.; Bonzon, C.; Hartmann, J.-M.; Chelnokov, A.; et al. Lasing in strained germanium microbridges. *Nat. Commun.* **2019**, *10*, 2724. [CrossRef] [PubMed]
12. Fadaly, E.M.T.; Dijkstra, A.; Suckert, J.R.; Ziss, D.; van Tilburg, M.A.J.; Mao, C.; Ren, Y.; van Lange, V.T.; Korzun, K.; Kölling, S.; et al. Direct-bandgap emission from hexagonal Ge and SiGe alloys. *Nature* **2020**, *580*, 205–209. [CrossRef] [PubMed]
13. Ji, Z.-M.; Luo, J.-W.; Li, S.-S. Interface–engineering enhanced light emission from Si/Ge quantum dots. *New J. Phys.* **2020**, *22*, 093037. [CrossRef]
14. Jannesari, R.; Schatzl, M.; Hackl, F.; Glaser, M.; Hingerl, K.; Fromherz, T.; Schäffler, F. Commensurate germanium light emitters in silicon–on–insulator photonic crystal slabs. *Opt. Express* **2014**, *21*, 25426–25435. [CrossRef] [PubMed]

15. Petykiewicz, J.; Nam, D.; Sukhdeo, D.S.; Gupta, S.; Buckley, S.; Piggott, A.Y.; Vuckovic, J.; Saraswat, K.C. Direct Bandgap Light Emission from Strained Germanium Nanowires Coupled with High-Q Nanophotonic Cavities. *Nano Lett.* **2016**, *16*, 2168–2173. [CrossRef] [PubMed]
16. Reboud, V.; Gassenq, A.; Hartmann, J.M.; Widiez, J.; Virot, L.; Aubin, J.; Guilloy, K.; Tardif, S.; Fédéli, J.M.; Pauc, N.; et al. Germanium based photonic components toward a full silicon/germanium photonic platform. *Prog. Cryst. Growth Charact. Mater.* **2017**, *63*, 1–24. [CrossRef]
17. Walters, R.J.; Bourianoff, G.I.; Atwater, H.A. Field-effect electroluminescence in silicon nanocrystals. *Nat. Mat.* **2005**, *4*, 143–146. [CrossRef]
18. Xu, X.; Tsuboi, T.; Chiba, T.; Usami, N.; Maruizumi, T.; Shiraki, Y. Silicon-based current-injected light emitting diodes with Ge self-assembled quantum dots embedded in photonic crystal nanocavities. *Opt. Express* **2012**, *20*, 14714–14721. [CrossRef]
19. Xu, X.; Usami, N.; Maruizumi, T.; Shiraki, Y. Enhancement of light emission from Ge quantum dots by photonic crystal nanocavities at room-temperature. *J. Crystal Growth* **2013**, *378*, 636. [CrossRef]
20. Xu, X.; Maruizumi, T.; Shiraki, Y. Waveguide-integrated microdisk light-emitting diode and photodetector based on Ge quantum dots. *Opt. Express* **2014**, *22*, 3902–3910. [CrossRef]
21. Zeng, C.; Ma, Y.; Zhang, Y.; Li, D.; Huang, Z.; Wang, Y.; Huang, Q.; Li, J.; Zhong, Z.; Yu, J.; et al. Single germanium quantum dot embedded in photonic crystal nanocavity for light emitter on silicon chip. *Opt. Exp.* **2015**, *23*, 22250–22261. [CrossRef] [PubMed]
22. Darma, Y.; Takaoka, R.; Murakami, H.; Miyazaki, S. Self-assembling formation of silicon quantum dots with a germanium core by low–pressure chemical vapor deposition. *Nanotechnology* **2003**, *14*, 413–415. [CrossRef]
23. Makihara, K.; Kondo, K.; Ikeda, M.; Ohta, A.; Miyazaki, S. Photoluminescence Study of Si Quantum Dots with Ge Core. *ECS Trans.* **2014**, *64*, 365–370. [CrossRef]
24. Yamada, K.; Kondo, K.; Makihara, K.; Ikeda, M.; Ohta, A.; Miyazaki, S. Effect of Ge Core Size on Photoluminescence from Si Quantum Dots with Ge Core. *ECS Trans.* **2016**, *75*, 695–700. [CrossRef]
25. Kondo, K.; Makihara, K.; Ikeda, M.; Miyazaki, S. Photoluminescence study of high density Si quantum dots with Ge core. *J. Appl. Phys.* **2016**, *119*, 033103. [CrossRef]
26. Makihara, K.; Ikeda, M.; Fujimura, N.; Yamada, K.; Ohta, A.; Miyazaki, S. Electroluminescence of superatom-like Ge-core/Si-shell quantum dots by alternate field–effect–induced carrier injection. *Appl. Phys. Exp.* **2018**, *11*, 011305. [CrossRef]
27. Tillack, B.; Yamamoto, Y.; Murota, J. Atomic Control of Doping during Si Based Epitaxial Layer Growth Processes. *ECS Trans.* **2010**, *33*, 603–614. [CrossRef]
28. Yamamoto, Y.; Zaumseil, P.; Arguirov, T.; Kittler, M.; Tillack, B. Low threading dislocation density Ge deposited on Si (100) using RPCVD. *Solid-State Electron.* **2011**, *60*, 2–6. [CrossRef]
29. Yamamoto, Y.; Zaumseil, P.; Murota, J.; Tillack, B. Phosphorus Profile Control in Ge by Si Delta Layers. *ECS J. Solid State Sci. Technol.* **2013**, *3*, P1–P4. [CrossRef]
30. Yamamoto, Y.; Zaumseil, P.; Schubert, M.A.; Tillack, B. Influence of annealing conditions on threading dislocation density in Ge deposited on Si by reduced pressure chemical vapor deposition. *Semicond. Sci. Technol.* **2018**, *33*, 124007. [CrossRef]

Disclaimer/Publisher's Note: The statements, opinions and data contained in all publications are solely those of the individual author(s) and contributor(s) and not of MDPI and/or the editor(s). MDPI and/or the editor(s) disclaim responsibility for any injury to people or property resulting from any ideas, methods, instructions or products referred to in the content.

Communication

Advantages of Ta-Doped Sb₃Te₁ Materials for Phase Change Memory Applications

Mingyue Shao [1,2], Yang Qiao [3], Yuan Xue [1,2,*], Sannian Song [1,2,*], Zhitang Song [1,2,*] and Xiaodan Li [1,2]

1. State Key Laboratory of Functional Materials for Informatics, Shanghai Institute of Microsystem and Information Technology, Chinese Academy of Sciences, Shanghai 200050, China
2. Center of Materials Science and Optoelectronics Engineering, University of Chinese Academy of Sciences, Beijing 100049, China
3. The Microelectronic Research & Development Center, Shanghai University, Shanghai 200444, China
* Correspondence: xueyuan@mail.sim.ac.cn (Y.X.); songsannian@mail.sim.ac.cn (S.S.); ztsong@mail.sim.ac.cn (Z.S.)

Abstract: Phase change memory (PCM), a typical representative of new storage technologies, offers significant advantages in terms of capacity and endurance. However, among the research on phase change materials, thermal stability and switching speed performance have always been the direction where breakthroughs are needed. In this research, as a high-speed and good thermal stability material, Ta was proposed to be doped in Sb_3Te_1 alloy to improve the phase transition performance and electrical properties. The characterization shows that Ta-doped Sb_3Te_1 can crystallize at temperatures up to 232 °C and devices can operate at speeds of 6 ns and 8×10^4 operation cycles. The reduction of grain size and the density change rate (3.39%) show excellent performances, which are both smaller than that of $Ge_2Sb_2Te_5$ (GST) and Sb_3Te_1. These properties conclusively demonstrate that Ta incorporation of Sb_3Te_1 alloy is a material with better thermal stability and faster crystallization rates for PCM applications.

Keywords: high speed; thermal stability; crystallization; PCM

1. Introduction

Since the 1950s, the rapid development of non-volatile memories has greatly contributed to faster data exchange in the age of the Internet. Phase change memory has attracted attention for its fast programming speed, excellent re-writable characteristics, good stability, and logical compatibility.

Breitwisch [1] showed that there is no clear physical limit to PCM so far, and that device can still phase change when the thickness of the phase change material is reduced to 2 nm. Therefore, PCM is considered to be the most likely solution to the storage technology problem and then to replace the current mainstream storage products as one of the new generations of non-volatile semiconductor memory devices that will be common in the future.

Raoux illustrates that phase change materials based on sulphur compounds have a transition between high and low resistance values when excited by an electric field. Moreover, sulphur-based materials can achieve fast nanosecond transitions between amorphous and crystalline states when thermally induced by a pulsed laser. The structural differences between these two material forms lead to a significant difference in their macroscopic optical reflectivity, and this difference enables the storage of two stable data states [2].

As a classical phase change material, $Ge_2Sb_2Te_5$ (GST) has the potential to be improved in terms of speed and stability when used in practical applications [3,4]. Since the Sb-Te system is a growth-dominated phase change material [5], the growth-dominated crystallization behaviour leads to a faster crystallization rate [6–8]. PCM devices based on Sb_3Te_1 can achieve higher operation speed on the premise of improving data retention capability compared with GST. Based on Ghosh's work [9], Stefan proposed a new version of the Sb-Te phase diagram, in which only the γ phase is used to determine all phase compositions between pure Sb and Sb_2Te_3, with the atomic percentage of Te in the phase ranges from 11.4 at.% to 56.9 at.% [10]. According to the study, it can be proved that the Te content in the Sb_3Te_1 is 25 at.%, which is a single phase.

In this study, the microstructural characterization and electrical properties of the Ta-doped Sb_3Te_1 films and devices were investigated. $Ta_{1.45}Sb_3Te_1$ has enhanced thermal stability, reduced grain size, and increased the switching speed of PCM devices. The improved performance of Ta-doped Sb_3Te_1 materials is analyzed in the paper through changes in the microstructure of the films.

2. Experiment

By using a Ta target and a Sb_3Te_1 target during magnetron co-sputtering, Sb_3Te_1 and $Ta_xSb_3Te_1$ (Ta_xST31) films were produced on SiO_2 substrates. The sputtering power controls the composition of the designed film. Production of pure Sb_3Te_1 films from Sb_3Te_1 targets are sputtered onto SiO_2 substrates using a 20 W RF magnetron. The RF magnetron power of the Ta target was set to 6 W and 8 W while the Sb_3Te_1 target was co-sputtered at 20 W to deposit two different ratios of Ta_xST31 films on the SiO_2 substrate. The ratios of the Ta_xST31 films were experimentally evaluated using energy dispersive spectroscopy (EDS) and recorded as $Ta_{1.08}Sb_3Te_1$ (TaST31-1), $Ta_{1.45}Sb_3Te_1$ (TaST31-2).

The experiment was then carried out with prepared 60 nm films of pure Sb_3Te_1, TaST31-1, and TaST31-2. The R-T measurements were performed at a 20 °C/min heating rate. The films were heated at a rate of 20 °C/min while under vacuum, and the crystallization temperature of the films was calculated for comparison and analysis.

Furthermore, the microstructure of 150 nm pure Sb_3Te_1, TaST31-1, and TaST31-2 films were explored using X-ray diffraction (XRD) for lattice information of the films. Furthermore, 30 nm thick films of pure Sb_3Te_1 and TaST31-2 were deposited on the Cu grid, and the transformation process was also studied by transmission electron microscopy (TEM) to investigate the structural evolution. Additionally to this, the 40 nm TaST31-1 and TaST31-2 films were characterized using X-ray reflectivity (XRR) to observe the density variation of the films. The 150 nm pure Sb_3Te_1 and TaST31-2 films were selected for analysis of their chemical bonding properties by applying X-ray photoelectron spectroscopy (XPS).

Finally, "T-shaped" PCM devices were fabricated to test their electrical properties. Pre-industrial delivery was carried out using 0.13 μm complementary metal-oxide semiconductor technology. A sputtering method was used to deposit a 100 nm thick phase change material and a 30 nm thick titanium nitride as an adhesion layer on a 190 nm diameter tungsten heated electrode. Eventually, resistance–voltage (R–V) measurements and endurance tests were carried out utilizing a Tektronix AWG5002B pulse generator and a Keithley 2400 C source meter.

3. Results and discussion

3.1. Effect of Ta Doping on Crystalline Properties and Microstructure

In this research, resistance–temperature (R-T) test experiments were first carried out on 60 nm films of pure Sb_3Te_1, TaST31-1, and TaST31-2.

Figure 1 shows the curve of resistance as a function of temperature for the three different components of the films. The resistivity of the films first reduces gradually with rising temperature. Before the crystallization temperature is attained, resistance rapidly decreases. The minimum value of the first-order inverse of the R-T curve is used to compute the crystallization temperature (T_c). The crystallization temperatures of pure Sb_3Te_1, TaST31-1, and TaST31-2 are 152 °C, 203 °C, and 230 °C, respectively. The crystallization temperature of the Sb_3Te_1 film can be increased by Ta doping. The substance GST crystallizes at a temperature of about 150 °C, according to studies [11,12]. In comparison, it can be found that the addition of the material Ta can increase the crystallization temperature by 50 °C to 80 °C, which greatly improves the thermal stability of the material.

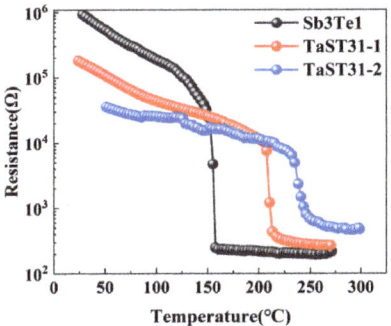

Figure 1. The R–T measurement results for Sb_3Te_1, TaST31-1, and TaST31-2.

The lattice information of the Sb_3Te_1 and TaST31-2 films were characterized by XRD measurements, as shown in Figure 2a,b. The Sb_3Te_1 and TaST21-2 films with a thickness of 150 nm were annealed in N_2 at 200 °C, 240 °C, and 280 °C for 5 min. The diffraction peaks of Sb_3Te_1 film appeared at 200 °C, which is corresponding to the hexagonal Sb_2Te phase. As observed in Figure 2b, the crystallization of TaST31-2 film is delayed, and the diffraction peak appeared at 240 °C. By comparing the crystal orientation indexes in Figure 2a,b, the TaST31-2 film shows an increase in the intensity of the 103 peak and a decrease in the intensity of the 004 and 005 peaks compared to the pure Sb_3Te_1 films, while the 114 and 023 peaks no longer appear. It can clearly be observed that the diffraction peaks of the Ta-doped Sb_3Te_1 film broaden, but there are no diffraction peaks of the metal Ta or Ta-containing metal compounds, and no isolated Sb [13]. The results of the XRD curves indicate that the main crystal structure of the Ta-doped Sb_3Te_1 film has not changed. Moreover, the half-peak width of the Ta-doped Sb_3Te_1 film has become larger, which means that the grains have been refined.

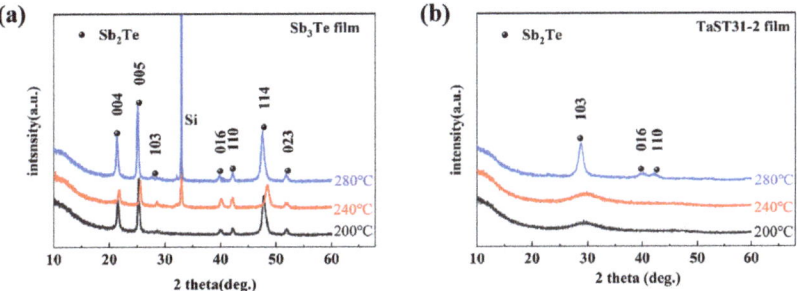

Figure 2. XRD curves of (**a**) Sb_3Te_1 and (**b**) TaST21-2 films at different temperatures.

Transmission electron microscopy (TEM) enables a more visual observation of the grain size and crystalline phase. The test samples for TEM are 30 nm thick films of Sb_3Te_1 and TaST31-2 deposited on copper grids, which can be directly used for TEM observation and analysis. All samples were annealed in N_2 at 280 °C for 5 min. BF TEM images of Sb_3Te_1 and TaST31-2 films and their associated SAED patterns are shown in Figure 3a,b, respectively. Figure 3c,d are HRTEM images of Sb_3Te_1 and TaST31-2 films, respectively. The comparison of BF images shows that the grain size of TaST31-2 film is significantly smaller than that of the Sb_3Te_1 film. It proves that after Ta doping, the crystallization behaviour of Sb_3Te_1 switches from being growth-dominated to nucleation-dominated. Meanwhile, the SAED pattern of the Sb_3Te_1 film shows single crystal diffraction spots, indicating that it has grown into large grains. In contrast, the intensity of diffraction rings that belong to 110, 103, and 016 crystallographic planes of hexagonal structure of Sb_2Te emerged in the corresponding SAED pattern of TaST31-2 film and showed a small degree of discontinuity, further explaining the inhibition effect of Ta on grains [14]. It also can be found that the grains become uniform on the HRTEM images of TaST31-2 films. Sb_3Te_1 films have a grain size of at least 100 nm, which is reduced to 5–10 nm after Ta doping. Therefore, it is concluded that the grain size decreases and no other phases in Ta doping Sb_3Te_1 film. The small size of the grain contributes to reducing the thermal conductivity and resistance drift as well as the formation of a homogeneous phase, which helps to improve the heating efficiency and the reliability of the device [15].

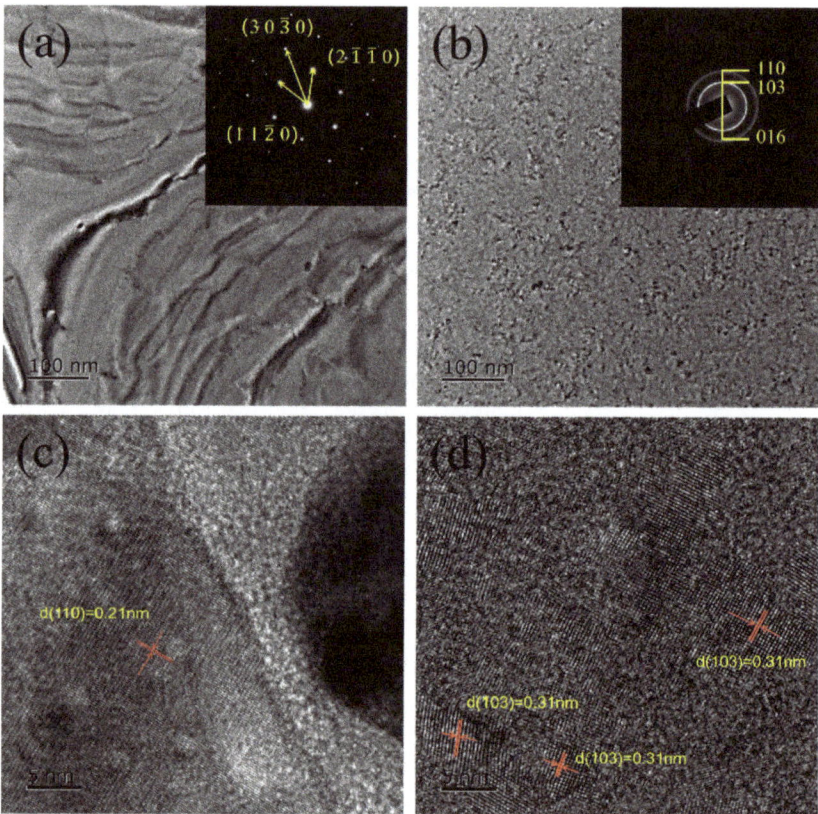

Figure 3. The BF TEM images and their corresponding SAED patterns of (**a**) Sb_3Te_1, (**b**) TaST31-2 films. The HRTEM images for (**c**) Sb_3Te_1 and (**d**) TaST31-2.

Bruker D8 Discover's X-ray reflectivity (XRR) investigations can be used to look at how the thickness and density of the films change both before and after crystallization. Figure 4a,c show the XRR diffraction images of the TaST31-1 and TaST31-2 films, respectively. The density change values for the amorphous and crystalline phases of the TaST31-1 and TaST31-2 films are shown in Figure 4b,d, respectively. The principle of operation of XRR is to take advantage of the reflection and refraction phenomena that occur on the surface of the sample when X-rays are reflected. The refracted light enters the interior of the sample and is reflected and refracted again at the Si substrate interface. The two reflected beams interfere to produce interference fringes [16]. Figure 4a,c show that the XRR curves of the Ta-doped Sb_3Te_1 films are wavy, with multiple peaks and troughs. The numbers 1–8 in the figures are the number of wave peaks. Based on the displacement of the peaks, fitting calculations are performed to show the thickness and density change corresponding to the amorphous to crystalline transition from Figure 4b,d. In the figures, the exact film thickness d can be deduced from the slope of the curve $[\left(\frac{\lambda}{2d}\right)^2]$ and the critical angle calculated from the intercept (α_m^2) on the y-axis. The density change of the film is calculated from Equation (1):

$$\nabla\rho = \frac{\rho_{mc} - \rho_{ma}}{\rho_{mc}} = \frac{\alpha_{cc}^2 - \alpha_{ca}^2}{\alpha_{cc}^2}, \qquad (1)$$

where ρ_{mc}, ρ_{ma} are the density of the crystalline and amorphous TaST31, respectively, and α_{ca}^2, α_{cc}^2 are the critical angles of the XRR curve for the amorphous and crystalline TaST31. It was calculated that the rate of change in thickness for the TaST31-1 film is 4.29%, compared to 1.01% for the TaST31-2 film. Meanwhile, the rate of change in density was 6.17% for film TaST31-1 and 3.39% for film TaST31-2. The density change of the film TaST31-2 (3.39%) is lower than that of GST (6.8%). From this result, it can be concluded that the change in density decreases as the Ta percentage increases. This value of density change improves the durability of the material [17,18].

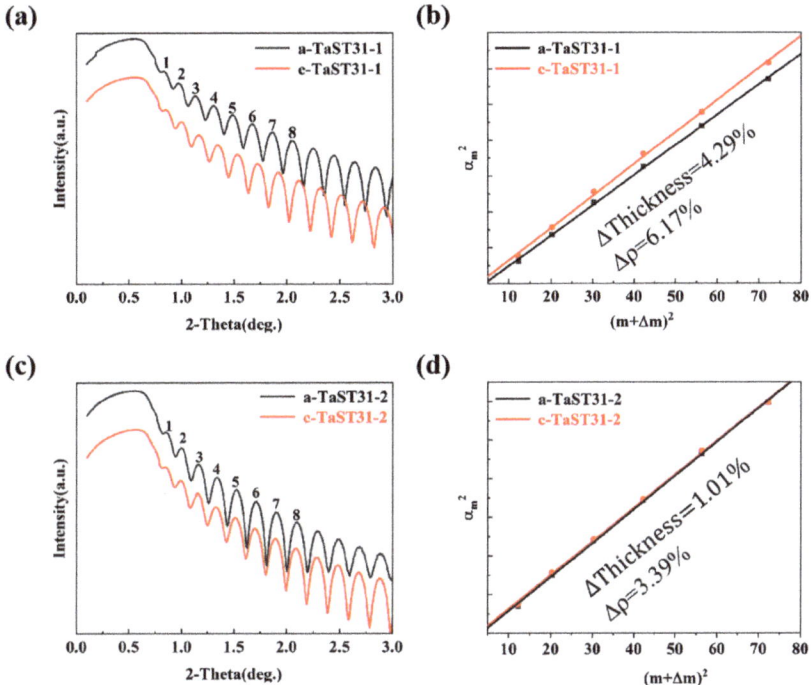

Figure 4. The density changes of the TaST31 films after crystallization. XRR curves of amorphous and crystalline (**a**) TaST31-1 and (**c**) TaST31-2 films. Bragg fitting curves of amorphous and crystalline films of (**b**) TaST31-1 and (**d**) TaST31-2 films.

The chemical bonding properties of Sb_3Te_1 and TaST31 films in the crystalline form were examined by the XPS method in order to examine the influence of Ta doping. Figure 5a,b show the XPS spectra of Te3d and Sb3d, respectively. In Figure 5a, it can be seen that the bond energies of 572.4 eV and 582.7 eV in the Sb_3Te_1 film correspond to Te 3d5/2 and 3d3/2, respectively, and the bond energies of 572.3 eV and 582.7 eV in the TaST31-2 film correspond to Te 3d5/2 and 3d3/2, respectively. The result indicates that by doping Sb_3Te_1 with Ta element, the binding energy of Te will shift towards a lower binding energy. As shown in Figure 5b, it is also observed that the bond energy of Sb 3d5/2 is reduced from 528.2 eV to 528.1 eV and the bond energy of Sb 3d3/2 is reduced from

537.6 eV to 537.5 eV with lower binding energy. The electronegativities are 1.4, 2.05, and 2.1 for Ta, Sb, and Te elements, respectively [19], making a stronger Ta-Te bond than an Sb-Te bond. Ta tends to combine with Te to form more Ta-Te bonds and extra Sb-Sb bonds. The bond energy of the Ta-Te bond is larger than that of the Sb-Te bond, indicating that more energy is required to break the Ta-Te bond, resulting in better thermal stability [20,21]. Therefore, the thermal stability of Sb_3Te_1 films was improved after doping with Ta elements.

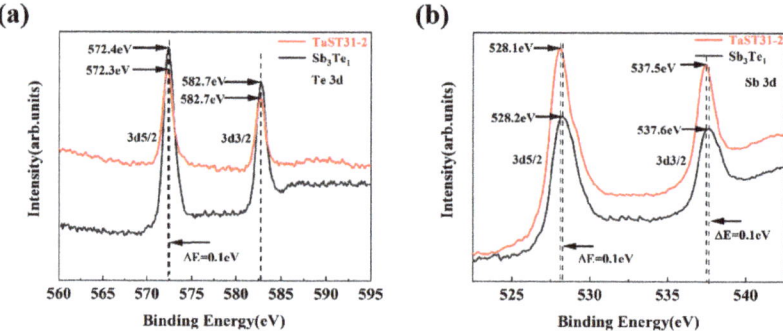

Figure 5. Sb_3Te_1 and TaST31-2 films' XPS spectra after 260 °C annealing (**a**) Te 3d and (**b**) Sb 3d.

3.2. Device Performance

The resistance–voltage (R-V) curve for a T-shaped device based on TaST31-2 material is depicted in Figure 6a. The devices based on TaST31-2 material can even perform SET and RESET at ultra-fast pulse widths of 6 ns, allowing for high-speed operation. Therefore, the Ta-doped Sb_3Te_1 material's electrical performance is more dominant than that of GST (10 ns) [22]. It is also evident by seeing the R-V curves that the Ta-doped Sb_3Te_1 material requires less voltage to operate in the SET/RESET state. The device performs best at a SET voltage of 2.5 V and a RESET voltage of 4 V for a pulse width of 100 ns. The highest and lowest resistances fall within a window of 1.7 orders of magnitude under these circumstances, which is sufficient for logical differentiation. Figure 6b shows the endurance performance of a T-shaped device based on TaST31-2 material. Set a 300 ns/1.8 V SET pulse and a 200 ns/2.8 V RESET pulse as the device's operating conditions. The device can be tested to hold for 8×10^4 cycles before a failure occurs. Consequently, the following conclusions are summarised from the above experiments, the switching speeds and SET/RESET voltages always have a huge advantage over GST materials in TaST31-2 device tests.

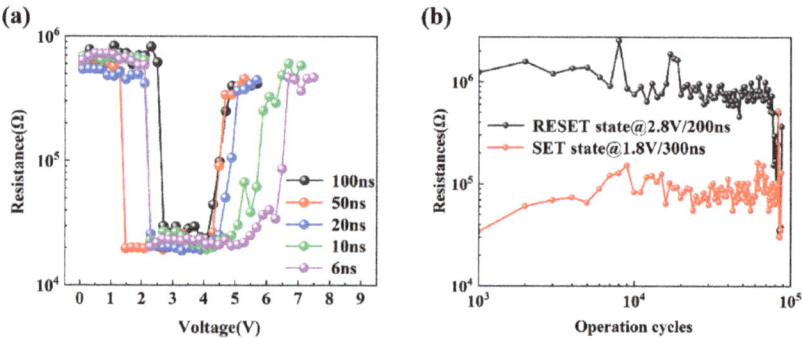

Figure 6. (**a**) The resistance–voltage characteristics of TaST31-2-based T-shaped PCM devices varying pulse widths. (**b**) Durability characteristics of TaST31-2-based PCM T-shaped devices.

4. Conclusions

The properties of Ta-doped Sb_3Te_1 materials for PCM applications are examined in this paper. The crystallization temperature increases with the amount of Ta doping, indicating that the element Ta improves thermal stability. The XRD results can be analyzed to show that Ta doping does not change the crystal structure of Sb_3Te_1 material and results in the refinement of the grains. Transmission electron microscopy images show more visually that the size of the crystal grains has become smaller in the TaST31 material, verifying the XRD test results. This reduction in grain size contributes to lower thermal conductivity and lower resistance drift, thus improving the heating efficiency and reliability of the device. X-ray reflectivity (XRR) experiments showed a thickness change (1.01%) and the density change (3.39%) for the TaST31-2 films, which is advantageous to the endurance of the material. XPS experiments illustrate that more Ta-Te bonds with higher bonding energy, verifying the conclusion that the doping of Ta enhances the thermal stability of the material. The PCM device based on TaST31-2 achieves 6 ns operation speed and 8×10^4 cycles. In summary, the $TaSb_3Te_1$ material holds great promise for phase change memory applications.

Author Contributions: Conceptualization, Y.X., S.S. and Z.S.; methodology, M.S., Y.X. and Y.Q.; formal analysis, M.S.; investigation, M.S., Y.Q., Y.X. and X.L.; data curation, M.S.; writing—original draft preparation, M.S.; writing—review and editing, M.S. and Y.X.; visualization, M.S. and Y.Q.; supervision, Y.X., S.S. and Z.S.; project administration, S.S. and Z.S.; funding acquisition, S.S. and Z.S. All authors have read and agreed to the published version of the manuscript.

Funding: This research was funded by the National Natural Science Foundation of China (62204251, 92164302, 61874129, 91964204, 61904186, 61904189, 61874178, 62274170), Strategic Priority Research Program of the Chinese Academy of Sciences (XDB44010200), Science and Technology Council of Shanghai (22DZ2229009, 19JC1416801, 20501120300).

Data Availability Statement: Not applicable.

Conflicts of Interest: The authors declare no conflict of interest.

References

1. Burr, G.W.; Breitwisch, M.J.; Franceschini, M.; Garetto, D.; Gopalakrishnan, K.; Jackson, B.; Kurdi, B.; Lam, C.; Lastras, L.A.; Padilla, A.; et al. Phase change memory technology. *J. Vac. Sci. Technol. B* 2010, *28*, 223. [CrossRef]
2. Raoux, S.; Wełnic, W.; Ielmini, D. Phase change materials and their application to nonvolatile memories. *Chem. Rev.* 2010, *110*, 240–267. [CrossRef]
3. Liu, B.; Song, Z.; Feng, S.; Chen, B. Chen, Characteristics of chalcogenide nonvolatile memory nano-cell-element based on Sb_2Te_3 material. *Microelectron. Eng.* 2005, *82*, 168–174. [CrossRef]
4. Sun, Z.; Zhou, J.; Pan, Y.; Song, Z.; Mao, H.K.; Ahuja, R. Pressure-induced reversible amorphization and an amorphous-amorphous transition in $Ge_2Sb_2Te_5$ phase change memory material. *Proc. Natl. Acad. Sci. USA* 2011, *108*, 10410–10414. [CrossRef] [PubMed]
5. Fujimori, S.; Yagi, S.; Yamazaki, H.; Funakoshi, N. Crystallization process of Sb-Te alloy films for optical storage. *J. Appl. Phys.* 1988, *64*, 1000–1004. [CrossRef]
6. Meinders, E.R.; Lankhorst, M.H.R. Determination of the crystallisation kinetics of fast-growth phase-change materials for mark-formation prediction. *Jpn. J. Appl. Phys.* 2003, *42*, 809–812. [CrossRef]
7. Lankhorst, M.H.R.; Van Pieterson, L.; Van Schijndel, M.; Jacobs, B.A.J.; Rijpers, J.C.N. Prospects of doped Sb–Te phase-change materials for high-speed recording. *Jpn. J. Appl. Phys.* 2003, *42*, 863–868. [CrossRef]
8. Van Pieterson, L.; Lankhorst, M.H.R.; van Schijndel, M.; Kuiper, A.E.T.; Roosen, J.H.J. Phase-change recording materials with a growth-dominated crystallization mechanism: A materials overview. *J. Appl. Phys.* 2005, *97*, 083520. [CrossRef]
9. Ghosh, G. The Sb-Te (antimony-tellurium) system. *J. Phase Equilibria* 1994, *15*, 349–360. [CrossRef]
10. Solé, S.; Schmetterer, C.; Richter, K.W. A Revision of the Sb-Te Binary Phase Diagram and Crystal Structure of the Modulated γ-Phase Field. *J. Phase Equilibria Diffus.* 2022, *43*, 648–659. [CrossRef]
11. Boniardi, M.; Redaelli, A.; Cupeta, C.; Pellizzer, F.; Crespi, L.; D'Arrigo, G.; Lacaita, A.L.; Servalli, G. Optimization metrics for phase change memory (PCM) cell architectures. In Proceedings of the 2014 IEEE International Electron Devices Meeting, San Francisco, CA, USA, 15–17 December 2014.
12. Rao, F.; Ding, K.Y.; Zhou, Y.X.; Zheng, Y.H.; Xia, M.J.; Lv, S.L.; Song, Z.T.; Feng, S.L.; Ronneberger, I.; Mazzarello, R.; et al. Reducing the stochasticity of crystal nucleation to enable subnanosecond memory writing. *Science* 2017, *358*, 1423–1426. [CrossRef] [PubMed]
13. Cho, J.-Y.; Kim, D.; Park, Y.-J.; Yang, T.-Y.; Lee, Y.-Y.; Joo, Y.-C. The phase-change kinetics of amorphous $Ge_2Sb_2Te_5$ and device characteristics investigated by thin-film mechanics. *Acta Mater.* 2015, *94*, 143–151. [CrossRef]

14. Zhu, M.; Wu, L.; Rao, F.; Song, Z.; Li, X.; Peng, C.; Zhou, X.; Ren, K.; Yao, D.; Feng, S. N-doped Sb2Te1 phase change materials for higher data retention. *J. Alloys Compd.* **2011**, *509*, 10105–10109. [CrossRef]
15. Wang, W.; Loke, D.; Shi, L.; Zhao, R.; Yang, H.; Law, L.-T.; Ng, L.-T.; Lim, K.-G.; Yeo, Y.-C.; Chong, T.-C.; et al. Enabling Universal Memory by overcoming the contradictory speed and stability nature of phase-change materials. *Sci. Rep.* **2010**, *2*, 360. [CrossRef] [PubMed]
16. Yan, S.; Cai, D.; Xue, Y.; Guo, T.; Song, S.; Song, Z. Sb-rich CuSbTe material: A candidate for high-speed and high-density phase change memory application. *Mater. Sci. Semicond. Process.* **2019**, *103*, 104625. [CrossRef]
17. Njoroge, W.K.; Wöltgens, H.-W.; Wuttig, M. Density changes upon crystallization of Ge2Sb2.04Te4.74 films. *J. Vac. Sci. Technol. B* **2002**, *20*, 230–233. [CrossRef]
18. Cheng, Y.; Song, Z.; Gu, Y.; Song, S.; Rao, F.; Wu, L.; Liu, B.; Feng, S. Influence of silicon on the thermally-induced crystallization process of Si-Sb4Te phase change materials. *Appl. Phys. Lett.* **2011**, *99*, 261914. [CrossRef]
19. Gordy, W.; Thomas, W.J.O. Electronegativities of the elements. *J. Chem. Phys.* **1956**, *24*, 439–444. [CrossRef]
20. Xia, Y.; Liu, B.; Wang, Q.; Zhang, Z.; Song, S.; Song, Z.; Yao, D.; Xi, W.; Guo, X.; Feng, S. Study on the phase change material Cr-doped Sb 3 Te 1 for application in phase change memory. *J. Non-Cryst. Solids* **2015**, *422*, 46–50. [CrossRef]
21. Zhu, M.; Wu, L.C.; Song, Z.T.; Rao, F.; Cai, D.L.; Peng, C.; Zhou, X.L.; Ren, K.; Song, S.N.; Liu, B.; et al. Ti10Sb60Te30 for phase change memory with high-temperature data retention and rapid crystallization speed. *Appl. Phys. Lett.* **2012**, *100*, 122101. [CrossRef]
22. Privitera, S.; Pimini, E.; Zonca, R. Amorphous-to-crystal transition of nitrogenand oxygen-doped $Ge_2Sb_2Te_5$ films studied by in situ resistance measurements. *Appl. Phys. Lett.* **2004**, *85*, 3044–3046. [CrossRef]

Disclaimer/Publisher's Note: The statements, opinions and data contained in all publications are solely those of the individual author(s) and contributor(s) and not of MDPI and/or the editor(s). MDPI and/or the editor(s) disclaim responsibility for any injury to people or property resulting from any ideas, methods, instructions or products referred to in the content.

Article

The Effect of Carbon Doping on the Crystal Structure and Electrical Properties of Sb₂Te₃

Jie Zhang [1], Ningning Rong [1], Peng Xu [1], Yuchen Xiao [1], Aijiang Lu [1], Wenxiong Song [2], Sannian Song [2], Zhitang Song [2], Yongcheng Liang [1] and Liangcai Wu [1,3,*]

[1] College of Science, Donghua University, Shanghai 201620, China
[2] State Key Laboratory of Functional Materials for Informatics, Shanghai Institute of Microsystem and Information Technology, Chinese Academy of Sciences, Shanghai 200050, China
[3] Shanghai Institute of Intelligent Electronics & Systems, Shanghai 200433, China
* Correspondence: lcwu@dhu.edu.cn

Abstract: As a new generation of non-volatile memory, phase change random access memory (PCRAM) has the potential to fill the hierarchical gap between DRAM and NAND FLASH in computer storage. Sb_2Te_3, one of the candidate materials for high-speed PCRAM, has high crystallization speed and poor thermal stability. In this work, we investigated the effect of carbon doping on Sb_2Te_3. It was found that the FCC phase of C-doped Sb_2Te_3 appeared at 200 °C and began to transform into the HEX phase at 25 °C, which is different from the previous reports where no FCC phase was observed in C-Sb_2Te_3. Based on the experimental observation and first-principles density functional theory calculation, it is found that the formation energy of FCC-Sb_2Te_3 structure decreases gradually with the increase in C doping concentration. Moreover, doped C atoms tend to form C molecular clusters in sp^2 hybridization at the grain boundary of Sb_2Te_3, which is similar to the layered structure of graphite. And after doping C atoms, the thermal stability of Sb_2Te_3 is improved. We have fabricated the PCRAM device cell array of a C-Sb_2Te_3 alloy, which has an operating speed of 5 ns, a high thermal stability (10-year data retention temperature 138.1 °C), a low device power consumption (0.57 pJ), a continuously adjustable resistance value, and a very low resistance drift coefficient.

Keywords: phase change random access memory; C-doped Sb_2Te_3; density functional theory; formation energy; continuously adjustable resistance value

Citation: Zhang, J.; Rong, N.; Xu, P.; Xiao, Y.; Lu, A.; Song, W.; Song, S.; Song, Z.; Liang, Y.; Wu, L. The Effect of Carbon Doping on the Crystal Structure and Electrical Properties of Sb_2Te_3. *Nanomaterials* 2023, 13, 671. https://doi.org/10.3390/nano13040671

Academic Editors: Shunri Oda, Kunji Chen, Linwei Yu and Sotirios Baskoutas

Received: 27 December 2022
Revised: 12 January 2023
Accepted: 21 January 2023
Published: 9 February 2023

Copyright: © 2023 by the authors. Licensee MDPI, Basel, Switzerland. This article is an open access article distributed under the terms and conditions of the Creative Commons Attribution (CC BY) license (https://creativecommons.org/licenses/by/4.0/).

1. Introduction

With the continuous development of an information society, people's demand for information storage and calculation continues to grow. Phase change random access memory (PCRAM) is greatly welcomed by researchers in industrial electronics, artificial intelligence, and other fields because of its simple process, high integration, non-volatility, multi-level storage, and other characteristics [1]. The non-volatile memory technology 3D-XPoint, developed by Intel and Micron and announced for the first time in August 2015, is a new type of non-volatile memory that can significantly reduce latency, so that more data can be stored near the central processing unit [2], and its essence is PCRAM. Intel claims that its speed and life span are 1000 times that of NAND Flash, its integration density is 10 times that of traditional memory, and its cost is half that of Dynamic Random Access Memory (DRAM) [3].

The key technology of PCRAM lies in phase change materials. PCRAM realizes data storage by using the resistance difference of the phase change material between the amorphous state and the crystalline state. The common $Ge_2Sb_2Te_5$ (GST) phase change material is the most widely used material at present, but it has the disadvantages of slow speed (~50 ns) and poor thermal stability (~82 °C) [4,5], which cannot meet the requirements of high-speed and high thermal stability PCRAM, thus limiting its application in electronic devices. In order to improve the performance of PCRAM, it is key to find phase change materials with fast reversible phase change speed and high thermal stability. Zuliani et al. explored the region rich in Ge element in the GST ternary diagram, which made the thermal stability of Ge-rich GST meet the specifications of automobile application, but the programming speed loss was about one-third of GST [6]. Diaz et al. improved the bottom electrode contact and thermal stability (1h retention is 230 °C) by adding a GST buffer layer under Ge-rich GST [7]. The high thermal stability of a Ge-rich GST alloy is due to the formation of local tetrahedral Ge-Ge bonds, which leads to a more disordered structure; that is, it is less prone to crystallization, which increases the resistivity of crystalline GST and reduces the RESET power consumption of PCRAM [6]. However, the operation speed of Ge-rich GST is still not satisfied.

In recent years, Sb_2Te_3 had great application potential in thermoelectrics [8], optoelectronics [9], PCRAM, and neuromorphic applications [10]. Sb_2Te_3 has a fast reversible phase transition, which is due to its rich Te and vacancies in the face-centered cubic (FCC) phase [11]. Sb_2Te_3 is considered one of the most advantageous candidate materials for PCRAM, but its thermal stability is poor [12–14]. In order to improve thermal stability while maintaining the advantages of its high crystallization speed, doping is a feasible means to improve the thermal stability of Sb_2Te_3 [15]. Although some dopants can improve the thermal stability of amorphous Sb_2Te_3, they will cause undesirable phase separation during the erasing and writing of PCRAM, which will seriously limit the crystallization speed and deteriorate the device's performance [16]. Dopant C is a relatively ideal dopant [17], and doping Sb_2Te_3 with element C can not only achieve the purpose of fast phase change, high thermal stability, and low power consumption of RESET [18], but also will not cause serious phase separation. Experimental results show that there is an FCC phase in C-doped Sb_2Te_3 (C-Sb_2Te_3). Combined with density functional theory, we found that the formation of an FCC-Sb_2Te_3 structure can gradually decrease with the increase in C doping concentration. The doped C atoms tend to form C molecular clusters in sp^2 hybridization at the grain boundary of Sb_2Te_3. Furthermore, PCRAM device cells based on C-Sb_2Te_3 were fabricated, which had an operating speed of 5 ns, high thermal stability, a low device power consumption, a continuously adjustable resistance, and an extremely low resistance drift coefficient.

2. Calculations and Experimental Section

2.1. Density Functional Theory (DFT) Methods

In the calculation work, we constructed a $3 \times 3 \times 3$ supercell model of an FCC-Sb_2Te_3, which contains 180 atoms (72 Sb atoms and 108 Te atoms), and randomly generated 36 Sb cation vacancies in the system. We also constructed an FCC-Sb_2Te_3 model with 160 atoms of Σ3 twin grain boundary, including 64 Sb atoms, 96 Te atoms, and 32 random Sb cation vacancies. The thickness of the vacuum layer between supercells is 10 Å.

We use the Vienna ab initio simulation package (VASP) for density functional theory calculation. We adopted the projector augmented wave (PAW) potentials for describing the ion–electron interaction, the generalized gradient approximation (GGA) of Perdew-Burke-Ernzerhof (PBE) for exchange–correlation interactions between electrons [15,19–23]. The calculated valence electrons include $2s^22p^2$ of C, $5s^25p^3$ of Sb, and $5s^25p^4$ of Te, and the plane wave cutoff energy is set to 550 eV. With Γ point as the origin, the Monkhorst–Pack method is used to generate $1 \times 1 \times 1$ and $3 \times 3 \times 1$ k-point grids, respectively, and the Gaussian smearing method is used to adjust each orbital occupation. The cutoff energy and k-point grids have been tested, and the atomic structures of these two models are fully optimized. The energy convergence criterion is 10^{-5} eV, while the atomic force is less than 0.05 eV/Å. We implemented the Crystal Orbital Hamiltonian Population (COHP) bonding analyses using the LOBSTER setup [24].

2.2. Experimental Methods

By magnetron sputtering, the C and Sb_2Te_3 were co-sputtered onto a SiO_2/Si (100) substrate, and the thickness of the film can be controlled by adjusting the sputtering time and sputtering power. The magnetron sputtering power of Sb$_2$Te$_3$ is 20W RF and the power of C is 40W DC. The deposition proceeds with Ar at a flow rate of 20 SCCM, with a background pressure of 3×10^{-4} Pa. The film was heated in situ at a heating rate of 60 °C/min in a self-made vacuum heating station, and the changes of resistance with time at various temperatures were recorded. Data retention for 10 years (and 100 years) was estimated by the Arrhenius Equation. To explore the electrical behavior of C-Sb$_2$Te$_3$ in memory cells, the PCRAM cells were fabricated with the traditional T-shaped (mushroom-type) structure. The bottom W heat electrode with a diameter of 190 nm was fabricated by 0.13 μm complementary metal oxide semiconductor technology. The bottom W heat electrode is covered with a C-Sb$_2$Te$_3$ film with a thickness of about 135 nm, and 40 nm TiN is deposited as the top electrode. The PCRAM cells were patterned using an etching process. The prototype PCRAM cells were annealed at 300 °C for 10 min in a N$_2$ atmosphere, and then the electrical properties, such as current voltage (I-V), resistance voltage (R-V), and resistance time (R-t), were tested by a self-made test system. The test system consists of an arbitrary waveform generator (Tektronix AWG5002B, Beaverton, OR, USA) and a digital source meter (Keithley-2400, Beaverton, OR, USA). The thin films were continuously annealed at 200–300 °C for 5 min in a N$_2$ atmosphere, and the lattice information of the thin films was explored by X-ray diffraction (XRD, Rigaku, Tokyo, Japan).

3. Results and Discussion

3.1. Atomic Configuration for C-Doped Sb$_2$Te$_3$

In order to determine the position of the C atom in Sb$_2$Te$_3$, we first consider four possible doping ways of the C atom in an FCC-Sb$_2$Te$_3$: replacing the Sb atom (C$_{Sb}$), replacing the Te atom (C$_{Te}$), occupying Sb cation vacancy (C$_V$), and interstitial doping (C$_I$). On this basis, the formation energy of C doping in an FCC-Sb$_2$Te$_3$ was calculated and compared. The Equation for calculating the formation energy of C doping is as follows [15,20,21,25]:

$$E^f[X] = E_{tot}[X] - E_{tot}[bulk] - \sum_i n_i \mu_i \quad (1)$$

where $E_{tot}[X]$ and $E_{tot}[bulk]$ are the total energies of a supercell with and without C doping, respectively, and n_i represents the number of doped atoms. $n_i > 0$ means adding atoms to the supercell, $n_i < 0$ means removing atoms from the supercell, and μ_i means the chemical potential of i substance. In this paper, the chemical potentials of C, Sb, and Te are calculated according to the simple substance trigonal phase.

The FCC-Sb$_2$Te$_3$ supercell used in the calculation contains 36 cation vacancies, 72 Sb atoms, and 108 Te atoms. The FCC-Sb$_2$Te$_3$ model is shown in Figure 1a. The calculation results show that the formation energy of C atoms in every position of the FCC-Sb$_2$Te$_3$ is very high, as shown in Figure 1b, which indicates that the doping system is not easy to form or unstable; that is, the substitution/occupation position of C atoms is unreasonable.

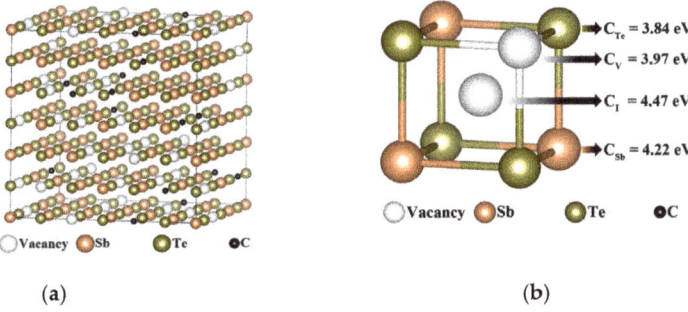

Figure 1. (a) The FCC-Sb$_2$Te$_3$ model with 180 atoms and (b) C atoms replace/occupy the formation energies of Sb, Te atoms, Sb vacancies, and intervals in the FCC-Sb$_2$Te$_3$, respectively.

It is found from the article [18,25–28] that C atoms are not simply doped in these four ways, but form C molecular clusters (such as C chain and/or C ring) on the crystal plane. The Σ3 twin grain boundary in an annealed C-Ge$_2$Sb$_2$Te$_5$ alloy accounts for 7.49% of the total polycrystalline structure [28]. Figure 2a shows the FCC-Sb$_2$Te$_3$ with 160 atoms of Σ3 twin grain boundary, and the thickness of the vacuum layer in the c-axis direction is 10 Å.

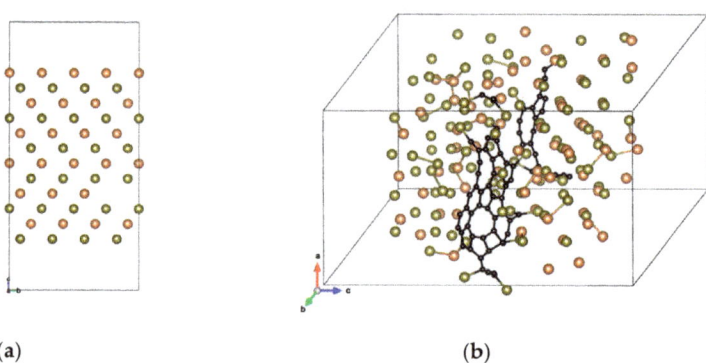

Figure 2. (a) The Σ3 twin grain boundary FCC-Sb$_2$Te$_3$ containing 160 atoms, and (b) the C atoms gradually converge at the crystal plane to form C molecular clusters.

After the structural relaxation, we found that the C atoms have a tendency to gradually converge together and form the C chain and/or C ring [29], as shown in Figure 2b. The longer the C chain, the more C rings, and the lower its formation energy. On the contrary, the more dispersed the C atom is, the higher its formation energy is. Figure 3 shows the formation energies of different C doping contents and forms calculated by Equation (1). As can be seen in Figure 3, comparing the formation energy of a single crystal structure (See Figure 1a) and a twin structure (See Figure 2a), the latter is lower, indicating that C atoms are more inclined to stay at the grain boundary than to replace Sb/Te atoms or occupy intervals/Sb cation vacancies in a single crystal. It is further analyzed that with the increase in C doping concentration, or with the increase in C chain and/or C ring, the formation energy of Sb$_2$Te$_3$ can gradually decrease, indicating that C atoms tend to converge to each other to form a C chain and/or C ring, which mainly exists in the form of a C chain and/or C ring at the twin grain boundary, as shown in Figure S1.

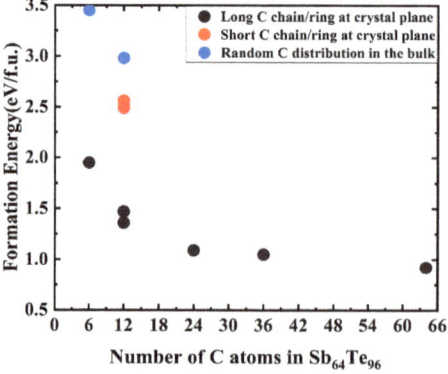

Figure 3. The formation energy of the Sb$_2$Te$_3$ structure doped with different contents and different forms of C atoms/molecular clusters. The percentages of C atoms are 3.61%, 6.98%, 13.04%, 18.37%, and 28.57%, respectively.

It can also be seen in Figure 3 that although the formation energy of C-Sb$_2$Te$_3$ gradually decreases with the increase in C doping concentration, its value is still greater than 0 (the formation energy of C$_{64}$Sb$_{64}$Te$_{96}$ is 0.92 eV/f.u.), indicating that the structure is not easy to form or unstable. Perhaps this is the reason why the FCC structure was not found in the previous reports of Sb$_2$Te$_3$ doped with carbon, yet the hexagonal (HEX) structure was found. Yin et al. reported that there was no FCC structure in the experiment of C-doped Sb$_2$Te$_3$, but the FCC structure was observed in the experiment of N-doped Sb$_2$Te$_3$ [30] and C-N co-doped Sb$_2$Te$_3$ [31]. The samples of pure Sb$_2$Te$_3$ and Sb$_2$Te$_3$ films doped with different carbon contents were prepared, and their crystal structures were analyzed by XRD, as shown in Figure 4. It can be seen in Figure 4 that when Sb$_2$Te$_3$ is at 225 °C, the FCC phase and the HEX phase coexist, which indicates that Sb$_2$Te$_3$ starts to change from the FCC phase to the HEX phase at this time. In C$_{40W}$Sb$_2$Te$_3$, there is no HEX phase at 200 and 225 °C, but the characteristic peak of the FCC phase appears, and the transition from the FCC phase to the HEX phase begins at 250 °C. However, the characteristic peak of the FCC phase was not observed at 225–250 °C in C$_{20W}$Sb$_2$Te$_3$, which indicated that with the increase in C doping concentration, the formation energy of C-Sb$_2$Te$_3$ decreased, so that the FCC phase can appear in C-doped Sb$_2$Te$_3$, and the FCC phase is likely to be stable in the C-Sb$_2$Te$_3$ structure as the C concentration increases. The formation energy of the metastable FCC-Sb$_2$Te$_3$ structure constructed by us is calculated to be 0.04eV/f.u., which is close to kT = 0.026 eV at room temperature. However, the formation energy of the FCC-Sb$_2$Te$_3$ structure is greater than 0, which also indicates that its stability is poor. In order to further understand the chemical stability of the Sb$_2$Te$_3$ structure, we performed the COHP analysis for Sb$_2$Te$_3$, as shown in Figure 5. The upper and lower portions of the -COHP curve indicate bonding (stable) and anti-bonding (unstable) interactions, respectively. From the -COHP of Sb$_2$Te$_3$ in Figure 5, the existence of the anti-bonding state of Sb-Te atoms below Fermi level (E$_f$) also indicates that the stability of the FCC-Sb$_2$Te$_3$ structure is poor [32], where the cutoff distance of an Sb-Te bond is 3.1 Å [18,28,33,34]. Kolobov et al. theoretically predicted and constructed the structure model of the FCC-Sb$_2$Te$_3$ in 2013 [35], but it was not until 2016 that Zheng et al. observed the structure of the FCC-Sb$_2$Te$_3$ in the experiment for the first time using TEM analysis [34].

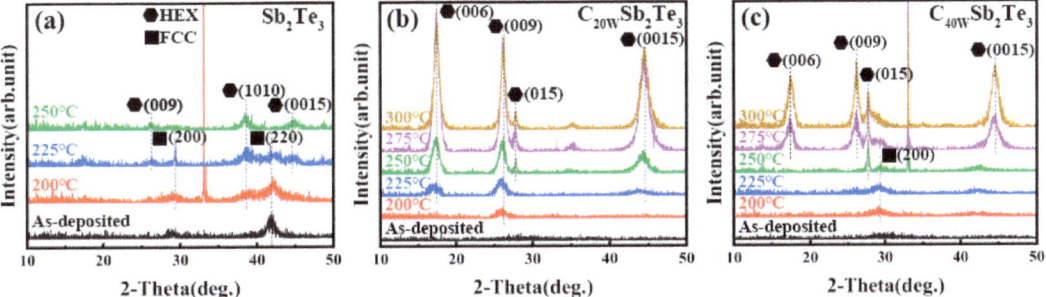

Figure 4. An XRD diagram of Sb$_2$Te$_3$ and C-Sb$_2$Te$_3$ in the experiment. (a) Sb$_2$Te$_3$; (b) C$_{20W}$Sb$_2$Te$_3$; and (c) C$_{40W}$Sb$_2$Te$_3$ were annealed for 5 min at different temperatures in a N$_2$ atmosphere.

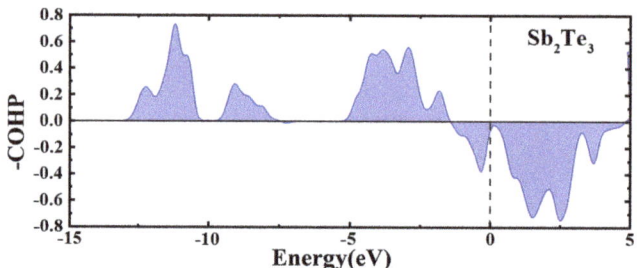

Figure 5. The -COHP curve of Σ3 twin boundary FCC-Sb_2Te_3.

3.2. Electronic Properties and Origin of Change of Crystalline C-doped Sb_2Te_3

To understand the mechanism of the thermal stability improvement of C-doped Sb_2Te_3, contour plots of electron localization function (ELF) projected on the same planes for Sb_2Te_3 and 64C-Sb_2Te_3 are shown in Figure 6, and the ELF maxima of various bonds are shown in Table 1. It is shown that the ELF maximum of the C-C bond is much higher than that of other bonds, indicating that the strength of the C-C bond is very high, which also proves that C atoms mainly exist in the form of C molecular clusters in Sb_2Te_3. The ELF maximum values of C-Te and C-Sb bonds are obviously much higher than 0.5, which indicates that C-Te and C-Sb bonds have high strength; that is, there are some molecular clusters containing C-Te and C-Sb. In addition, it also shows that the doping of the C atoms changes the local environment of each element in Sb_2Te_3 and increases the strength of the Sb-Te covalent bond, thus obviously improving the stability of Sb_2Te_3 after C doping [36,37].

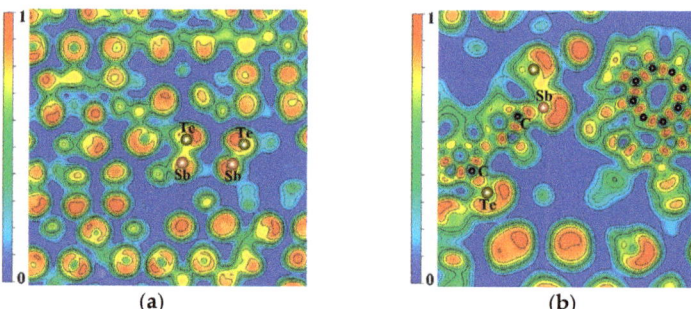

(a) (b)

Figure 6. Two-dimensional electronic local function contour plots of Sb_2Te_3 and C-Sb_2Te_3. (a) Σ3 twin grain boundary FCC-Sb_2Te_3 without doping C atoms and (b) Σ3 twin grain boundary FCC-Sb_2Te_3-doped C atoms.

Table 1. The maximum electronic local function value of each covalent bond in the FCC-Sb_2Te_3 with Σ3 twin grain boundary of undoped/doped C atoms in the direction of bond length.

Bonds	Sb_2Te_3	C-Sb_2Te_3
Sb-Te	0.78	0.80
C-C	/	0.94
C-Sb	/	0.91
C-Te	/	0.89

In order to represent the effect of C element doping on the amorphous thermal stability of Sb_2Te_3 materials, we tested the failure time of thin film materials at different temperatures. The failure time is considered to be that the film resistance drops to half of the initial resistance at this set temperature T. As shown in Figure 7, the 10-year (and 100-year) data retention is estimated according to the Arrhenius equation:

$$t = \tau \exp\left(\frac{E_a}{k_b T}\right), \quad (2)$$

where t is the failure time of the film at a set temperature T, τ is the preexponential factor, E_a is the activation energy, and K_b is the Boltzmann constant. It can be seen in Figure 7 that the addition of C atoms obviously improves the 10-year (or 100-year) data retention of Sb_2Te_3.

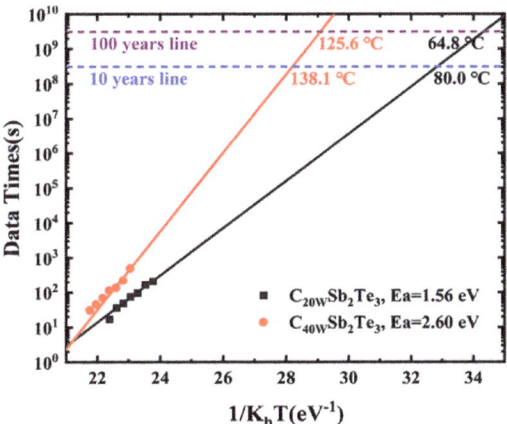

Figure 7. The 10-year (or 100-year) data retention temperature and activation energy of crystallization are deduced based on the Arrhenius equation, according to the failure time versus reciprocal temperature.

Figure 8a shows the pair correlation function (PCF) of the C-C bond and the Sb-Te bond in crystal Sb_2Te_3 and $C-Sb_2Te_3$ after structural relaxation. For the first peak, we found that the peak value of the C-C bond is far greater than that of the Sb-Te bond, C-Sb bond, C-Te bond, etc., indicating that C atoms are more inclined to combine with C atoms in Sb_2Te_3 to form C molecular clusters. To our surprise, the position of the first peak of the C-C bond is 1.406 Å, which is very close to the inter-layer atomic spacing of 1.42 Å in graphite structure [38,39]. In addition, the maximum bond angle distribution of the C-C-C configuration in C molecular clusters is about 105–125°, and the coordination number of C atoms is mainly 3-coordinate, as shown in Figure 8b,c. These indicate that the doped C atoms in Sb_2Te_3 are not randomly formed when forming C molecular clusters, but tend to form graphite-like layered structures by sp^2 hybridization [18,28]. The low PCF peaks of the C-Sb and C-Te bonds in Figure S2 indicate that C atoms bond less with Sb and Te atoms. It is observed that after doping the C atom in Sb_2Te_3, the position of the first peak value of the Sb-Te bond decreases and the bond length becomes shorter, indicating that the binding between Sb and Te atoms is strengthened, thus making the structure more stable. Meanwhile, the extremely unstable Sb-Sb homobonds are reduced, which also contributes to the enhancement of structural stability, as shown in Figure S2.

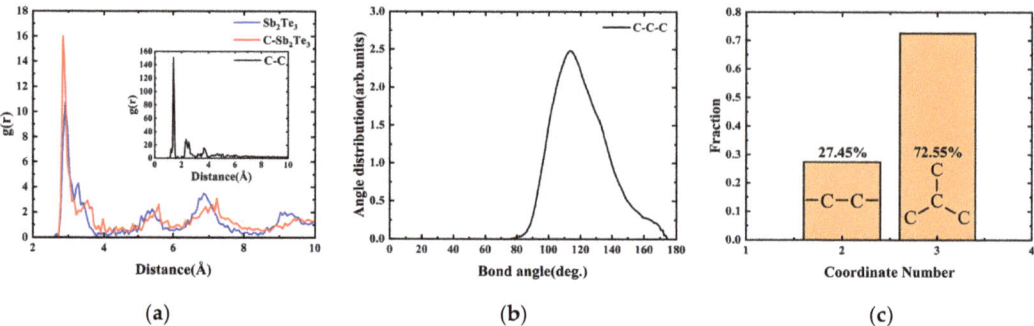

Figure 8. (**a**) Pair correlation function (PCF) of the Sb-Te bond and the C-C bond in crystal Sb_2Te_3 and C-Sb_2Te_3 after structural relaxation; (**b**) bond angle distribution of the C-C-C configuration; and (**c**) coordination number statistics of C atoms.

3.3. Electrical Performance Test of a Prototype PCRAM Device Based on C-Sb_2Te_3 Material

The electrical programming characteristics of the C-Sb_2Te_3 prototype PCRAM device are shown in Figure 9a. The illustration is the voltage pulse we applied to the PCRAM cell and the pulse width is fixed and the voltage amplitude step is set to 0.1 V. Starting from the first pulse amplitude of 0.1 V until the PCRAM cell stops after the SET and the RESET, as illustrated in Figure 9a, a reading voltage of 0.1 V is applied between every two pulses to record the resistance value of the PCRAM test cell. The initial state resistance value and the final state resistance value are controlled to be equal as much as possible. Obviously, the final state resistance value before adjusting the voltage pulse width is taken as the initial state resistance value after adjusting the voltage pulse width, and the initial state resistance value will affect the SET voltage value and even the RESET voltage value of this electrical programming. It can be seen from Figure 9a that the resistance resolution of the PCRAM cells exceeds two orders of magnitude, which meets the requirements of PCRAM. Our PCRAM device cells can be programmed at a very low SET/RESET voltage with a pulse width of 500 ns–5 ns, and the SET voltage and RESET voltage are as low as 1.5 V and 2.2 V when the voltage pulse width is 6 ns, which indicates that our PCRAM device has an operating speed of 5 ns and a low device power consumption (0.57 pJ), as shown in Figure 9b. Further observation in Figure 9b shows that the RESET power consumption of the PCRAM device cell decreases with the decrease in the width of the programming pulse. At the same time, it was noted that the PCRAM device cell could not be completely SET operated, as the width of the programming pulse decreased, resulting in the resistance value of its low resistance state to increase. This explains the behavior of the RESET power consumption change of the PCRAM device cell. However, the PCRAM device cell cannot be completely SET, which may be caused by the FCC phase appearing before the amorphous phase is transformed into the HEX phase due to the high C doping concentration. The multistage storage function can be realized by setting different voltage pulse widths and voltage pulse amplitudes, such as "0" at 10^7 Ω, "1" at 10^5 Ω, and "2" at 10^4 Ω [40]. We also noticed that during the SET/RESET process of our PCRAM device, there was a continuous resistance change. In this case, we used a voltage pulse with a pulse width of 500 ns to RESET the PCRAM in a low resistance state, and the continuously adjustable resistance value can be obtained as shown in Figure 10a. With this change, maybe we can realize several basic synaptic functions at the cell level, including long-term plasticity (LTP) [41,42], short-term plasticity (STP) [41,42], spike timing-dependent plasticity (STDP) [43,44], and spike rate-dependent plasticity (SRDP) [44,45], and maybe can also realize more complex or higher-order learning behaviors at the network level, such as supervised learning [46] and associative learning [47], as well as non-von Neumann architecture of in-memory computing [48,49]. In general, for this phenomenon of continuous resistance change, the resistance drift caused by the widening of the band gap due to the structural relaxation

(SR) of amorphous Sb_2Te_3 is a great obstacle to multilevel storage, neuromorphic learning and in-memory computing. Li et al. greatly reduced this resistance drift phenomenon by bipolar pulse operation on the PCRAM cell [50], which provides an effective means to improve the stability of the phase-change neuromorphic applications. We adjusted the resistance values of the PCRAM cell to high and low resistance states and each intermediate resistance state by controlling the electrical signal applied to the PCRAM cell, and fitting the resistance drift coefficient by Equation (3):

$$R(t) = R_0 \left(\frac{t}{t_0}\right)^\nu, \quad (3)$$

where R_0 is the resistance value of the PCRAM cell at time t_0; that is, the initial resistance. ν is the resistance drift coefficient, indicating the change in the resistance value of the PCRAM cell with time. The results measured at room temperature are shown in Figure 10b.

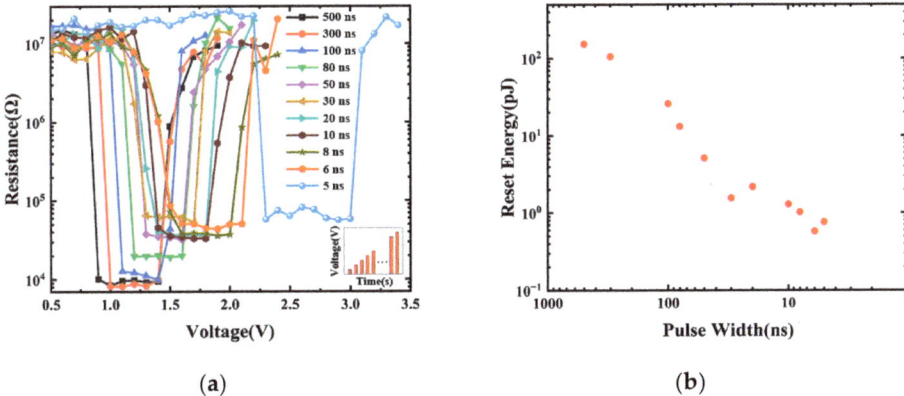

Figure 9. (a) The R-V electrical test of the prototype PCRAM device based on C-Sb_2Te_3 material, and (b) the relationship between power consumption and pulse width.

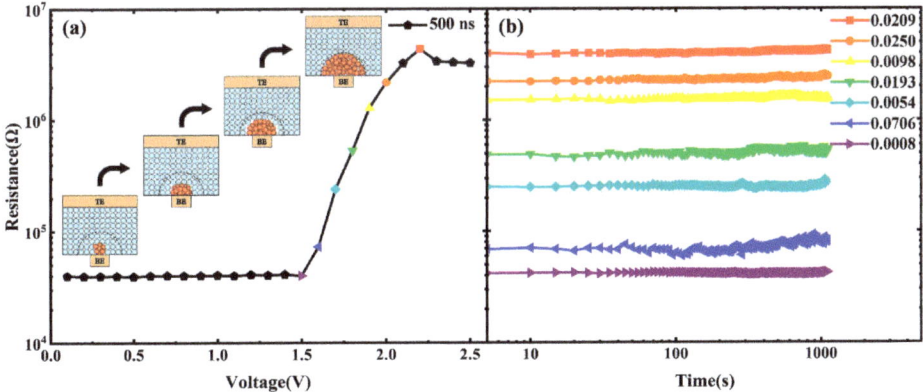

Figure 10. (a) The R-V curve is obtained by resetting the voltage pulse with a pulse width of 500 ns, and (b) the R-t curve and the resistance drift coefficient ν is obtained by fitting with Equation (2).

4. Conclusions

We have obtained an understanding of the doping position and existence form of C atoms in the FCC-Sb_2Te_3 and the improved device performance after C doping by performing density functional theory calculations for different concentrations and forms of C doping in single crystal FCC-Sb_2Te_3 and Σ3 twin boundary FCC-Sb_2Te_3. The results show that the formation energy of C-Sb_2Te_3 decreases with the increase in C doping concentration, which is consistent with the appearance of the FCC phase in high-concentration C-doped Sb_2Te_3 in our experiment. In addition, C atoms prefer to form C molecule clusters by sp^2 hybridization at the grain boundary of Sb_2Te_3, similar to the layered structure of graphite, which changes the local environment of each element in Sb_2Te_3, resulting in the improvement of thermal stability of Sb_2Te_3. We fabricated the prototype device cells of PCRAM, which had an operating speed of 5 ns, a high thermal stability (10-year data retention temperature 138.1 °C), a low device power consumption of 0.57 pJ, and a resistance drift coefficient as low as 0.025, showing a continuously adjustable resistance. These performances all indicate that C-Sb_2Te_3-based PCRAM devices have great potential in applications, such as multilevel storage and spiking neural networks.

Supplementary Materials: The following supporting information can be downloaded at https://www.mdpi.com/article/10.3390/nano13040671/s1, Figure S1: After structural relaxation of different C doping contents and C existing forms; Figure S2: The pair correlation function after structure relaxation; Figure S3: XRD diagram of $C_{40W}Sb_2Te_3$ at 225 °C; Figure S4: The optical image of the fabricated PCRAM cells; Figure S5: I-V curve of PCRAM unit set from high resistance state to low resistance state by DC current; Figure S6: SET the PCRAM in a low resistance state with a large/small set pulse, and then RESET it with a 500 ns pulse to obtain a continuously adjustable resistance value.

Author Contributions: Conceptualization, L.W. and Z.S.; methodology, J.Z., L.W. and P.X.; software, validation, J.Z., Y.X., A.L., W.S. and Y.L.; formal analysis, J.Z.; investigation, J.Z. and N.R.; resources, L.W., Z.S. and S.S.; data curation, writing—original draft preparation, J.Z.; writing—review and editing, L.W., N.R. and A.L.; visualization, J.Z.; supervision, project administration, funding acquisition, L.W. and Z.S. All authors have read and agreed to the published version of the manuscript.

Funding: This research was funded by the National Natural Science Foundation of China (grant No. 61874151), the Science and Technology Council of Shanghai (grant. No. 19JC1416802, 22ZR1402200), and the Fundamental Research Funds for the Central Universities (grant. No. 2232022A-06).

Institutional Review Board Statement: Not applicable.

Informed Consent Statement: Not applicable.

Data Availability Statement: Not applicable.

Acknowledgments: The authors acknowledge the financial support from the State Key Laboratory of Functional Materials for Informatics, the Shanghai Institute of Microsystem and Information, the Chinese Academy of Sciences, and the College of Science, Donghua University.

Conflicts of Interest: The authors declare no conflict of interest.

References

1. Fong, S.W.; Neumann, C.M.; Wong, H.S.P. Phase-Change Memory—Towards a Storage-Class Memory. *IEEE Trans. Electron Devices* **2017**, *64*, 4374–4385. [CrossRef]
2. Wong, H.S.P.; Salahuddin, S. Memory leads the way to better computing. *Nat. Nanotechnol.* **2015**, *10*, 191–194. [CrossRef]
3. Ung, G.M. Intel, Micron Announce New 3D XPoint Memory Type That's 1,000 Times Faster than NAND, Really. Available online: https://www.pcworld.com/article/422680/intel-micron-announce-new-3dxpoint-memory-type-thats-1000-times-faster-than-nand.html (accessed on 1 March 2022).
4. Rao, F.; Ding, K.; Zhou, Y.; Zheng, Y.; Xia, M.; Lv, S.; Song, Z.; Feng, S.; Ronneberger, I.; Mazzarello, R.; et al. Reducing the stochasticity of crystal nucleation to enable subnanosecond memory writing. *Science* **2017**, *358*, 1423–1427. [CrossRef] [PubMed]
5. Zhao, J.; Yuan, Z.H.; Song, W.X.; Song, Z.T. High performance of Er-doped Sb_2Te material used in phase change memory. *J. Alloy. Compd.* **2021**, *889*, 161701. [CrossRef]
6. Zuliani, P.; Varesi, E.; Palumbo, E.; Borghi, M.; Tortorelli, I.; Erbetta, D.; Libera, G.D.; Pessina, N.; Gandolfo, A.; Prelini, C.; et al. Overcoming Temperature Limitations in Phase Change Memories With Optimized $Ge_xSb_yTe_z$. *IEEE Trans. Electron Devices* **2013**, *60*, 4020–4026. [CrossRef]

7. Diaz Fattorini, A.; Cheze, C.; Lopez Garcia, I.; Petrucci, C.; Bertelli, M.; Righi Riva, F.; Prili, S.; Privitera, S.M.S.; Buscema, M.; Sciuto, A.; et al. Growth, Electronic and Electrical Characterization of Ge-Rich Ge-Sb-Te Alloy. *Nanomaterials* **2022**, *12*, 1340. [CrossRef]
8. Xu, B.; Zhang, J.; Yu, G.; Ma, S.; Wang, Y.; Wang, Y. Thermoelectric properties of monolayer Sb_2Te_3. *J. Appl. Phys.* **2018**, *124*, 15104. [CrossRef]
9. Verma, S.K.; Kandpal, K.; Kumar, P.; Kumar, A.; Wiemer, C. Performance of Topological Insulator (Sb_2Te_3)-Based Vertical Stacking Photodetector on n-Si Substrate. *IEEE Trans. Electron Devices* **2022**, *69*, 4342–4348. [CrossRef]
10. Bryja, H.; Gerlach, J.W.; Prager, A.; Ehrhardt, M.; Rauschenbach, B.; Lotnyk, A. Epitaxial layered Sb_2Te_3 thin films for memory and neuromorphic applications. *2D Mater.* **2021**, *8*, 045027. [CrossRef]
11. Sun, Z.; Zhou, J.; Blomqvist, A.; Johansson, B.; Ahuja, R. Formation of large voids in the amorphous phase-change memory $Ge_2Sb_2Te_5$ alloy. *Phys. Rev. Lett.* **2009**, *102*, 075504. [CrossRef]
12. Li, L.; Song, S.N.; Xue, Y.; Chen, X.; Zhao, J.; Shen, J.B.; Song, Z.T. Thermally stable tungsten and titanium doped antimony tellurium films for phase change memory application. *J. Mater. Sci.-Mater. Electron.* **2020**, *31*, 10912–10918. [CrossRef]
13. Ren, K.; Rao, F.; Song, Z.T.; Lv, S.L.; Cheng, Y.; Wu, L.C.; Peng, C.; Zhou, X.L.; Xia, M.J.; Liu, B.; et al. Pseudobinary Al_2Te_3-Sb_2Te_3 material for high speed phase change memory application. *Appl. Phys. Lett.* **2012**, *100*, 052105. [CrossRef]
14. Zhang, T.; Zhang, K.; Wang, G.; Greenberg, E.; Prakapenka, V.B.; Yang, W.E. Multiple phase transitions in Sc doped Sb_2Te_3 amorphous nanocomposites under high pressure. *Appl. Phys. Lett.* **2020**, *116*, 021903. [CrossRef]
15. Hu, S.W.; Liu, B.; Li, Z.; Zhou, J.; Sun, Z.M. Identifying optimal dopants for Sb_2Te_3 phase-change material by high-throughput ab initio calculations with experiments. *Comput. Mater. Sci.* **2019**, *165*, 51–58. [CrossRef]
16. Zhu, M.; Wu, L.; Song, Z.; Rao, F.; Cai, D.; Peng, C.; Zhou, X.; Ren, K.; Song, S.; Liu, B.; et al. $Ti_{10}Sb_{60}Te_{30}$ for phase change memory with high-temperature data retention and rapid crystallization speed. *Appl. Phys. Lett.* **2012**, *100*, 122101. [CrossRef]
17. Cheng, Y.; Cai, D.; Zheng, Y.; Yan, S.; Wu, L.; Li, C.; Song, W.; Xin, T.; Lv, S.; Huang, R.; et al. Microscopic Mechanism of Carbon-Dopant Manipulating Device Performance in CGeSbTe-Based Phase Change Random Access Memory. *ACS Appl. Mater. Interfaces* **2020**, *12*, 23051–23059. [CrossRef]
18. Zhou, X.; Xia, M.; Rao, F.; Wu, L.; Li, X.; Song, Z.; Feng, S.; Sun, H. Understanding phase-change behaviors of carbon-doped $Ge_2Sb_2Te_5$ for phase-change memory application. *ACS Appl. Mater. Interfaces* **2014**, *6*, 14207–14214. [CrossRef]
19. Sun, Z.; Zhou, J.; Mao, H.K.; Ahuja, R. Peierls distortion mediated reversible phase transition in GeTe under pressure. *Proc. Natl. Acad. Sci. USA* **2012**, *109*, 5948–5952. [CrossRef]
20. Li, Z.; Si, C.; Zhou, J.; Xu, H.; Sun, Z. Yttrium-Doped Sb_2Te_3: A Promising Material for Phase-Change Memory. *ACS Appl. Mater. Interfaces* **2016**, *8*, 26126–26134. [CrossRef]
21. Zhao, Z.Y.; Peng, S.; Tan, Z.L.; Wang, C.J.; Wen, M. Doping effects of Ru on Sb_2Te and Sb_2Te_3 as phase change materials studied by first-principles calculations. *Mater. Today Commun.* **2022**, *31*, 103669. [CrossRef]
22. Lee, T.H.; Elliott, S.R. The Relation between Chemical Bonding and Ultrafast Crystal Growth. *Adv. Mater.* **2017**, *29*, 1700814. [CrossRef] [PubMed]
23. Lee, T.H.; Loke, D.; Elliott, S.R. Microscopic Mechanism of Doping-Induced Kinetically Constrained Crystallization in Phase-Change Materials. *Adv. Mater.* **2015**, *27*, 5477–5483. [CrossRef] [PubMed]
24. Maintz, S.; Deringer, V.L.; Tchougréeff, A.L.; Dronskowski, R. Analytic projection from plane-wave and PAW wavefunctions and application to chemical-bonding analysis in solids. *J. Comput. Chem.* **2013**, *34*, 2557–2567. [CrossRef] [PubMed]
25. Wang, G.J.; Zhou, J.; Elliott, S.R.; Sun, Z.M. Role of carbon-rings in polycrystalline $GeSb_2Te_4$ phase-change material. *J. Alloy. Compd.* **2019**, *782*, 852–858. [CrossRef]
26. Zhou, X.; Wu, L.; Song, Z.; Rao, F.; Zhu, M.; Peng, C.; Yao, D.; Song, S.; Liu, B.; Feng, S. Carbon-doped $Ge_2Sb_2Te_5$ phase change material: A candidate for high-density phase change memory application. *Appl. Phys. Lett.* **2012**, *101*, 142104. [CrossRef]
27. Guo, T.; Song, S.; Zheng, Y.; Xue, Y.; Yan, S.; Liu, Y.; Li, T.; Liu, G.; Wang, Y.; Song, Z.; et al. Excellent thermal stability owing to Ge and C doping in Sb_2Te-based high-speed phase-change memory. *Nanotechnology* **2018**, *29*, 505710. [CrossRef]
28. Song, W.X.; Cheng, Y.; Cai, D.L.; Tang, Q.Y.; Song, Z.T.; Wang, L.H.; Zhao, J.; Xin, T.J.; Liu, Z.P. Improving the performance of phase-change memory by grain refinement. *J. Appl. Phys.* **2020**, *128*, 075101. [CrossRef]
29. Zheng, L.; Song, W.X.; Zhang, S.F.; Song, Z.T.; Zhu, X.Q.; Song, S.N. Designing artificial carbon clusters using $Ge_2Sb_2Te_5$/C superlattice-like structure for phase change applications. *J. Alloy. Compd.* **2021**, *882*, 160695. [CrossRef]
30. Yin, Y.; Higano, N.; Ohta, K.; Miyachi, A.; Asai, M.; Niida, D.; Sone, H.; Hosaka, S. Characteristics of N-doped Sb_2Te_3 films by X-ray diffraction and resistance measurement for phase-change memory. In Proceedings of the Symposium on Materials and Processes for Nonvolatile Memories II Held at the 2007 MRS Spring Meeting, San Francisco, CA, USA, 10–13 April 2007; p. 287.
31. Yin, Y. C–N-codoped Sb_2Te_3 chalcogenides for reducing writing current of phase-change devices. *J. Appl. Phys.* **2021**, *117*, 153502. [CrossRef]
32. Zhao, J.; Song, W.X.; Xin, T.; Song, Z. Rules of hierarchical melt and coordinate bond to design crystallization in doped phase change materials. *Nat. Commun.* **2021**, *12*, 6473. [CrossRef]
33. Sun, Z.; Zhou, J.; Pan, Y.; Song, Z.; Mao, H.K.; Ahuja, R. Pressure-induced reversible amorphization and an amorphous-amorphous transition in $Ge_2Sb_2Te_5$ phase-change memory material. *Proc. Natl. Acad. Sci. USA* **2011**, *108*, 10410–10414. [CrossRef]
34. Zheng, Y.; Xia, M.; Cheng, Y.; Rao, F.; Ding, K.; Liu, W.; Jia, Y.; Song, Z.; Feng, S. Direct observation of metastable face-centered cubic Sb_2Te_3 crystal. *Nano Res.* **2016**, *9*, 3453–3462. [CrossRef]

35. Kolobov, A.V.; Fons, P.; Tominaga, J.; Ovshinsky, S.R. Vacancy-mediated three-center four-electron bonds in GeTe-Sb$_2$Te$_3$ phase-change memory alloys. *Phys. Rev. B* **2013**, *87*, 165206. [CrossRef]
36. Zhou, W.Y.; Wu, L.C.; Zhou, X.L.; Rao, F.; Song, Z.T.; Yao, D.N.; Yin, W.J.; Song, S.N.; Liu, B.; Qian, B.; et al. High thermal stability and low density variation of carbon-doped Ge$_2$Sb$_2$Te$_5$ for phase-change memory application. *Appl. Phys. Lett.* **2014**, *105*, 243113. [CrossRef]
37. Li, T.; Shen, J.B.; Wu, L.C.; Song, Z.T.; Lv, S.L.; Cai, D.L.; Zhang, S.F.; Guo, T.Q.; Song, S.N.; Zhu, M. Atomic-Scale Observation of Carbon Distribution in High-Performance Carbon-Doped Ge$_2$Sb$_2$Te$_5$ and Its Influence on Crystallization Behavior. *J. Phys. Chem. C* **2019**, *123*, 13377–13384. [CrossRef]
38. Bernal, J.D. The structure of graphite. *Proc. R. Soc. London. Ser. A Contain. Pap. A Math. Phys. Character* **1924**, *106*, 749–773. [CrossRef]
39. Chung, D. Review graphite. *J. Mater. Sci.* **2002**, *37*, 1475–1489. [CrossRef]
40. Jin, S.M.; Kang, S.Y.; Kim, H.J.; Lee, J.Y.; Nam, I.H.; Shim, T.H.; Song, Y.H.; Park, J.G. Sputter-grown GeTe/Sb$_2$Te$_3$ superlattice interfacial phase change memory for low power and multi-level-cell operation. *Electron. Lett.* **2022**, *58*, 38–40. [CrossRef]
41. Chee, M.Y.; Dananjaya, P.A.; Lim, G.J.; Du, Y.; Lew, W.S. Frequency-Dependent Synapse Weight Tuning in 1S1R with a Short-Term Plasticity TiOx-Based Exponential Selector. *ACS Appl. Mater. Interfaces* **2022**, *14*, 35959–35968. [CrossRef]
42. Sarwat, S.G.; Kersting, B.; Moraitis, T.; Jonnalagadda, V.P.; Sebastian, A. Phase-change memtransistive synapses for mixed-plasticity neural computations. *Nat. Nanotechnol.* **2022**, *17*, 507–513. [CrossRef]
43. Kuzum, D.; Jeyasingh, R.G.; Lee, B.; Wong, H.S. Nanoelectronic programmable synapses based on phase change materials for brain-inspired computing. *Nano Lett.* **2012**, *12*, 2179–2186. [CrossRef] [PubMed]
44. Milo, V.; Pedretti, G.; Laudato, M.; Bricalli, A.; Ambrosi, E.; Bianchi, S.; Chicca, E.; Ielmini, D. Resistive switching synapses for unsupervised learning in feed-forward and recurrent neural networks. In Proceedings of the 2018 IEEE International Symposium on Circuits and Systems (ISCAS), Florence, Italy, 27–30 May 2018.
45. Milo, V.; Pedretti, G.; Carboni, R.; Calderoni, A.; Ramaswamy, N.; Ambrogio, S.; Ielmini, D. A 4-Transistors/1-Resistor Hybrid Synapse Based on Resistive Switching Memory (RRAM) Capable of Spike-Rate-Dependent Plasticity (SRDP). *IEEE Trans. Very Large Scale Integr. (VLSI) Syst.* **2018**, *26*, 2806–2815. [CrossRef]
46. Nandakumar, S.R.; Boybat, I.; Le Gallo, M.; Eleftheriou, E.; Sebastian, A.; Rajendran, B. Experimental Demonstration of Supervised Learning in Spiking Neural Networks with Phase-Change Memory Synapses. *Sci. Rep.* **2020**, *10*, 8080. [CrossRef] [PubMed]
47. Cheng, Y.K.; Lin, Y.; Zeng, T.; Shan, X.Y.; Wang, Z.Q.; Zhao, X.N.; Ielmini, D.; Xu, H.Y.; Liu, Y.C. Pavlovian conditioning achieved via one-transistor/one-resistor memristive synapse. *Appl. Phys. Lett.* **2022**, *120*, 133503. [CrossRef]
48. Sebastian, A.; Le Gallo, M.; Eleftheriou, E. Computational phase-change memory: Beyond von Neumann computing. *J. Phys. D Appl. Phys.* **2019**, *52*, 443002. [CrossRef]
49. Wright, C.D.; Hosseini, P.; Diosdado, J.A.V. Beyond von-Neumann Computing with Nanoscale Phase-Change Memory Devices. *Adv. Funct. Mater.* **2012**, *23*, 2248–2254. [CrossRef]
50. Li, X.; He, Q.; Tong, H.; Miao, X.S. Resistance Drift-Reduced Multilevel Storage and Neural Network Computing in Chalcogenide Phase Change Memories by Bipolar Operation. *IEEE Electron Device Lett.* **2022**, *43*, 565–568. [CrossRef]

Disclaimer/Publisher's Note: The statements, opinions and data contained in all publications are solely those of the individual author(s) and contributor(s) and not of MDPI and/or the editor(s). MDPI and/or the editor(s) disclaim responsibility for any injury to people or property resulting from any ideas, methods, instructions or products referred to in the content.

 nanomaterials

Communication

The Relationship between Electron Transport and Microstructure in Ge$_2$Sb$_2$Te$_5$ Alloy

Cheng Liu [1], Yonghui Zheng [1,2,*], Tianjiao Xin [1], Yunzhe Zheng [1], Rui Wang [1] and Yan Cheng [1]

[1] Key Laboratory of Polar Materials and Devices (MOE), Department of Electronics, East China Normal University, Shanghai 200241, China
[2] Chongqing Key Laboratory of Precision Optics, Chongqing Institute of East China Normal University, Chongqing 401120, China
* Correspondence: yhzheng@phy.ecnu.edu.cn

Abstract: Phase-change random-access memory (PCRAM) holds great promise for next-generation information storage applications. As a mature phase change material, Ge$_2$Sb$_2$Te$_5$ alloy (GST) relies on the distinct electrical properties of different states to achieve information storage, but there are relatively few studies on the relationship between electron transport and microstructure. In this work, we found that the first resistance dropping in GST film is related to the increase of carrier concentration, in which the atomic bonding environment changes substantially during the crystallization process. The second resistance dropping is related to the increase of carrier mobility. Besides, during the cubic to the hexagonal phase transition, the nanograins grow significantly from ~50 nm to ~300 nm, which reduces the carrier scattering effect. Our study lays the foundation for precisely controlling the storage states of GST-based PCRAM devices.

Keywords: phase change memory; Ge$_2$Sb$_2$Te$_5$; phase transition; electron transport

Citation: Liu, C.; Zheng, Y.; Xin, T.; Zheng, Y.; Wang, R.; Cheng, Y. The Relationship between Electron Transport and Microstructure in Ge$_2$Sb$_2$Te$_5$ Alloy. *Nanomaterials* **2023**, *13*, 582. https://doi.org/10.3390/nano13030582

Academic Editor: Jeremy Sloan

Received: 15 December 2022
Revised: 20 January 2023
Accepted: 26 January 2023
Published: 31 January 2023

Copyright: © 2023 by the authors. Licensee MDPI, Basel, Switzerland. This article is an open access article distributed under the terms and conditions of the Creative Commons Attribution (CC BY) license (https://creativecommons.org/licenses/by/4.0/).

1. Introduction

Phase-change random-access memory (PCRAM) is one of the most promising and mature new memory technologies due to its high operating speed, cycle life comparable to that of dynamic random-access memory, and large operation window [1–3]. As the core of PCRAM, phase-change materials (PCMs) store information by the huge difference in resistivity between amorphous (high resistance) and crystalline state (low resistance) [4,5]. Currently, the Ge$_2$Sb$_2$Te$_5$ (GST) alloy is one of the most mature PCMs due to its excellent electrical properties, and it is also the parent phase for constructing low-power superlattice or the doped material [6,7]. There are two crystalline structures in GST materials: one is the metastable face-centered cubic (f-), and the other is the stable hexagonal (h-) structure [8,9]. In the f-phase, Te atoms occupy the anion lattice, and the Ge/Sb/20% vacancies occupy the cation lattice. With vacancy aggregation, the f-phase will transform into the h-phase at high temperatures [10,11].

Although literature investigates the microscopic storage mechanism of GST from experiments and theoretical methods, the relationship between the electronic transport property and the microstructure is not fully elucidated currently [12–15]. In GeSb$_2$Te$_4$, a similar alloy to GST, electronic measurement showed that it undergoes a metal–insulator transition mechanism, accompanied by the disorder-to-order process [16,17], while the corresponding structure variation is not well explained.

In this work, we investigated the relationship between the electric property and microstructure through resistance–temperature (R-T) curves, in-situ transmission electron microscope (TEM) heating, Hall test, and the electron backscattering diffraction (EBSD) technique. With the increasing temperature, the amorphous GST film first transformed into the f-phase, and then the h-phase, accompanied by the two sharp resistance drops. At the first drop, the carrier concentration increased substantially due to the variation of the atomic bonding environment. At the second drop, the nanograins grew significantly with an obvious [0001] texture, increasing the carrier mobility.

2. Materials and Methods

2.1. Sample Preparation

GST films (~200 nm thick) were deposited on a Si (100)/SiO$_2$ substrate by a magnetron sputtering stoichiometric Ge$_2$Sb$_2$Te$_5$ target in magnetron sputtering equipment (ACS-4000-C4, ULVAC, Kanagawa, Japan), and the films for TEM observation were deposited on TEM grids with a thickness of 30 nm. The base pressure of the vacuum system is about 1.5×10^{-4} Pa, the Ar flow rate is 50 sccm, the sputtering DC power is 50 W and the corresponding sputtering pressure is about 0.2 Pa. With the aid of energy dispersive spectroscopy (EDS) accessory (EDAX, Oxford, UK), the average concentrations of Ge, Sb, and Te atoms were estimated to be 20.7%, 29.4%, and 49.9%, respectively. Then an amorphous SiO$_2$ film (~5 nm in thickness) was deposited on top of GST films as the anti-oxidation layer.

2.2. The Hall Test

The electrical property of the film was obtained by the Hall measurement system (H-50, MMR, Baton Rouge, LA, USA) at room temperature. First of all, the sample film was cut into small squares (1×1 cm^2 to 1.3×1.3 cm^2). Each sample was annealed in Ar gas at different temperatures. The annealing temperature range was from 100 °C to 300 °C, and the interval step was 25 °C. Each sample was tested 20 times for accuracy.

2.3. Microscopic Characterization

The microstructure of the film was studied using X-ray diffraction (XRD) in an X-ray diffractometer (PANalytical X'Pert PRO, PANalytical B.V., Almelo, The Netherlands) with Cu Kα radiation source. The in-situ crystallization process of the GST film was carried out in a Gatan 628 heating holder (Gatan 628, Gatan, Pleasanton, CA, USA) with a heating rate of 20 °C/min. The bright field (BF) images and selected area electron diffraction (SAED) patterns were obtained in JEOL 2100F TEM (JEM-2100F, JEOL, Tokyo, Japan) under 200 kV. The phase structure at 180 °C and 270 °C was obtained in the focused ion beam system (Helios G4 UX, FEI, Hillsboro, OR, USA) using EBSD technology. The voltage was 30 kV and the step size was 10 nm and 50 nm, respectively.

3. Results

Figure 1a shows the resistance–temperature (R-T) curves of GST films annealed for different temperatures (100–300 °C). These samples went through a constant heating process with a rate of 5 °C/min from room temperature by the two-probe method. Before 150 °C, the resistance gradually decreased from 10^9 Ω to 10^6 Ω. Around 150 °C, a sharp resistance decrease could be observed (from 10^6 Ω to 10^4 Ω), corresponding to the amorphous-to-f phase transition. Further increasing the temperature to 250 °C, the resistance continued to decrease to 10^3 Ω. Around 250 °C, the second resistance decrease could be observed (from 10^3 Ω to 10^4 Ω), corresponding to the f-to-h phase transition. The structure of GST films at different temperatures was also investigated through XRD experiments. The results shown in Figure 1b confirmed an amorphous state at room temperature, the f-phase at 200 °C, and the h-phase at 300 °C.

The R-T curves during the cooling process are also investigated to explore the metal–insulator transition in GST films. When the annealing temperature was lower than 150 °C, the resistance of GST films backtracked to 10^9 Ω at room temperature. When the films crystallized into the f-phase, the cooling R-T curves no longer coincided with the heating R-T curve, and the temperature coefficient of resistance (TCR, yellow lines) was less than zero, demonstrating a semiconductor behavior. The higher the annealing temperature, the larger the TCR is. At 250 °C, the TCR is ~0.64, indicating that metal–insulator transition (MIT) occurs.

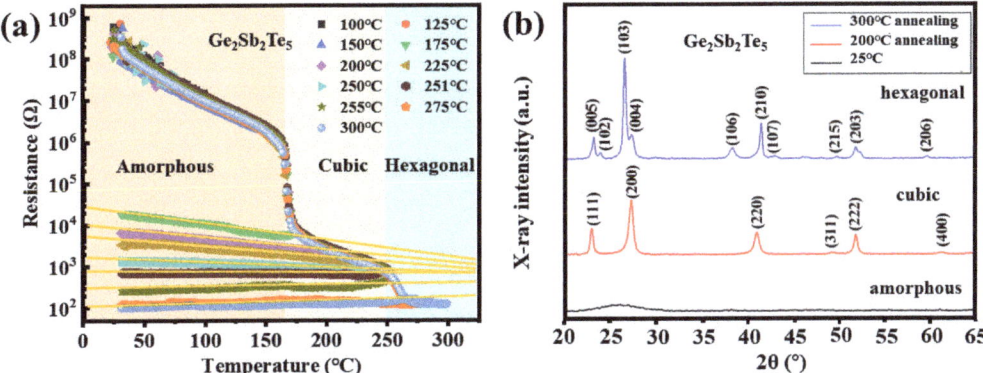

Figure 1. (a) The resistance–temperature curves of GST films annealed from 100 °C to 300 °C. (b) XRD patterns at 25 °C, 200 °C, and 300 °C, respectively.

To explore the electronic transport property in GST films, Hall experiments were conducted. Figure 2a shows the schematic of the Hall test [18], when the current, magnetic induction, and sample thickness are known, it is only necessary to measure the Hall voltage to obtain the carrier concentration and mobility. Here, we did not use the in-situ heating Hall test method because of a large experiment error. Instead, we carried out the Hall test with squared samples directly. After annealing, the structure of the sample was stable, and the obtained Hall effect parameters were relatively accurate, which reflects the relationship between the electronic transport properties and the structure. Here, the films annealed from 25 °C to 350 °C with an interval step of 25 °C were investigated. Figure 2b shows that the resistivity was ~5×10^3 $\Omega \cdot$cm at low temperatures. When the temperature rose to 150 °C, the resistivity sharply decreased to 10^0 $\Omega \cdot$cm, corresponding to the amorphous-f phase transition. As the annealing temperature increased to 225 °C, the resistivity dropped to 10^{-3} $\Omega \cdot$cm, corresponding to the f-to-h phase transition. The mismatch of the f-to-h phase transition temperature in Figures 1a and 2b should be ascribed to the two different thermal treatments, while both of them maintained a low-level resistivity at higher temperatures. Figure 2c,d shows the carrier concentration and carrier mobility of the film at different temperatures. Before 150 °C, the carrier concentration was ~5×10^{15} $\Omega \cdot$cm^{-3}. At 150 °C, the carrier concentration sharply increased to ~1×10^{19} $\Omega \cdot$cm^{-3}, which shows the same variation orders of the magnitude observed in Figure 1a. Therefore, the decrease in resistance during crystallization is mainly attributed to the rapid increase in carrier concentration. When the temperature continued to increase, the carrier concentration was maintained at a high level. Interestingly, the carrier mobility suddenly increased from 1 cm^2/(V·s) to ~50 cm^2/(V·s) at 225 °C; thus, the decrease in resistance during the f-to-h phase transition is related to the increase of carrier mobility. At high temperatures, the value maintained around ~10 cm^2/(V·s) to ~100 cm^2/(V·s).

To understand the electronic transport property from a microscopic point of view, an in-situ heating experiment of GST film inside TEM was performed as shown in Figure 3. At room temperature, the morphology of the film was uniform, and the corresponding SAED pattern showed a diffused ring (Figure 3a), demonstrating a typical amorphous state. At 140 °C, some nanograins appeared and (220) a lattice plane (a continued and sharp ring) of the f-phase was detected, demonstrating the start of crystallization. As the temperature rose to 240 °C, the nanograins and other f-phase lattice planes were more and more obvious, but note that the sizes of the nanograins were still small. Therefore, the increase of carrier concentration is mainly related to amorphous to f-phase transition. During the crystallization process, the tetrahedral Ge atoms jumped to the octahedral coordination, and the change in the constituent atom environment increased the carrier concentration. With increasing temperature, the vacancies gradually aggregated, while the variation of carrier concentration or carrier mobility was small. At 278 °C, a large grain suddenly occupied the entire top left area (bright contrast, Figure 3e). As the temperature increased to 300 °C, the grain quickly merged the nanograins throughout the whole area as shown in Figure 3f,g [19]. The corresponding SAED patterns (Figure 3) showed a typical single crystal structure, which belongs to the [0001] orientation of the h-phase.

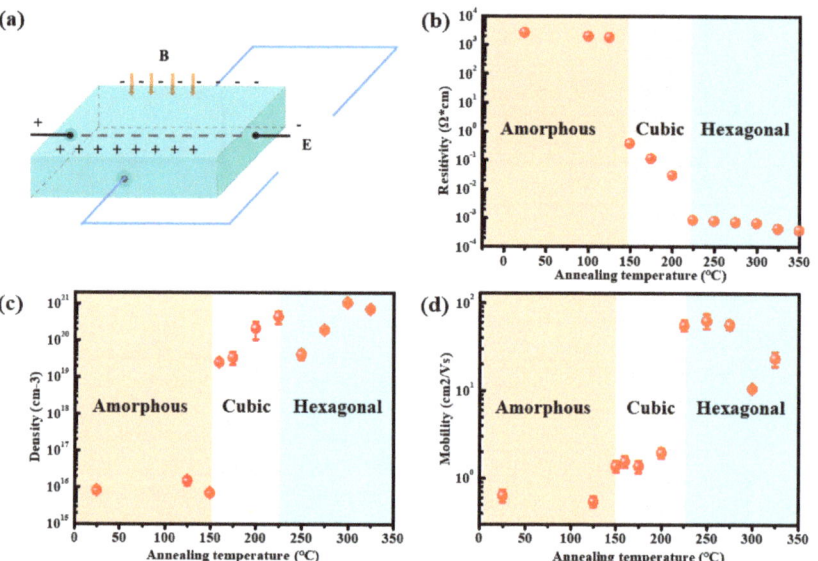

Figure 2. (a) The Hall test schematic. (b) The resistivity, (c) carrier density, and (d) carrier mobility at different annealing temperatures.

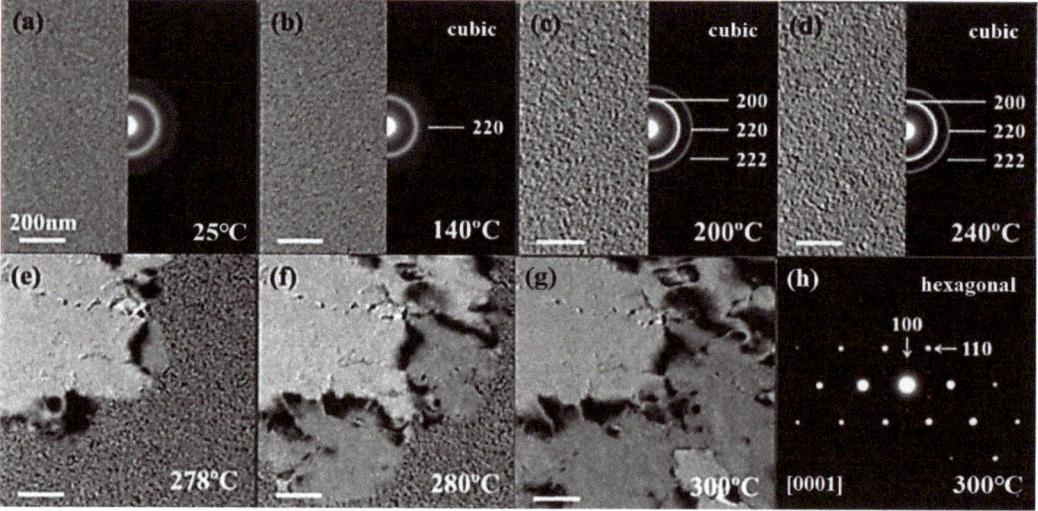

Figure 3. (a–d) The morphology and SAED patterns of GST films at 25 °C, 140 °C, 160 °C and 200 °C, respectively. (e–g) The morphology of GST films at 278 °C, 280 °C, and 300 °C, respectively. (h) The SAED patterns at 300 °C, which is along the [0001] axis.

Through an in-situ heating experiment inside TEM, we observed the phase transition process in a small area. To study the crystal structure in a larger area, we used the EBSD technique to analyze the GST film that was annealed at 180 °C and 270 °C, respectively. Figure 4a,c show the inverse polar figure (IPF) of the film at 180 °C and 270 °C, respectively. At 180 °C, the orientation difference of the IPF was very inconsistent, while the orientation of the grains was very similar at 270 °C, demonstrating that the film had a preferred orientation, which corresponds to the [0001] orientation in the h-phase as observed in Figure 3h. Further, the statistical grain size distribution [20–22] at two temperatures are also shown in Figure 4b,d. The calculated average grain size was 48.5 nm for 180 °C, and 292.3 nm for 270 °C, indicating that the nanograins grew significantly during the f-to-h phase transition.

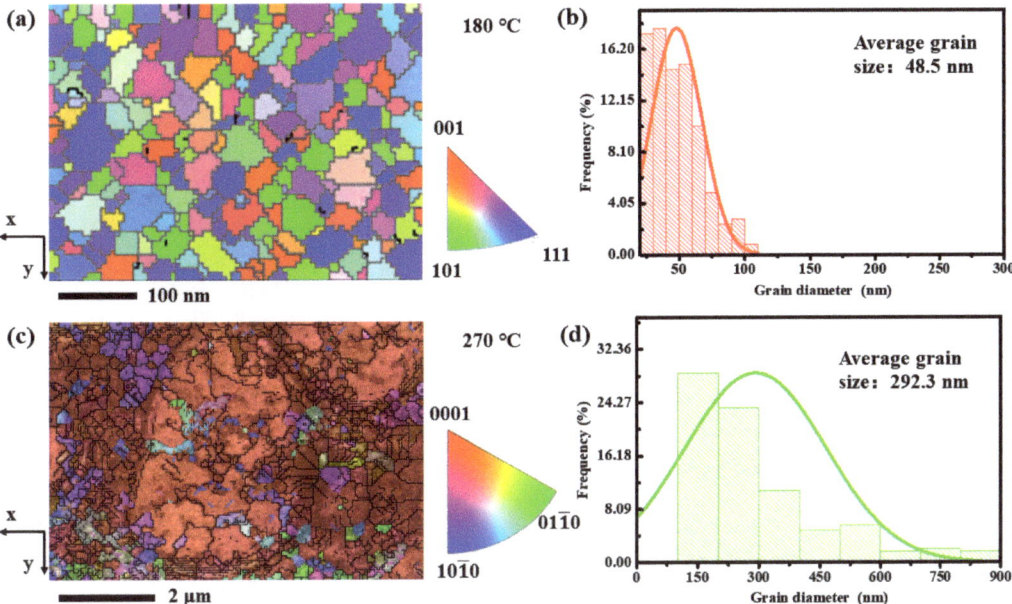

Figure 4. (**a**,**b**) are the inverse polar figure (IPF) of the film and the statistical grain size distribution at 180 °C. (**c**,**d**) are the IPF of the film and the statistical grain size distribution at 270 °C.

To explore the preferred orientation relationship between the f-phase and the h-phase, we analyzed the EBSD pole figures (PF) for {100}, {110}, and {111} crystallographic planes for the f-phase (180 °C). As can be seen from Figure 5a, the maximum density of orientation difference was 2.97; thus, there was no consistent preferred orientation in the f-phase films at 180 °C. As for the PF in the h-phase films (270 °C), there was no preferred orientation in {11$\bar{2}$0}, and {10$\bar{1}$0} planes, while in the {0001} plane, the maximum orientation difference density was 65.15 at the central position, which is along the [0001] orientation. Therefore, the large h-phase grains in GST film indeed had a preferred orientation as observed in Figure 3h. During the f-to-h phase transition, the carrier concentration did not increase significantly. However, due to the growth of the nanograins, the scattering affected by the crystal lattice was weakened, hence increasing carrier mobility. When the structural transformation was completed, the grain size did not change, and thus the mobility reached a stable value at this time.

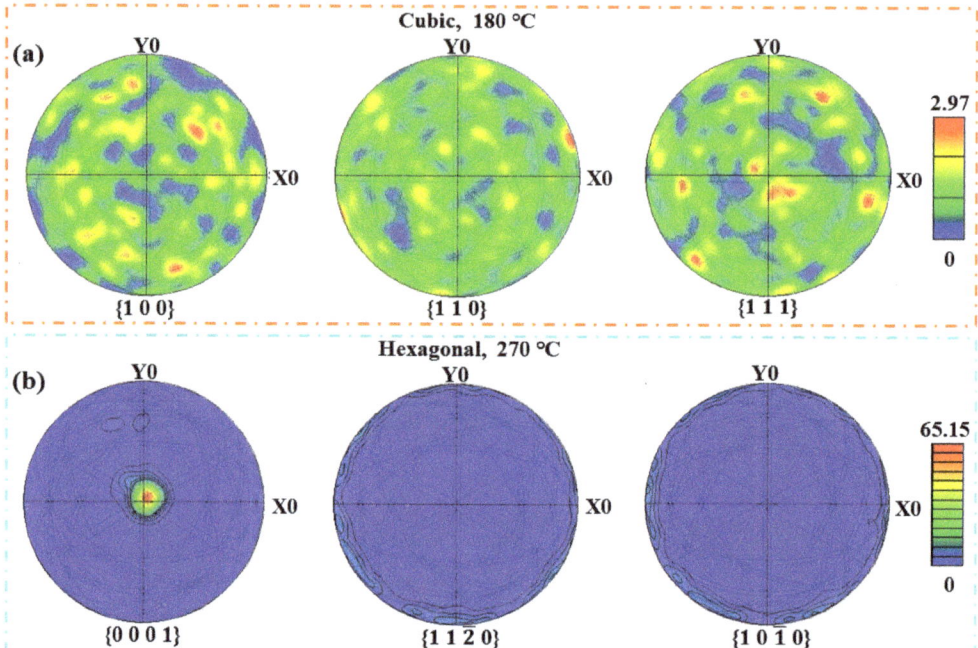

Figure 5. (a) EBSD pole figures for {100}, {110}, and {111} crystallographic planes for the f-phases film (180 °C). (b) EBSD pole figures for {0001}, {11$\bar{2}$0}, and {10$\bar{1}$0} crystallographic planes for the h-phase films (270 °C).

4. Conclusions

In summary, the relationship between electronic transport property and microstructure in GST alloy was studied. Induced by thermal treatment, GST alloy first crystallized into the f-phase, with a sharp resistance drop. At the same time, the carrier concentration increased substantially due to the variation of the atomic bonding environment during the crystallization process. At higher temperatures, the carrier concentration maintained a high level. When the temperature was high enough, a second resistance drop was observed, which is related to the increase in carrier mobility. This is because the f-phase grains transformed into the h-phase, and significantly grew up from ~50 nm to ~300 nm, which could reduce the carrier scattering effect. Our study lays the foundation for establishing the relationship between electron transport and microstructure in GST materials, and also provides an optimization direction for precisely controlling the storage states of GST-based PCRAM devices.

Author Contributions: C.L., Y.Z. (Yonghui Zheng) and Y.C. carried out the experiments, analysis of the data and wrote the manuscript. T.X., Y.Z. (Yunzhe Zheng) and R.W. provided valuable suggestion for the analysis of the data. The project was initiated and conceptualized by Y.C. All authors have read and agreed to the published version of the manuscript.

Funding: This work was financially supported by the National Natural Science Foundation of China (62104071, 62174054, 92064003, 12134003), Shanghai Sailing Program (21YF1410900), Natural Science Foundation of Chongqing (cstc2021jcyj-msxmX0750).

Data Availability Statement: All data that support the findings of this study are available from the corresponding authors upon reasonable request.

Conflicts of Interest: The authors declare no conflict of interest.

References

1. Raoux, S.; Xiong, F.; Wuttig, M.; Pop, E. Phase change materials and phase change memory. *MRS Bull.* **2014**, *39*, 703–710. [CrossRef]
2. Wuttig, M.; Ma, E. Designing crystallization in phase-change materials for universal memory and neuro-inspired computing. *Nat. Rev. Mater.* **2019**, *4*, 150–168.
3. Wong, H.S.P.; Salahuddin, S. Memory leads the way to better computing. *Nat. Nanotechnol.* **2015**, *10*, 191–194. [CrossRef] [PubMed]
4. Wuttig, M.; Yamada, N. Phase-change materials for rewriteable data storage. *Nat. Mater.* **2007**, *6*, 824–832. [CrossRef]
5. Ovshinsky, S.R. Reversible electrical switching phenomena in disordered structures. *Phys. Rev. Lett.* **1968**, *21*, 1450. [CrossRef]
6. Simpson, R.E.; Fons, P.; Kolobov, A.V.; Fukaya, T.; Krbal, M.; Yagi, T.; Tominaga, J. Interfacial phase-change memory. *Nat. Nanotechnol.* **2011**, *6*, 501–505. [CrossRef]
7. Cheong, B.-K.; Lee, S.; Jeong, J.-H.; Park, S.; Han, S.; Wu, Z.; Ahn, D.-H. Fast and scalable memory characteristics of Ge-doped SbTe phase change materials. *Phys. Status Solidi B* **2012**, *249*, 1985–1991. [CrossRef]
8. Zheng, Y.; Cheng, Y.; Huang, R.; Qi, R.; Rao, F.; Ding, K.; Yin, W.; Song, S.; Liu, W.; Song, Z.; et al. Surface energy driven cubic-to-hexagonal grain growth of $Ge_2Sb_2Te_5$ thin film. *Sci. Rep.* **2017**, *7*, 5915. [CrossRef]
9. Nonaka, T.; Ohbayashi, G.; Toriumi, Y.; Mori, Y.; Hashimoto, H. Crystal structure of GeTe and $Ge_2Sb_2Te_5$ meta-stable phase. *Thin Solid Film.* **2000**, *370*, 258–261. [CrossRef]
10. Zhang, B.; Zhang, W.; Shen, Z.; Chen, Y.; Li, J.; Zhang, S.; Zhang, Z.; Wuttig, M.; Mazzarello, R.; Ma, E.; et al. Element-resolved atomic structure imaging of rocksalt $Ge_2Sb_2Te_5$ phase-change material. *Appl. Phys. Lett.* **2016**, *108*, 191902. [CrossRef]
11. Siegrist, T.; Jost, P.; Volker, H.; Woda, M.; Merkelbach, P.; Schlockermann, C.; Wuttig, M. Disorder-induced localization in crystalline phase-change materials. *Nat. Mater.* **2011**, *10*, 202–208. [CrossRef] [PubMed]
12. Zheng, Y.; Wang, Y.; Xin, T.; Cheng, Y.; Huang, R.; Liu, P.; Luo, M.; Zhang, Z.; Lv, S.; Song, Z.; et al. Direct atomic identification of cation migration induced gradual cubic-to-hexagonal phase transition in $Ge_2Sb_2Te_5$. *Commun. Chem.* **2019**, *2*, 13. [CrossRef]
13. Zhang, B.; Wang, X.-P.; Shen, Z.-J.; Li, X.-B.; Wang, C.-S.; Chen, Y.-J.; Li, J.-X.; Zhang, J.-X.; Zhang, Z.; Zhang, S.-B.; et al. Vacancy structures and melting behavior in rock-salt GeSbTe. *Sci. Rep.* **2016**, *6*, 25453. [CrossRef] [PubMed]
14. Zhang, W.; Thiess, A.; Zalden, P.; Zeller, R.; Dederichs, P.; Raty, J.-Y.; Wuttig, M.; Blügel, S.; Mazzarello, R. Role of vacancies in metal–insulator transitions of crystalline phase-change materials. *Nat. Mater.* **2012**, *11*, 952–956. [CrossRef] [PubMed]
15. Xu, L.; Tong, L.; Geng, L.; Yang, F.; Xu, J.; Su, W.; Liu, D.; Ma, Z.; Chen, K. A comparative study on electrical transport properties of thin films of $Ge_1Sb_2Te_4$ and $Ge_2Sb_2Te_5$ phase-change materials. *J. Appl. Phys.* **2011**, *110*, 013703. [CrossRef]
16. Jost, P.; Volker, H.; Poitz, A.; Poltorak, C.; Zalden, P.; Schäfer, T.; Lange, F.R.L.; Schmidt, R.M.; Holländer, B.; Wirtssohn, M.R.; et al. Disorder-Induced Localization in Crystalline Pseudo-Binary $GeTe–Sb_2Te_3$ Alloys between $Ge_3Sb_2Te_6$ and GeTe. *Adv. Funct. Mater.* **2015**, *25*, 6399–6406. [CrossRef]
17. Wang, J.-J.; Xu, Y.-Z.; Mazzarello, R.; Wuttig, M.; Zhang, W. A Review on Disorder-Driven Metal–Insulator Transition in Crystalline Vacancy-Rich GeSbTe Phase-Change Materials. *Materials* **2017**, *10*, 862. [CrossRef]
18. Woodland, M.V. Data of rock analyses—VI: Bibliography and index of rock analyses in the periodical and serial literature of Scotland. *Geochim. Cosmochim. Acta* **1959**, *17*, 136–147. [CrossRef]
19. Liu, C.; Tang, Q.; Zheng, Y.; Zhang, B.; Zhao, J.; Song, W.; Cheng, Y.; Song, Z. The origin of hexagonal phase and its evolution process in $Ge_2Sb_2Te_5$ alloy. *APL Mater.* **2022**, *10*, 021102. [CrossRef]
20. Zhao, P.; Chen, B.; Kelleher, J.; Yuan, G.; Guan, B.; Zhang, X.; Tu, S. High-cycle-fatigue induced continuous grain growth in ultrafine-grained titanium. *Acta Mater.* **2019**, *174*, 29–42. [CrossRef]
21. Zhao, P.; Yuan, G.; Wang, R.; Guan, B.; Jia, Y.; Zhang, X.; Tu, S. Grain-refining and strengthening mechanisms of bulk ultrafine grained CP-Ti processed by L-ECAP and MDF. *J. Mater. Sci. Technol.* **2021**, *83*, 196–207.
22. Zhao, P.; Chen, B.; Zheng, Z.; Guan, B.; Zhang, X.; Tu, S. Microstructure and Texture Evolution in a Post-dynamic Recrystallized Titanium During Annealing, Monotonic and Cyclic Loading. *Metall. Mater. Trans. A Phys. Metall. Mater. Sci.* **2020**, *52*, 394–412. [CrossRef]

Disclaimer/Publisher's Note: The statements, opinions and data contained in all publications are solely those of the individual author(s) and contributor(s) and not of MDPI and/or the editor(s). MDPI and/or the editor(s) disclaim responsibility for any injury to people or property resulting from any ideas, methods, instructions or products referred to in the content.

Communication

Great Potential of Si-Te Ovonic Threshold Selector in Electrical Performance and Scalability

Renjie Wu [1,2], Yuting Sun [1,2], Shuhao Zhang [1,2], Zihao Zhao [1,2,*] and Zhitang Song [1,*]

[1] State Key Laboratory of Functional Materials for Informatics, Shanghai Institute of Micro-System and Information Technology, Chinese Academy of Sciences, Shanghai 200050, China
[2] University of Chinese Academy of Sciences, Beijing 100029, China
* Correspondence: zzh@mail.sim.ac.cn (Z.Z.); ztsong@mail.sim.ac.cn (Z.S.)

Abstract: The selector is an indispensable section of the phase change memory (PCM) chip, where it not only suppresses the crosstalk, but also provides high on-current to melt the incorporated phase change material. In fact, the ovonic threshold switching (OTS) selector is utilized in 3D stacking PCM chips by virtue of its high scalability and driving capability. In this paper, the influence of Si concentration on the electrical properties of Si-Te OTS materials is studied; the threshold voltage and leakage current remain basically unchanged with the decrease in electrode diameter. Meanwhile, the on-current density (J_{on}) increases significantly as the device is scaling down, and 25 MA/cm^2 on-current density is achieved in the 60-nm SiTe device. In addition, we also determine the state of the Si-Te OTS layer and preliminarily obtain the approximate band structure, from which we infer that the conduction mechanism conforms to the Poole-Frenkel (PF) model.

Keywords: OTS material; high scalability; high on-current density; PF model

Citation: Wu, R.; Sun, Y.; Zhang, S.; Zhao, Z.; Song, Z. Great Potential of Si-Te Ovonic Threshold Selector in Electrical Performance and Scalability. *Nanomaterials* 2023, 13, 1114. https://doi.org/10.3390/nano13061114

Academic Editors: Antonio Di Bartolomeo and Filippo Giubileo

Received: 2 March 2023
Revised: 16 March 2023
Accepted: 17 March 2023
Published: 21 March 2023

Copyright: © 2023 by the authors. Licensee MDPI, Basel, Switzerland. This article is an open access article distributed under the terms and conditions of the Creative Commons Attribution (CC BY) license (https://creativecommons.org/licenses/by/4.0/).

1. Introduction

The fact that the digital data are exponentially increasing at every moment has put forward a series of higher requirements for memory—more reliability with less cost, greater memory capacity with smaller sizes, and faster read/write speed with lower power consumption [1,2]. Compared to NAND Flash devices, PCM has been one of the most promising competitors of the next-generation memories on the strength of structural transition of chalcogenides between crystalline and amorphous states [3,4] which has shown extraordinary advantages in operating speed, storage density and endurance [5]. Moreover, the successful commercialization of PCM [6] is also inseparable from the help of ovonic threshold switching (OTS), which tends to suppress leakage current in case of operating errors. Compared with traditional Si-based switches, OTS materials acting as just nm-scale films dispense with the share of Si substrate and possess highly potential in scalability [7].

However, loads of materials with superior performance emerging from the OTS research in full swing contain eco-unfriendly elements such as sulfur [8–10], arsenic [11,12] etc.; additionally, most OTS materials are ternary, quaternary or even more [13,14], which means they inevitably suffer from element segregation and inhomogeneous composition. At the same time, OTS switches are required to maintain an amorphous state at the high temperature of 400–450 °C, which is also called back-end-of-line (BEOL) process [1,15], posing a serious challenge to binary Se- and Te-based materials [16,17]. In 2021, a major breakthrough was made by Shen et al., in that tellurium was discovered to be the simplest threshold switching material with an endurance of 10^8 cycles. The selectivity is higher than 10^3 and the on current density is more than 11 MA/cm^2, which is enough to drive the incorporated PCM. Different from previous OTS materials, the Te layer stays in a crystalline state before and after pulse operation; therefore, Shen et al. creatively raised a brand-new principle called the 'Crystalline-Liquid-Crystalline Mechanism', which corresponds the off state of the switch to the crystalline state of Te and then Te melts into liquid under the external voltage, which means the switch turns on. At the removal of applied voltage, Te rapidly recrystallizes and the device turns off. In addition, the single element also avoids the limitations from the high temperature of the BEOL process and the element/phase segregation [1].

Unluckily, the leakage current at 1/2 V_{th}, i.e., I_{off}, of 60 nm Te device is only 0.5 µA according to Shen's research, which indicates the subthreshold region of Te is composed of two parts; one is dominated by the Schottky barrier between Te and TiN electrodes in low electric field, and the other depends on the intrinsic excitation of Te in higher field. Therefore, the ways to optimize the I_{off} lie in two perspectives: one is to increase the Schottky barrier by changing the electrode material, such as Pt; the other is to increase the band gap of the material including Si [18–20] or B [21] doping.

In our work, we carry out electrical tests on SiTe-based devices to explore the influence of different Si contents on device performance and their potential to scale down. Then we study the conduction mechanism of SiTe devices through characterization.

2. Materials and Methods

2.1. Film Preparation and Testing

Four compositions are designed to be SiTe, $SiTe_2$, $SiTe_4$ and $SiTe_8$. All the films were deposited by co-sputtering of Si and Te targets, while the power values of Si target are 65 W, 40 W, 30 W and 20 W with 10 W of Te target. The exact components were determined by energy-dispersive spectroscopy (EDS). 100 nm-thick films were deposited on SiO_2 substrates for in situ resistance-temperature (R-T) tests with 20 °C/min heating rate and X-ray diffraction (XRD) tests ranging from 10–67°. The Fermi level of different films were measured by X-ray photoelectron spectroscopy (XPS), and 400-nm Si-Te films were deposited on quartz glass as the ultraviolet-visible Spectrophotometer (UV-Vis) samples.

2.2. Device Fabrication

The 10 nm Si-Te OTS layers were deposited on TiN bottom electrodes whose diameters are 200, 150, 120 and 60 nm, respectively. Then, TiN was sputtered as the top electrode with 40-nm thickness. Keithley 4200A-SCS parameter analyzer was used to measure the electrical performance. Tektronix MSO54 mixed signal oscilloscope was used to capture the input pulses and device responses.

3. Results

3.1. Variations of Electrical Performance with Different Si Concentrations

T-shaped device structure is shown in Figure 1a where cylindrical TiN serves as the bottom electrode and Si_3N_4 acts as the isolation. Then current-voltage (I-V) curves of 200 nm-electrode devices with all four Si-Te compositions under 10 continuous triangular pulses (Figure 1b) are shown in Figure 1c. The red lines represent the response of device under the first pulse, which is called first fire (FF). During the process, SiTe device suddenly turns to low resistance state at 2.83 V, while under the second pulse, the threshold voltage (V_{th}) goes down to 1.71 V as described by the blue lines. With the decrease of Si content, the fire voltage (V_{fire}) of $SiTe_2$ drops to 2.09 V. After FF process, only 1.11–1.31 V V_{th} is needed to operate the cell. The downward trend continues with $SiTe_4$ and $SiTe_8$, whose V_{fire} were 1.79 V and 1.63 V, respectively. Moreover, V_{th} of $SiTe_4$ is ~0.8 V lower than its V_{fire} which fluctuates between 0.99 and 1.13 V, while the V_{th} of $SiTe_8$ ranges from 0.89 V to 1.05 V. The amplitude of the first pulse applied to the SiTe device is 4 V and the height of the remaining nine pulses is 3 V. Meanwhile, the pulse amplitude of all ten triangular pulses used for other Si-Te devices is 4 V. The rising edges, falling edges and pulse intervals of these pulses are 1 µs. Clearly, both V_{fire} and V_{th} keep decreasing with the continuous decrease of Si contents.

Afterwards, we carried out DC I-V tests and the circuit is shown in Figure 2a. From Figure 2b, we can obviously find that the change trend of V_{th} is consistent with the results of pulsed I-V, that is, the addition of silicon will lead to the rise of V_{th}. As can be seen in the figure, all Si-Te devices show >10 mV/dec nonlinear characteristics and more than 10^3 selective ratio. In addition, the leakage current of SiTe at ~1/2 V_{th}, namely I_{off}, which is about 0.12 μA, and the I_{off} of SiTe$_2$ is situated at ~93 nA. It continued to decrease to 53 nA by the concentration of Si declines to 20 at. %, but contrary to the down trend of Si concentration, I_{off} increases to 0.15 μA in SiTe$_8$. However, after summarizing and comparing the DC curves of the four components as shown in Figure 2c, we clearly observed that - the leakage current decreases with the increase of Si content.

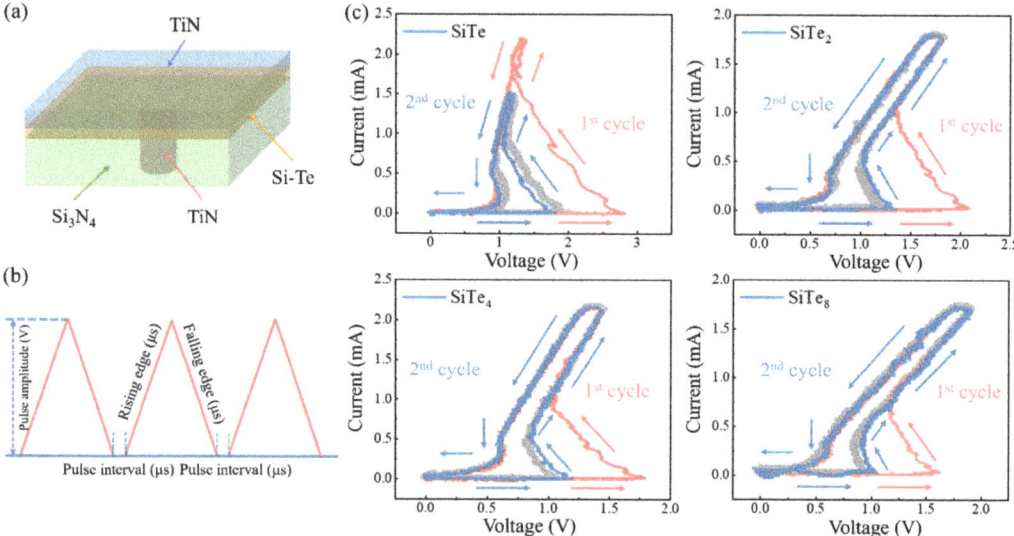

Figure 1. Pulse *I-V* curves of devices with different Si contents. (**a**) The schematic diagram of device structure. (**b**) Triangular pulses applied to the devices with modifiable rising edge time, falling edge time, pulse amplitude and interval. (**c**) *I-V* curves of devices with different compositions under 10 consecutive triangular pulses. Dotted lines correspond to FF, and the dashed lines refer to the second operation.

Other electrical performances are shown in Figure 3. Instantaneous responses of the device are observed through the oscilloscope, from which we are able to capture the moment when the device is turned on and off, as shown in Figure 3a. The abrupt rise of current signifies that the device is switched on, while the time span is considered to be the on-speed. Similarly, a sudden drop in current refers to the off-speed. As shown in Figure 3b, the content of silicon seems independent of the switching speeds. Moreover, the lifetime of the devices was measured with the help of the oscilloscope. A fixed number of continuous pulses is applied to the device, and whether the device still works normally is judged from the screen. In case the device responds to each pulse, a DC test is conducted next to measure its I_{off} in order to verify whether the off-state is in a high-resistance state as before. Only if both conditions are met, the endurance with that fixed number is proved. Then the number of input pulses gradually increases until the device keeps silent to the input or the I_{off} of the device goes too large. In this way, we find the endurance of both SiTe and SiTe$_8$ are merely 10^5, but the reason for that low value is slightly different. SiTe is unable to be switched on after 10^5 pulses. As for SiTe$_8$, it could be turned on but its I_{off} is too large, which may be due to the low crystallization temperature. As shown in Figure 3c, SiTe2 device enables to operate for 108 cycles which shows the longest lifetime among these four contents.

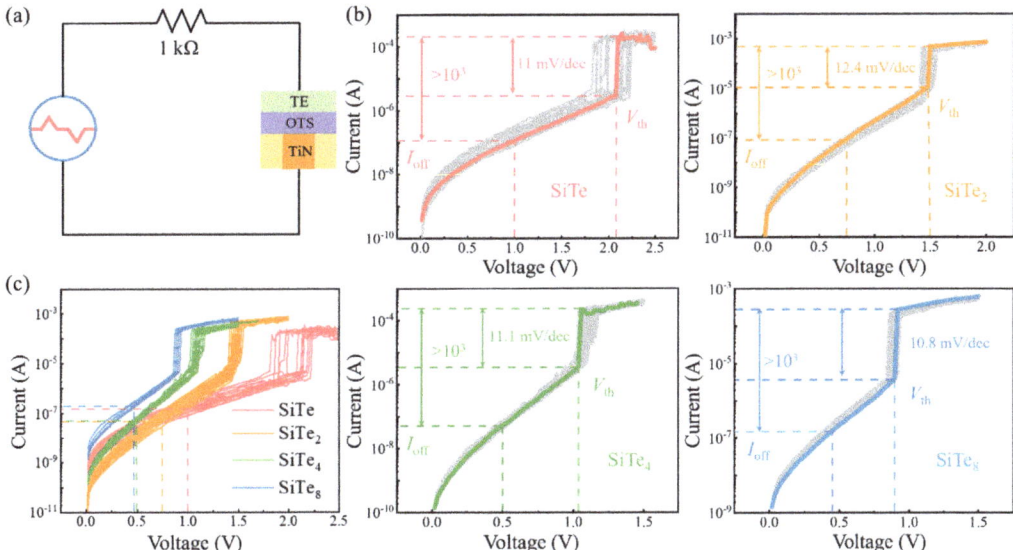

Figure 2. DC curves of devices with different Si contents. (**a**) A 1 kΩ resistor is in series with the device in the DC testing circuit. (**b**) DC I-V curves of Si–Te devices. (**c**) The synthesized DC curves of four compositions.

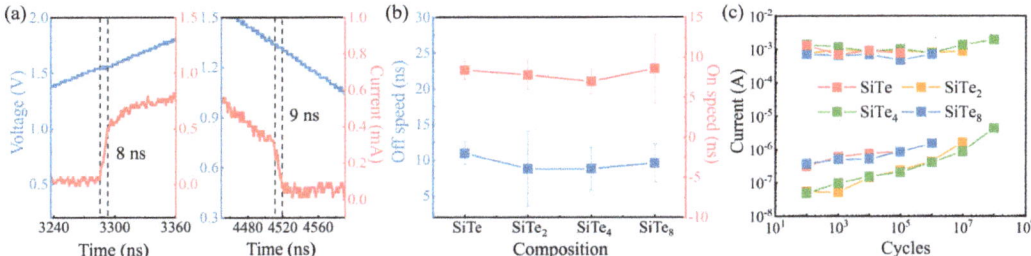

Figure 3. Speed and endurance distribution of Si–Te devices. (**a**) The on–speed and off–speed are 8 ns and 9 ns, respectively. (**b**) Statistical operating speeds distribution of different Si–Te devices. (**c**) Device lifetime of devices with all four compositions.

3.2. Effect of Device Size on Electrical Performance

From the perspective of market demand, device shrinkage has been focused, and whether the device maintains the performance with scaling down has been a major problem. First, we explored the relationships between V_{fire}, V_{th}, holding voltage (V_h) and electrode sizes. In fact, V_{fire} significantly affects the device lifetime and power consumption, while V_{th} and V_h are directly related to the read margin [22], that is, a slight shift of V_{th} or V_h may cause serious operation errors. Luckily, Si-Te series devices show high potential in scalability from Figure 4a. According to the statistics of more than 50 independent cells, V_{fire}, V_{th} and V_h are almost irrelevant to the device size. Clearly shown in Figures 4a and 1c, the read margin becomes narrow with the decrease of Si content, but the good news is that V_{fire} also becomes smaller. Therefore, the key to application is how to balance these two points. In addition, the reduction of device size brings the increase of on-current density as well. Generally, >10 MA/cm² is required to drive the PCM. From Figure 4b, the on-current density of 60 nm-SiTe device reaches 25 MA/cm² at most, and J_{on} of the other components also exceed 10 MA/cm². As for I_{off}, it hardly changes with the electrode size according to Figure 4c. The discrepancy of SiTe$_8$ results from the number of devices participating in statistics. Under comprehensive consideration, SiTe$_2$ is equipped with >15 mA/cm^2 on-current density, 10^8 cycles of endurance, and moderate I_{off} among these compositions, which is the best choice of the four components.

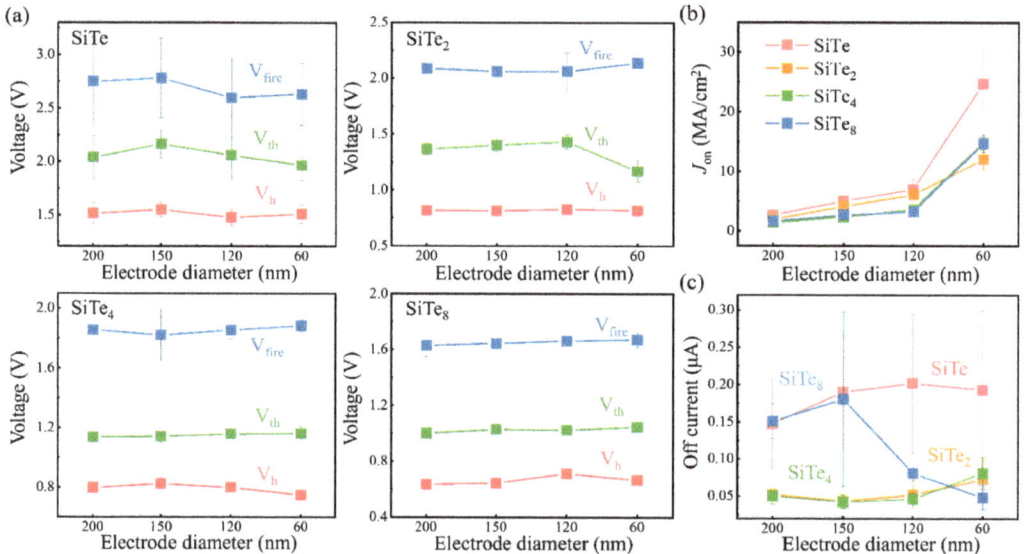

Figure 4. Device performances along with the device size. (**a**) V_{fire}, V_{th} and V_h variations of SiTe, SiTe$_2$, SiTe$_4$ and SiTe$_8$ with different electrode diameters. (**b**) Change trend of average on-current density with electrode sizes. (**c**) Tendency of average I_{off} with different electrodes.

3.3. Characterization of Si-Te Films

We first carried out XRD tests on the as-deposited films of each component, as shown in Figure 5a. As seen from the figure, Si-Te films of the four components show no peak, but there are obvious crystallization peaks in the Te sample, indicating that the deposited Te is in crystalline state, while the state of film turns to amorphous after Si incorporation. At the same time, the state of the material has also been proved by R-T test in Figure 5b. Interestingly, the crystallization temperature of SiTe is only 78 °C, and the crystallization temperature reaches the peak of ~235 °C at SiTe$_2$. However, with the further reduction of Si content, the crystallization temperature decreases, and that of SiTe$_8$ decreases to 128 °C, the tendency which is very similar to Ge-Te system [23]. Moreover, the different states directly indicate that the conduction mechanism of Si-Te is different from that of Te, while the Si-Te device is more consistent with the traditional PF model [24], where the switching characteristic strongly depends on the amorphous state. Then the location of the valence band maximums (VBM, E_v) are determined by linear extrapolation of the linear part of the XPS valence band spectrum, they are situated at −0.48 eV, −0.45 eV, −0.41 eV and −0.39 eV for SiTe, SiTe$_2$, SiTe$_4$ and SiTe$_8$, respectively, in Figure 5c. Meanwhile, the band gap is obtained by linear fitting according to $\alpha h\nu = C\,(h\nu - E_g)^2$, where C is the constant, and α is the absorption coefficient, as shown in Figure 5d. The band gap value of SiTe is 1.02 eV. With the decrease of Si concentration, the band gap slightly increases to 1.07 eV, and further drops to 0.88 eV and 0.87 eV for SiTe$_4$ and SiTe$_8$. The Fermi level is basically located in the center of the band gap, which is generally believed that the high concentration defect states would pin the Fermi level in the middle of the band gap. Si-Te series materials basically conform to this rule, that is to say, their conduction mechanism may be closely related to the defect state, which requires further research.

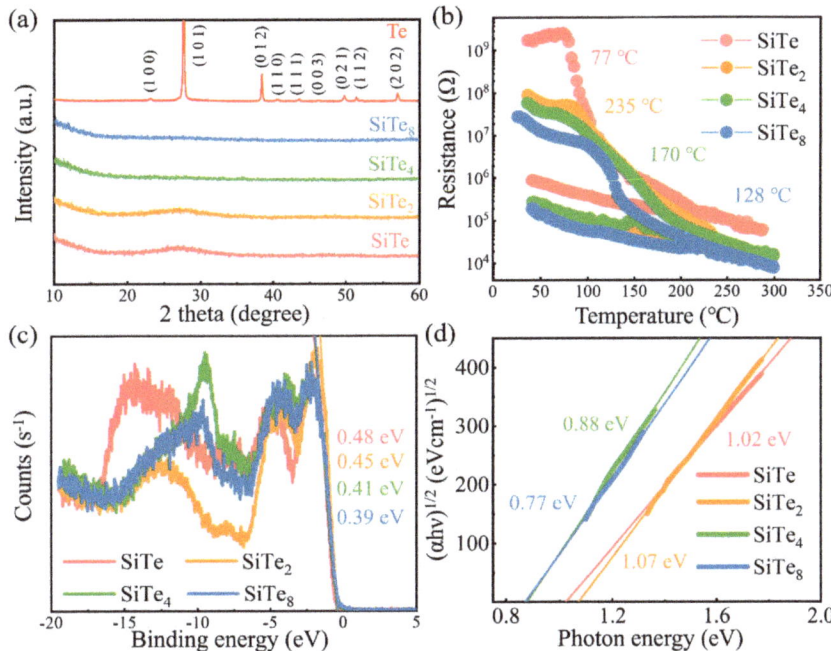

Figure 5. Characterization of Si−Te films. (a) XRD results of SiTe, SiTe$_2$, SiTe$_4$, SiTe$_8$ and Te films. (b) R-T results of films with four Si−Te compositions. (c) XPS valence band spectrum of SiTe, SiTe$_2$, SiTe$_4$, SiTe$_8$. (d) $(\alpha h\nu)^{1/2}$ versus hν interprets the bandgap of four compositions.

4. Conclusions

In this paper, we comprehensively studied the relationships between the electrical properties of Si-Te devices and Si concentration. Although the operation speed of the Si-Te device is maintained at ~10 ns, V_{fire} and V_{th} gradually rise and I_{off} drops as the Si concentration goes up. However, the steady increase of Si concentration also leads to the lifetime degradation when the Si concentration exceeds 33 at. %, where the SiTe$_2$ device exhibits a superior endurance of 10^8 cycles to other contents. In addition, V_{th}, I_{on} and I_{off} of all components hardly change with the electrode size, and the J_{on} of 60-nm SiTe device even increases to 25 MA/cm^2, indicating the great potential of Si-Te devices in scaling down. In terms of the conduction mechanism of Si-Te, we firstly find that Si-Te materials are amorphous by XRD, and observe the change in trend whereby the crystallization temperature increases and then decreases with the decrease of Si contents through in situ RT experiment. Through UV-vis, we also find that the root of I_{off} variation with Si concentration originates from the band gap, that is, the larger the content of Si, the wider the band gap, and the smaller the I_{off}. Afterwards, the Fermi level is found to be pinned at the center of the band gap through XPS, indicating that the conduction mechanism is more consistent with the traditional PF model associated with trap state which is different from Te switching. The specific relationship between the switching performance of Si-Te materials and the defect state is also worthy of further study.

Author Contributions: R.W. deposited the films, prepared the devices, measured device performances and wrote the paper. Y.S., S.Z. and Z.Z. revised the paper. The project was initiated and conceptualized by Z.S. All authors have read and agreed to the published version of the manuscript.

Funding: Financial support was provided by the Strategic Priority Research Program of the Chinese Academy of Sciences (XDB44010200) and the National Natural Science Foundation of China (61904046).

Data Availability Statement: Not applicable.

Acknowledgments: The authors acknowledge the financial support from the State Key Laboratory of Functional Materials for Informatics, Shanghai Institute of Microsystem and Information, Chinese Academy of Sciences.

Conflicts of Interest: The authors declare no conflict of interest.

References

1. Shen, J.; Jia, S.; Shi, N.; Ge, Q.; Gotoh, T.; Lv, S.; Liu, Q.; Dronskowski, R.; Elliott, S.R.; Song, Z.; et al. Elemental Electrical Switch Enabling Phase Segregation–Free Operation. *Science* **2021**, *374*, 1390–1394. [CrossRef] [PubMed]
2. Calarco, R.; Arciprete, F. Keep It Simple and Switch to Pure Tellurium. *Science* **2021**, *374*, 1321–1322. [CrossRef] [PubMed]
3. Wuttig, M.; Yamada, N. Phase-Change Materials for Rewriteable Data Storage. *Nat. Mater.* **2007**, *6*, 824–832. [CrossRef] [PubMed]
4. Zhu, M.; Song, W.; Konze, P.M.; Li, T.; Gault, B.; Chen, X.; Shen, J.; Lv, S.; Song, Z.; Wuttig, M.; et al. Direct Atomic Insight into the Role of Dopants in Phase-Change Materials. *Nat. Commun.* **2019**, *10*, 3525. [CrossRef] [PubMed]
5. Lai, S.; Lowrey, T. OUM—A 180 nm Nonvolatile Memory Cell Element Technology for Stand Alone and Embedded Applications. In Proceedings of the International Electron Devices Meeting. Technical Digest, Washington, DC, USA, 2–5 December 2001; IEEE: Piscataway, NJ, USA, 2001; pp. 36.5.1–36.5.4. [CrossRef]
6. Choe, J. Memory Technology 2021: Trends & Challenges. In Proceedings of the 2021 International Conference on Simulation of Semiconductor Processes and Devices (SISPAD), Dallas, TX, USA, 27–29 September 2021; IEEE: Piscataway, NJ, USA, 2021; pp. 111–115. [CrossRef]
7. Kau, D.; Tang, S.; Karpov, I.V.; Dodge, R.; Klehn, B.; Kalb, J.A.; Strand, J.; Diaz, A.; Leung, N.; Wu, J.; et al. A Stackable Cross Point Phase Change Memory. In Proceedings of the Technical Digest—International Electron Devices Meeting, IEDM, Baltimore, MD, USA, 7–9 December 2009; IEEE: Piscataway, NJ, USA, 2009; pp. 1–4. [CrossRef]
8. Jia, S.; Li, H.; Gotoh, T.; Longeaud, C.; Zhang, B.; Lyu, J.; Lv, S.; Zhu, M.; Song, Z.; Liu, Q.; et al. Ultrahigh Drive Current and Large Selectivity in GeS Selector. *Nat. Commun.* **2020**, *11*, 4636. [CrossRef] [PubMed]
9. Wu, R.; Jia, S.; Gotoh, T.; Luo, Q.; Song, Z.; Zhu, M. Screening Switching Materials with Low Leakage Current and High Thermal Stability for Neuromorphic Computing. *Adv. Electron. Mater.* **2022**, *8*, 2200150. [CrossRef]
10. Jia, S.; Li, H.; Liu, Q.; Song, Z.; Zhu, M. Scalability of Sulfur-Based Ovonic Threshold Selectors for 3D Stackable Memory Applications. *Phys. Status Solidi Rapid Res. Lett.* **2021**, *15*, 2100084. [CrossRef]
11. Yuan, Z.; Li, X.; Song, S.; Song, Z.; Zha, J.; Han, G.; Yang, B.; Jimbo, T.; Suu, K. The Enhanced Performance of a Si–As–Se Ovonic Threshold Switching Selector. *J. Mater. Chem. C* **2021**, *9*, 13376–13383. [CrossRef]
12. Yuan, Z.; Li, X.; Wang, H.; Xue, Y.; Song, S.; Song, Z.; Zhu, S.; Han, G.; Yang, B.; Jimbo, T.; et al. Characteristic of As3Se4-Based Ovonic Threshold Switching Device. *J. Mater. Sci. Mater. Electron.* **2021**, *32*, 7209–7214. [CrossRef]
13. Cheng, H.Y.; Chien, W.C.; Kuo, I.T.; Lai, E.K.; Zhu, Y.; Jordan-Sweet, J.L.; Ray, A.; Carta, F.; Lee, F.M.; Tseng, P.H.; et al. An Ultra High Endurance and Thermally Stable Selector Based on TeAsGeSiSe Chalcogenides Compatible with BEOL IC Integration for Cross-Point PCM. In Proceedings of the 2017 IEEE International Electron Devices Meeting (IEDM), San Francisco, CA, USA, 2–6 December 2017; IEEE: Piscataway, NJ, USA, 2017; Volume 1, pp. 2.2.1–2.2.4. [CrossRef]
14. Garbin, D.; Devulder, W.; Degraeve, R.; Donadio, G.L.; Clima, S.; Opsomer, K.; Fantini, A.; Cellier, D.; Kim, W.G.; Pakala, M.; et al. Composition Optimization and Device Understanding of Si-Ge-As-Te Ovonic Threshold Switch Selector with Excellent Endurance. In Proceedings of the 2019 IEEE International Electron Devices Meeting (IEDM), San Francisco, CA, USA, 7–11 December 2019; IEEE: Piscataway, NJ, USA, 2019; pp. 35.1.1–35.1.4. [CrossRef]
15. Zhu, M.; Ren, K.; Song, Z. Ovonic Threshold Switching Selectors for Three-Dimensional Stackable Phase-Change Memory. *MRS Bull.* **2019**, *44*, 715–720. [CrossRef]
16. Keukelier, J.; Opsomer, K.; Devulder, W.; Clima, S.; Goux, L.; Sankar Kar, G.; Detavernier, C. Tuning of the Thermal Stability and Ovonic Threshold Switching Properties of GeSe with Metallic and Non-Metallic Alloying Elements. *J. Appl. Phys.* **2021**, *130*, 165103. [CrossRef]
17. Anbarasu, M.; Wimmer, M.; Bruns, G.; Salinga, M.; Wuttig, M. Nanosecond Threshold Switching of GeTe$_6$ Cells and Their Potential as Selector Devices. *Appl. Phys. Lett.* **2012**, *100*, 143505. [CrossRef]
18. Li, X.; Yuan, Z.; Lv, S.; Song, S.; Song, Z. Extended Endurance Performance and Reduced Threshold Voltage by Doping Si in GeSe-Based Ovonic Threshold Switching Selectors. *Thin Solid Films* **2021**, *734*, 1–9. [CrossRef]
19. Koo, Y.; Lee, S.; Park, S.; Yang, M.; Hwang, H. Simple Binary Ovonic Threshold Switching Material SiTe and Its Excellent Selector Performance for High-Density Memory Array Application. *IEEE Electron Device Lett.* **2017**, *38*, 568–571. [CrossRef]
20. Velea, A.; Opsomer, K.; Devulder, W.; Dumortier, J.; Fan, J.; Detavernier, C.; Jurczak, M.; Govoreanu, B. Te-Based Chalcogenide Materials for Selector Applications. *Sci. Rep.* **2017**, *7*, 8103. [CrossRef] [PubMed]
21. Yoo, J.; Lee, D.; Park, J.; Song, J.; Hwang, H. Steep Slope Field-Effect Transistors with B-Te-Based Ovonic Threshold Switch Device. *IEEE J. Electron Devices Soc.* **2018**, *6*, 821–824. [CrossRef]

22. Cheng, H.Y.; Chien, W.C.; Kuo, I.T.; Yeh, C.W.; Gignac, L.; Kim, W.; Lai, E.K.; Lin, Y.F.; Bruce, R.L.; Lavoie, C.; et al. Ultra-High Endurance and Low I_{OFF} Selector Based on AsSeGe Chalcogenides for Wide Memory Window 3D Stackable Crosspoint Memory. In Proceedings of the 2018 IEEE International Electron Devices Meeting (IEDM), San Francisco, CA, USA, 1–5 December 2018; IEEE: Piscataway, NJ, USA, 2018; pp. 37.3.1–37.3.4. [CrossRef]
23. Chen, M.; Rubin, K.A.; Barton, R.W. Compound Materials for Reversible, Phase-change Optical Data Storage. *Appl. Phys. Lett.* **1986**, *49*, 502–504. [CrossRef]
24. Ielmini, D.; Zhang, Y. Analytical Model for Subthreshold Conduction and Threshold Switching in Chalcogenide-Based Memory Devices. *J. Appl. Phys.* **2007**, *102*, 054517. [CrossRef]

Disclaimer/Publisher's Note: The statements, opinions and data contained in all publications are solely those of the individual author(s) and contributor(s) and not of MDPI and/or the editor(s). MDPI and/or the editor(s) disclaim responsibility for any injury to people or property resulting from any ideas, methods, instructions or products referred to in the content.

Article

Amorphous As₂S₃ Doped with Transition Metals: An *Ab Initio* Study of Electronic Structure and Magnetic Properties

Vladimir G. Kuznetsov [1], Anton A. Gavrikov [2,*], Milos Krbal [3], Vladimir A. Trepakov [1] and Alexander V. Kolobov [2,*]

[1] Ioffe Institute, 26 Polytechnicheskaya Str., 194021 St. Petersburg, Russia
[2] Institute of Physics, Herzen State Pedagogical University of Russia, 48 Moika Emb., 191186 St. Petersburg, Russia
[3] Center of Materials and Nanotechnologies, Faculty of Chemical Technology, University of Pardubice, Nam. Čs. Legii 565, 530 02 Pardubice, Czech Republic
* Correspondence: antongavr@gmail.com (A.A.G.); akolobov@herzen.spb.ru (A.V.K.)

Abstract: Crystalline transition-metal chalcogenides are the focus of solid state research. At the same time, very little is known about amorphous chalcogenides doped with transition metals. To close this gap, we have studied, using first principle simulations, the effect of doping the typical chalcogenide glass As₂S₃ with transition metals (Mo, W and V). While the undoped glass is a semiconductor with a density functional theory gap of about 1 eV, doping results in the formation of a finite density of states (semiconductor-to-metal transformation) at the Fermi level accompanied by an appearance of magnetic properties, the magnetic character depending on the nature of the dopant. Whilst the magnetic response is mainly associated with *d*-orbitals of the transition metal dopants, partial densities of spin-up and spin-down states associated with arsenic and sulphur also become slightly asymmetric. Our results demonstrate that chalcogenide glasses doped with transition metals may become a technologically important material.

Citation: Kuznetsov, V.G.; Gavrikov, A.A.; Krbal, M.; Trepakov, V.A.; Kolobov, A.V. Amorphous As₂S₃ Doped with Transition Metals: An *Ab Initio* Study of Electronic Structure and Magnetic Properties. *Nanomaterials* **2023**, *13*, 896. https://doi.org/10.3390/nano13050896

Academic Editors: Kunji Chen, Shunri Oda and Linwei Yu

Received: 27 January 2023
Revised: 19 February 2023
Accepted: 21 February 2023
Published: 27 February 2023

Copyright: © 2023 by the authors. Licensee MDPI, Basel, Switzerland. This article is an open access article distributed under the terms and conditions of the Creative Commons Attribution (CC BY) license (https:// creativecommons.org/licenses/by/ 4.0/).

Keywords: chalcogenides glasses; As₂S₃; transition metal doping; electronic structure and magnetism; density functional theory simulations

1. Introduction

The discovery of semiconducting properties of chalcogenide glasses by Goryunova and Kolomiets has laid the foundation of a new class of semiconductors, viz. amorphous semiconductors [1]. Due to the presence of lone-pair electrons, i.e., paired *p*-electrons that do not participate in the formation of covalent bonds but occupy the top of the valence band [2], these materials exhibit a number of photo-induced phenomena, such as reversible photostructural changes, photo-induced anisotropy, etc. (for reviews, see [3,4]). In these processes, photo-induced bond modification is caused by bond switching that involves excitation of lone-pair electrons and subsequent formation of a new bonding configuration [5].

On the other hand, transition-metal chalcogenides contain Ch-TM (Ch = chalcogen, TM = transition metal) bonds that are formed by sharing chalcogen lone-pair electrons with empty *d*-orbitals of a transition metal, which can lead to the appearance of hybridized states [6]. This process results in (i) a decrease in the ability of chalcogenide glasses to undergo photostructural changes and should also result in (ii) an appearance of magnetism in doped glasses. It would also be extremely interesting if one could change the local bonding configuration of TM-doped chalcogenide glasses using electronic excitation in a way that would affect its magnetic properties. This would allow one to design glasses with optically switchable magnetic response. As the first step to addressing this issue, we studied, using first principles simulations, the electronic structure and magnetic properties of a prototypical chalcogenide glass a-As$_2$S$_3$ (amorphous As$_2$S$_3$) doped with transition metals such as molybdenum, tungsten and vanadium.

Transition metals have the following general electronic configuration: $(n-1)\mathrm{d}^{1\div 10}n\mathrm{s}^2$. As one goes across a row from left to right in the Periodic Table, electrons are generally added to the $(n-1)\mathrm{d}$ shell that is filled according to the aufbau principle, which states that electrons fill the lowest available energy levels before filling higher levels, and the Hund rule. The aufbau principle correlates well with an empirical so called $(n+l)$-rule, which is also known as the Madelung rule [7,8], according to which (i) electrons are assigned to subshells in order of increasing value of $n+l$, (ii) for subshells with the same value of $n+l$, electrons are assigned first to the subshell with lower n. Later, Klechkovskii proposed a theoretical explanation of the Madelung rule based on the Thomas–Fermi model of the atom [9,10]. Therefore, the $(n+l)$-rule is also cited as the Klechkovskii rule.

However, the Klechkovskii energy ordering rule applies only to neutral atoms in their ground state and has twenty exceptions (eleven in the d-block and nine in the f-block), for which this rule predicts an electron configuration that differs from that determined experimentally, although the rule-predicted electron configurations are at least close to the ground state even in those cases. For example, in molybdenum $_{42}$Mo, according to the Klechkovskii rule, the 5s subshell ($n+l=5+0=5$) is occupied before the 4d subshell ($n+l=4+2=6$). The rule then predicts the electron configuration [Kr] 4d^{4}5s^2 where [Kr] denotes the configuration of krypton, the preceding noble gas. However, the measured electron configuration of the molybdenum atom is [Kr] 4d^{5}5s^1, despite the fact that the tungsten atom, which is isoelectronic to the molybdenum atom and is an atom of the same group, does have the configuration [Xe] 4f^{14}5d^{4}6s^2 predicted by the Klechkovskii rule. By filling the 5d subshell, molybdenum can be in a lower energy state.

The choice of Mo, W and V as transition-metal dopants was determined by the great interest in layered crystalline chalcogenides based on these elements. The interest in transition metal (di)chalcogenides was triggered by the discovery of the fact that in the monolayer limit MoS$_2$, which is an indirect-gap semiconductor in the bulk form, becomes a direct-gap semiconductor, which opens up a plethora of possible applications of these materials. The most studied crystalline materials of this class are Mo- and W-dichalcogenides, while vanadium is interesting because its crystalline dichalcogenides possess so called charge-density waves [11–19]. At the same time, very little is known about the structure and properties, magnetic in particular, of amorphous chalcogenides containing TMs.

2. Simulation Details

The aim of this work was to study theoretically the amorphous magnetism [20–23] of a prototypical chalcogenide glass a-As$_2$S$_3$ doped with transition metals (Mo, W and V) by means of computer simulations of electronic and magnetic properties within density functional theory (DFT).

The amorphous phase generation procedure used in this study is a "standard" procedure to obtain in silico a melt-quenched amorphous phase of chalcogenides, which has been used in various publications [24,25]. The idea behind this approach is (i) to randomize the structure at a high temperature, (ii) to equilibrate the structure above the melting point of a material and (iii) to quench the melt in order to obtain the amorphous (glassy) phase.

The amorphous phase was generated with the help of *ab initio* molecular dynamics (AIMD) simulations via the following procedure. As the initial structure we used, the $2\times 2\times 3$ supercell of the orpiment [26] (a = 22.95 Å; b = 19.15 Å; c = 12.77 Å; $\alpha = \gamma = 90°$; $\beta = 90.35°$) consisting of 240 atoms (48 formula units As$_2$S$_3$) with a density of 3.49 g/cm^3. The initial structure was randomized using an NVT-ensemble with a Nosé thermostat by heating the structure to 3000 K during 20 ps followed by cooling to 900 K (a temperature just above the melting point) over a period of 15 ps. The amorphous phase was generated by quenching from 900 K to 300 K over a duration of 15 ps. The times for each stage we used are similar to the times used by others in the field [24,25,27]. A similar procedure was repeated for Mo doped a-As$_2$S$_3$. A total of 5 Mo atoms were included in the 240 atom Mo-doped cell corresponding to the Mo concentration of 2%.

The generalized gradient approximation (GGA) for the exchange-correlation functional by Perdew–Burke–Ernzerhof (PBE) [28] along with the Projected Augmented Wave (PAW) [29,30] method to describe the electron–ion interaction as implemented in the plane wave pseudopotential code VASP [31–33] were used. A value of 260 eV was chosen for plane wave kinetic energy cutoff E_{cutoff}. Integration over Brillouin zone (BZ) was accomplished with only Γ-point taking into account the relatively large size of the simulation cell.

To address the issue of the effect of different transition metals on the structure and properties of the glass, we used a simplified approach adopting the following strategy. Since the glass formation process is stochastic and the obtained a-As$_2$S$_3$:Mo structure is just one of the many possible structures, we replaced Mo species by W and V (assuming that such local structures will statistically be formed), after which the structures were relaxed at 0 K. Full geometry optimization (GOpt) at 0 K of all melt-quenched (mq) amorphous structures, both pure and doped, was accomplished via the Broyden–Fletcher–Goldfarb–Shanno (BFGS) [34–37] optimizer using spin-polarized DFT within the GGA with the exchange-correlation functional PW91 by Perdew and Wang [38] and the van der Waals (vdW) dispersion correction by Tkatchenko and Scheffler (TS) [39], as implemented in the plane-wave pseudopotential code CASTEP [40,41]. The Vanderbilt ultrasoft pseudopotentials (USPs) [42] were chosen to describe the electron–ion interactions. As: $4s^24p^3$, S: $3s^23p^4$; Mo: $4s^24p^64d^55s^1$; V: $3s^23p^63d^34s^2$; and W: $5s^25p^65d^46s^2$; these electrons were assigned to the valence space. Geometry optimization (full relaxation) at 0 K was performed at the Γ-point of BZ with E_{cutoff} = 330 eV. Optimization was carried out until the energy difference per atom, the Hellman–Feynman forces on the atoms and all the stress components did not exceed the values of 5×10^{-7} eV/atom, 1×10^{-2} eV/Å and 2×10^{-2} GPa, respectively. The convergence of the SCF-energy was achieved with a tolerance of 5×10^{-8} eV/atom.

While amorphous structures under consideration are not layered and in this sense are not van der Waals (vdW) solids, the presence of strongly polarizable lone-pair electrons requires the use of vdW corrections in calculations. In the CASTEP code, various types of vdW corrections are implemented for different chemical elements and different exchange-correlation functionals, but not all of them are compatible with all types of GGA-functionals and implemented for all Mo, W and V. The only combination of the GGA functional and the vdW correction implemented in the CASTEP code, available for all three TMs (Mo, W and V), is the combination PW91+TS (Perdew-Wang + Tkatchenko-Scheffler).

Finally, for pure a-As$_2$S$_3$ and doped a-As$_2$S$_3$:TM structures, both melt-quenched and melt-quenched with subsequent geometry optimization (mq-GOpt) at 0 K, we calculated different electronic and magnetic properties, including band structure, density of states (DOS), partial and local densities of states (PDOS and LDOS), interatomic distances, atomic, bond and spin Mulliken populations [43,44], total and absolute magnetizations. The Monkhorst–Pack [45] 3 × 3 × 5 k-mesh (23 k-points in the irreducible BZ) was used for calculating DOS and PDOS.

The nature of magnetic ordering of a-As$_2$S$_3$:TM was determined by comparing the calculated values of total and absolute magnetizations. The total magnetization, or in other words the doubled integrated spin density (ISD), gives the doubled total spin of the system. The magnitude of absolute magnetization, or in other words the doubled integrated modulus spin density (IMSD), is a measure of the local unbalanced spin.

Of course, the rather small 240-atom model with just 5 TM atoms may miss some details associated with certain atomic configurations formed in a real glass. Nevertheless, we believe that our model is able to capture the general features of amorphous magnetism in TM-doped chalcogenides glasses.

3. Results and Discussion

The mass densities of the simulated equilibrated amorphous structures were verified against the experimental data (Table 1). The presented data demonstrate that they are all slightly lower (3.095–3.321 g/cm^3) than that of the crystalline phase, which agrees with the experimental value reported (3.193 g/cm^3) [46]. The data presented in Table 1 demonstrate that the discrepancy between the theoretical and experimental a-As$_2$S$_3$ density values is 3%, which is a very reasonable value. Indeed, its magnitude is comparable to the experimental spread of density data reported by different groups for orpiment (crystalline As$_2$S$_3$). Obviously for a glass this spread will be even larger.

When doped with TM impurities, the mass densities of amorphous structures increase from 3.095 g/cm^3 for undoped a-As$_2$S$_3$ to 3.159, 3.202 and 3.321 g/cm^3 for a-As$_2$S$_3$ doped with the V, Mo and W impurities, respectively. In this case, we see that the order of increasing mass density upon doping the a-As$_2$S$_3$ with impurity TM-atoms correlates well with an increase in the atomic number of the impurity TM-atom.

Table 1. Densities of As$_2$S$_3$ and As$_2$S$_3$:TM structures.

Type of Structure	Density (g/cm^3)
c-As$_2$S$_3$ (experiment) [26]	3.494
a-As$_2$S$_3$ (experiment) [46]	3.193
a-As$_2$S$_3$ mq-GOpt	3.095
a-As$_2$S$_3$:V mq-GOpt	3.159
a-As$_2$S$_3$:Mo mq-GOpt	3.202
a-As$_2$S$_3$:W mq-GOpt	3.321

We also note that undoped a-As$_2$S$_3$ is a semiconductor with a DFT gap of around 1.0 eV as shown in Figure 1(left panel), which is in reasonable agreement with the experimental value of 2.4 eV considering that DFT usually underestimates the gap value by about 50% due to incomplete exclusion of electron self-interaction when using LDA and GGA approximations [47,48].

Figure 2(left) shows the a-As$_2$S$_3$:Mo melt-quenched structure. At the right of Figure 2, we show fragments of the structure around the Mo atoms. The following observations can be made. One can see that some of the S atoms are three-fold coordinated, i.e., their lone-pair electrons are consumed to form covalent (donor-acceptor) bonds with Mo. Analysis of the obtained structure shows that of the 25 S atoms that form covalent bonds with Mo species 13 S are three-fold coordinated. In other words, doping with transition metals results in a decreased concentration of lone-pair electrons.

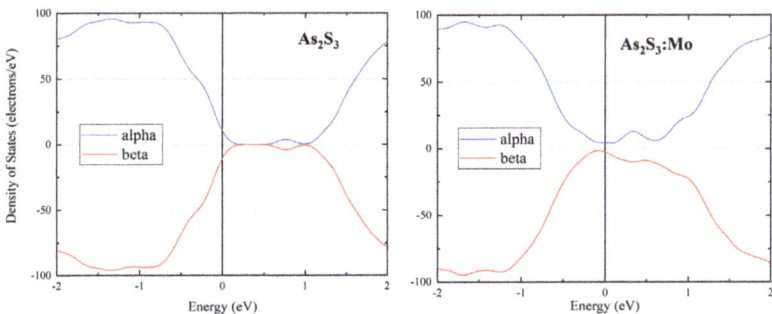

Figure 1. DOS of mq-GOpt structures of a-As$_2$S$_3$ (**left**) and a-As$_2$S$_3$:Mo (**right**).

A detailed analysis of the short-range order around molybdenum atoms is difficult as the structure used in the calculations is rather large. That is why to accomplish this analysis fragments of the structure were taken, in which broken bonds were saturated with hydrogen atoms. The differential electron density (charge density difference, CDD), allowing visualization of both covalent bonds (CBs) and lone-pair (LP) electrons, was calculated. CDD is the difference in electron density in the structure under study and the sum of isolated atoms. Consequently the CDD clouds, i.e., an increase in atomic density between atoms, correspond to covalent bonds. Lone pairs are also associated with an increased electron density. An example of electronic distribution visualization is shown in Figure 3. It can be seen that molybdenum atoms are connected with surrounding atoms with covalent bonds. Additionally, lone-pairs formed by s-electrons of arsenic atoms are also visible. At the same time, it is interesting to note that CDD corresponding to p-lone-pairs of sulfur atoms in some cases are not observed in agreement with the formation of three-fold coordinated sulfur atoms mentioned above.

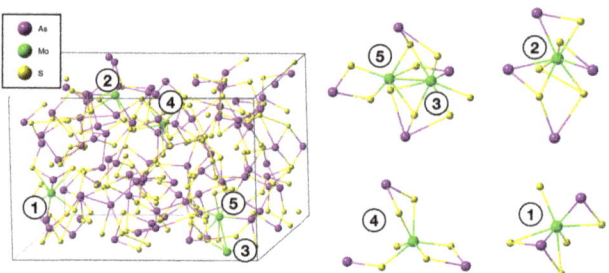

Figure 2. Melt-quenched a-As_2S_3:Mo obtained from AIMD simulations (**left**) and fragments of the structure centered around the Mo atoms (**right**). Mo atoms in the a-As_2S_3 structure are numbered from 1 to 5 (left panel). In the right panel, local structures around these atoms are shown. As atoms—violet; S atoms—yellow; Mo atoms—green.

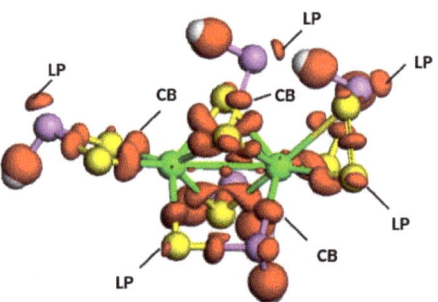

Figure 3. A fragment of Mo-doped a-As_2S_3. Dangling bonds are saturated with hydrogen atoms (grey). As atoms—violet; S atoms—yellow; Mo atoms—green. Charge Density Difference is shown in red for covalent bonds (CBs) and lone-pairs (LPs).

The effect of Mo-doping on the electronic structure can be seen from Figure 1, where densities of states for the undoped and Mo-doped mq-GOpt structures are given. One can see that spin-up and spin-down DOSs in pure a-As_2S_3 are mirror-symmetric in agreement with the experiment (a-As_2S_3 is diamagnetic). Here it is important to note that spin-up and spin-down DOSs become different in the doped glass, which is an indication that the material became magnetic. Another apparent difference is the formation of a strong tail at the bottom of the conduction band extending down to the valence band and effectively closing the band gap in the Mo-doped glass.

Molybdenum doping also results in the formation of a finite density of states at the Fermi level, i.e., the material becomes a metal. This is in line with the fact that amorphous MoS_2 is metallic [49] while the corresponding crystal in its stable 2H form is a semiconductor. The metallic conductivity of amorphous MoS_2 is due to the formation of a large concentration of metallic Mo-Mo bonds in the amorphous phase. It is remarkable that despite a rather low concentration of Mo atoms in our model, a Mo-Mo dimer is formed, which, again, correlates with the strong tendency of molybdenum to form Mo-Mo bonds in the amorphous phase.

It is not unnatural to assume that isolated Mo atoms and a Mo-Mo dimer contribute differently to the density of states. To address this issue, we removed one of the atoms in the molybdenum dimer, after which the structure was additionally relaxed. In the structure without a dimer, the tail states disappear, while some states in the gap remain. This result allows us to draw a conclusion that the conduction band tail is associated with Mo dimers.

Since Mo atoms all possess different local structures, it is interesting to see which of them (and how) contribute to the magnetic properties. This was done via DFT-method calculating the contributions from individual molybdenum atoms, as well as vanadium and tungsten atoms, to the total spin of the a-As_2S_3:TM structures using the well-known Mulliken approximation for the analysis of atomic, bond and spin populations. The results of the Mulliken analysis of spin populations of individual atoms, along with the atomic and bond populations, are presented in Appendix A. The performed DFT calculations showed that due to the different local structure of impurity TM-atoms, not all of them contribute to the total spin of the system. In particular, as can be seen from Tables A1 and A2, only Mo1 and Mo3 atoms contribute noticeably to the total spin of a-As_2S_3:Mo. Similar conclusions can also be drawn for vanadium atoms, where four out of five V atoms make a significant contribution to the total spin for mq a-As_2S_3:V (see Table A3) and only two out of five—for mq-GOpt a-As_2S_3:V (see Table A4). As regards the antiferromagnetic structure mq a-As_2S_3:W with zero total spin, the main noticeable contributions to the total spin come from the projections of spins of opposite signs of atoms W1 and W3 (see Table A5). For equilibrated paramagnetic mq-GOpt a-As_2S_3:W structure, all tungsten atoms have zero spins (see Table A6). It should be noted that, in addition to the Mulliken atomic and spin populations, Appendix A also lists the bond populations along with the bond lengths between impurity TM atoms and ligand atoms (see Tables A7–A12). An analysis of these data allows us to conclude that in the glassy (amorphous) structures a-As_2S_3:TM there are mainly two types of hybridization of the orbitals of impurity TM atoms, namely, trigonal-pyramidal (d^1sp^3) and octahedral (d^2sp^3). The first hybridization type involves s, p_x, p_y, p_z and d_{z^2} orbitals in the combination $(s+p_x+p_y)+(p_z+d_{z^2})$. The other one contains another $d_{x^2-y^2}$ in addition to the same orbitals in the combination $(s+p_x+p_y+d_{x^2-y^2})+(p_z+d_{z^2})$. Of course, in glass hybridization is distorted due to the spread in bond lengths and angles.

While the calculation of the contributions of individual atoms to the total spin of the system using the Mulliken approach is very approximate, the total spin can also be calculated more accurately by integrating the total spin density. Indeed, the CASTEP code enables one to calculate two quantities, which qualitatively determine the type of magnetic ordering, namely the integrated spin density (total spin of the system) and the integrated modulus spin density (a measure of the local unbalanced spin), which allow one to characterize magnetism in materials. Depending on the obtained values, a material can be characterized as being paramagnetic, ferromagnetic, ferrimagnetic or antiferromagnetic. Thus, if both quantities are equal to zero, the material is paramagnetic. Both non-zero quantities and the same magnitude determine ferromagnetic ordering. In the case of non-zero value for the first number and a bigger value for the second, the material is ferrimagnetic. Finally, a zero value for the first number and non-zero for the second characterizes antiferromagnetic ordering of material (the total spin is zero, but the local spin density varies).

In order to verify the appearance of magnetism, we performed spin-up and spin-down PDOS (alpha and beta) calculations. As can be seen from Figure 4, spin-up and spin-down contributions from Mo d-electrons are clearly different. At the same time, PDOS of spin-up and spin-down states associated with As and S atoms also become slightly asymmetric which indicates the presence of some contribution to the magnetic properties of the a-As_2S_3:TM from As and S atoms when doping with a TM impurity. It is interesting to note that a similar result, namely, an appearance of magnetism on normally diamagnetic S atoms was also observed for edge states of WS_2 nanosheets [50].

Of significant interest is the observation that the contribution of d-electrons substantially changes upon relaxation of the structure. Since the relaxation is done at 0 K when bond breaking is very unlikely to happen and only the existing bond lengths and bond angles slightly change, the result suggests that magnetism in doped a-As_2S_3:TM is rather "fragile". In Figure 5, we compare spin-up and spin-down contributions to PDOS from d-electrons of Mo, W and V atoms before and after relaxation. One can see that—as in the case of Mo doping—PDOS shape changes after relaxation. Of special interest is the fact that in the relaxed W-doped a-As_2S_3 spin-up and spin-down contributions become identical (paramagnetic state). We believe that the observed fragility of magnetism in doped a-As_2S_3 is related to the absence of dark ESR in chalcogenide glasses, which is due to the absence of dangling bonds with unpaired spins caused by the fact that a melt-quenched relaxed glassy structure adjusts to have all bonds saturated [51]. It should be noted that magnetism is also not observed in single crystal transition-metal chalcogenides but appears at sample edges and grain boundaries [6]. Similarly, magnetism in the studied glasses may be associated with the presence of soft modes [52], where some of the bonds may be electron deficient. The latter may be easily affected by slight changes in the local environment.

Analysis of the integrated spin density and integrated modulus spin density shows that the nature of magnetic ordering in a-As_2S_3:TM is different for different dopants, namely, the melt-quenched a-As_2S_3:Mo and melt-quenched a-As_2S_3:V are ferrimagnetic, while the melt-quenched a-As_2S_3:W structure is antiferromagnetic. After subsequent relaxation at 0 K of the melt-quenched structure geometry, the Mo- and V-doped structures remain ferrimagnetic, while the W-doped structure becomes paramagnetic. A summary of the results is given in Table 2, which shows the integrated spin densities (total spin of the system) and integrated modulus spin densities (a measure of the local unbalanced spin) for the materials studied. From Table 2, one can see that despite the observed differences in PDOS of as-quenched and 0K-optimized structures the type of magnetism is preserved (ferrimagnetic) in Mo- and V-doped a-As_2S_3, while in W-doped a-As_2S_3 structural optimization (relaxation) changes the type of magnetic ordering from antiferromagnetic to paramagnetic.

We attribute this unusual behavior of the W-doped amorphous structure to the fact that when the Mo atom is substituted by a bigger W atom, the lattice becomes locally stressed and when this stress is removed during the subsequent optimization of the geometry at 0 K, the nature of the magnetic ordering changes significantly, as a result of which the optimized W-doped a-As_2S_3 structure loses magnetic properties. This opens up the interesting possibility of controlling the magnetic properties of the TM-doped amorphous structure by applying external pressure. The study of pressure effect on the magnetic properties of a-As_2S_3 is currently underway.

Table 2. Type of magnetic ordering of a-As_2S_3:TM structure.

	Melt-Quenched			Melt-Quenched and Relaxed		
Dopant	ISD ($\hbar$/2)	IMSD ($\hbar$/2)	Magnetic Ordering	ISD ($\hbar$/2)	IMSD ($\hbar$/2)	Magnetic Ordering
Mo ($4s^2 4p^6 4d^5 5s^1$)	2.00	2.58	ferrimagnetic	2.00	2.54	ferrimagnetic
V ($3s^2 3p^6 3d^3 4s^2$)	2.97	4.36	ferrimagnetic	1.00	1.88	ferrimagnetic
W ($5s^2 5p^6 5d^4 6s^2$)	0.00	1.45	antiferromagnetic	0.00	0.00	paramagnetic

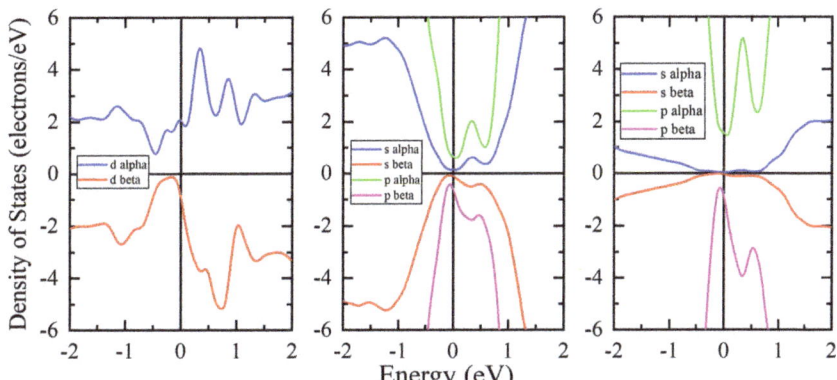

Figure 4. Partial densities of Mo d-states (**left**), As sp-states (**center**) and of S sp-states (**right**) for mq-GOpt a-As$_2$S$_3$:Mo structure.

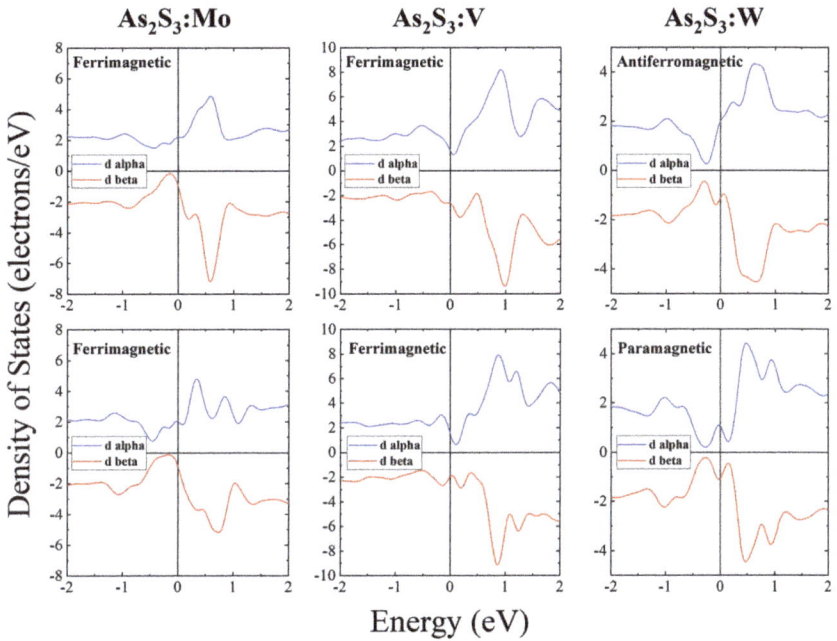

Figure 5. Partial densities of Mo d-states, V d-states and W d-states for mq structures (**upper panel**) and for mq-GOpt structures (**lower panel**).

4. Conclusions

The electronic configuration of sulphur, $s^2p_x^1p_y^1p_z^2$, means that two of the four p-electrons are used to form covalent bonds and two are left as a lone-pair. The accomplished DFT simulations indeed showed that for both structures of a pure glass and a-As$_2$S$_3$:Mo the majority of As and S atoms satisfy the 8-N rule, i.e., As atoms are three-fold coordinated and S atoms are two-fold coordinated. In the doped glass, some lone-pair electrons are consumed to form Ch-TM bonds.

We demonstrate a strong effect of TM-dopants on the electronic structure of the a-As_2S_3 as well as an appearance of a magnetic response. Of special interest is the fact that doping with different transition metals results in the formation of materials with different magnetic properties. Our results suggest that chalcogenide glasses doped with transition metals may become a technologically important material.

Author Contributions: Conceptualization—A.V.K. and M.K.; methodology—A.V.K. and V.G.K.; investigation—V.G.K., A.A.G. and M.K.; writing original draft—V.G.K. and A.V.K.; writing review and editing—V.G.K., A.A.G., M.K., V.A.T. and A.V.K. All authors have read and agreed to the published version of the manuscript.

Funding: This work has been performed within a joint Russian-Czech project partly funded by the Russian Foundation for Basic Research (grant 19-53-26017) and the Czech Science Foundation (grant 20-23392J).

Data Availability Statement: Additional data can obtained from the authors upon reasonable request.

Acknowledgments: The authors are grateful to Yuta Saito and Paul Fons for their involvement during the initial stage of this project.

Conflicts of Interest: The authors declare no conflict of interest.

Abbreviations

The following abbreviations are used in this manuscript:

DFT	Density Functional Theory
AIMD	Ab initio molecular dynamics
GGA	Generalized Gradient Approximation
PW91	Perdew-Wang
PBE	Perdew-Burke-Ernzerhof
vdW	van der Waals
TS	Tkatchenko and Scheffler
PAW	Projected Augmented Wave
VASP	Vienna Ab initio Simulation Package
CASTEP	CAmbridge Serial Total Energy Package
USP	Ultrasoft pseudopotentials
BFGS	Broyden-Fletcher-Goldfarb-Shanno
BZ	Brillouin zone
Ch	Chalcogen
TM	Transition metal
DOS	Density of States
PDOS	Partial Density of States
LDOS	Local Density of States
SCF	Self-Consistent Field
ISD	Integrated Spin Density
IMSD	Integrated Modulus Spin Density
mq	Melt-quenched
GOpt	Geometry Optimization
CB	Covalent bond
LP	Lone pair
CDD	Charge Density Difference

Appendix A. Atomic and Bond Populations

Table A1. Atomic populations (Mulliken) for mq As$_2$S$_3$:Mo.

Species	Spin	s	p	d	Total	Charge (e)	Spin ($\hbar$/2)
Mo1	up	1.261	3.332	2.840	7.433	−0.150	0.715
	down	1.240	3.304	2.174	6.717		
Mo2	up	1.258	3.313	2.541	7.112	−0.206	0.017
	down	1.257	3.313	2.524	7.095		
Mo3	up	1.268	3.308	2.843	7.420	−0.004	0.836
	down	1.235	3.256	2.093	6.584		
Mo4	up	1.248	3.325	2.456	7.029	−0.056	0.001
	down	1.248	3.325	2.454	7.027		
Mo5	up	1.238	3.240	2.537	7.014	−0.019	0.010
	down	1.240	3.245	2.519	7.004		

Table A2. Atomic populations (Mulliken) for mq-GOpt As$_2$S$_3$:Mo.

Species	Spin	s	p	d	Total	Charge (e)	Spin ($\hbar$/2)
Mo1	up	1.248	3.302	2.811	7.362	−0.084	0.640
	down	1.230	3.275	2.216	6.722		
Mo2	up	1.242	3.301	2.540	7.083	−0.158	0.009
	down	1.242	3.301	2.532	7.074		
Mo3	up	1.272	3.278	2.856	7.405	0.062	0.873
	down	1.234	3.225	2.073	6.533		
Mo4	up	1.237	3.297	2.459	6.993	0.014	0.001
	down	1.237	3.297	2.458	6.993		
Mo5	up	1.246	3.226	2.540	7.012	0.014	0.039
	down	1.247	3.230	2.497	6.974		

Table A3. Atomic populations (Mulliken) for mq As$_2$S$_3$:V.

Species	Spin	s	p	d	Total	Charge (e)	Spin ($\hbar$/2)
V1	up	1.234	3.351	1.893	6.478	0.051	0.007
	down	1.234	3.351	1.886	6.471		
V2	up	1.242	3.358	2.223	6.824	0.007	0.654
	down	1.224	3.332	1.614	6.170		
V3	up	1.244	3.340	2.167	6.751	0.176	0.678
	down	1.224	3.306	1.542	6.073		
V4	up	1.229	3.345	1.927	6.501	0.136	0.139
	down	1.226	3.341	1.796	6.363		
V5	up	1.229	3.317	2.294	6.840	0.154	0.834
	down	1.206	3.287	1.513	6.006		

Table A4. Atomic populations (Mulliken) for mq-GOpt As$_2$S$_3$:V.

Species	Spin	s	p	d	Total	Charge (e)	Spin ($\hbar/2$)
V1	up	1.214	3.315	1.909	6.439	0.115	−0.008
	down	1.214	3.316	1.917	6.447		
V2	up	1.210	3.329	1.951	6.490	0.041	0.022
	down	1.210	3.328	1.930	6.469		
V3	up	1.231	3.305	2.164	6.700	0.214	0.614
	down	1.214	3.282	1.590	6.086		
V4	up	1.212	3.317	1.879	6.408	0.182	−0.001
	down	1.212	3.317	1.880	6.410		
V5	up	1.220	3.290	2.226	6.735	0.184	0.654
	down	1.203	3.269	1.609	6.081		

Table A5. Atomic populations (Mulliken) for mq As$_2$S$_3$:W.

Species	Spin	s	p	d	Total	Charge (e)	Spin ($\hbar/2$)
W1	up	1.327	3.375	2.227	6.930	−0.160	−0.301
	down	1.337	3.391	2.503	7.230		
W2	up	1.340	3.385	2.372	7.097	−0.200	−0.007
	down	1.340	3.385	2.378	7.104		
W3	up	1.338	3.367	2.487	7.192	0.008	0.393
	down	1.320	3.333	2.147	6.800		
W4	up	1.333	3.383	2.291	7.007	−0.015	0.000
	down	1.334	3.383	2.291	7.008		
W5	up	1.315	3.310	2.393	7.018	−0.004	0.031
	down	1.314	3.309	2.363	6.987		

Table A6. Atomic populations (Mulliken) for mq-GOpt As$_2$S$_3$:W.

Species	Spin	s	p	d	Total	Charge (e)	Spin ($\hbar/2$)
W1	up	1.321	3.362	2.366	7.049	−0.098	0.000
	down	1.321	3.362	2.366	7.049		
W2	up	1.329	3.377	2.373	7.079	−0.158	0.000
	down	1.329	3.377	2.373	7.079		
W3	up	1.330	3.327	2.312	6.969	0.062	0.000
	down	1.330	3.327	2.313	6.969		
W4	up	1.325	3.361	2.290	6.976	0.047	0.000
	down	1.325	3.361	2.290	6.976		
W5	up	1.321	3.305	2.361	6.987	0.026	0.000
	down	1.321	3.305	2.361	6.987		

Table A7. Bond populations for mq As$_2$S$_3$:Mo.

Bond	Population	Length (Å)
S 62 – Mo 1	0.42	2.35153
S 97 – Mo 1	0.70	2.36834
S 43 – Mo 1	0.55	2.45819
S 96 – Mo 1	0.50	2.55119
As 2 – Mo 1	0.10	2.62421
S 117 – Mo 1	0.19	2.66196
S 37 – Mo 2	0.51	2.39509
S 71 – Mo 2	0.55	2.44170
S 104 – Mo 2	0.34	2.51202
S 51 – Mo 2	0.25	2.51837
S 48 – Mo 2	0.29	2.54396
S 50 – Mo 3	0.65	2.28939
S 136 – Mo 3	0.39	2.40071
S 18 – Mo 3	0.45	2.41271
S 103 – Mo 3	0.42	2.47104
S 66 – Mo 3	0.37	2.48602
S 100 – Mo 3	0.47	2.54992
S 21 – Mo 4	0.83	2.27300
S 5 – Mo 4	0.57	2.35213
S 65 – Mo 4	0.69	2.36482
S 132 – Mo 4	0.52	2.40702
S 138 – Mo 4	0.57	2.44573
S 102 – Mo 4	0.26	2.67791
S 61 – Mo 5	0.55	2.31356
S 98 – Mo 5	0.31	2.38004
S 103 – Mo 5	0.53	2.44358
S 136 – Mo 5	0.33	2.47183
S 139 – Mo 5	0.14	2.51564
S 66 – Mo 5	0.38	2.59185

Table A8. Bond populations for mq-GOpt As$_2$S$_3$:Mo.

Bond	Population	Length (Å)
S 97 – Mo 1	0.69	2.34497
S 62 – Mo 1	0.40	2.38299
S 43 – Mo 1	0.52	2.44055
S 96 – Mo 1	0.56	2.44842
S 117 – Mo 1	0.17	2.60819
As 2 – Mo 1	0.03	2.67550
S 37 – Mo 2	0.49	2.36059
S 71 – Mo 2	0.53	2.42611
S 104 – Mo 2	0.37	2.49311
S 51 – Mo 2	0.27	2.49964
S 48 – Mo 2	0.22	2.58459
As 69 – Mo 2	0.05	2.65684
S 50 – Mo 3	0.61	2.34939
S 136 – Mo 3	0.41	2.41678
S 18 – Mo 3	0.48	2.42998
S 103 – Mo 3	0.42	2.43346
S 66 – Mo 3	0.35	2.45545
S 100 – Mo 3	0.50	2.49639

Table A8. *Cont.*

Bond	Population	Length (Å)
S 21 – Mo 4	0.81	2.24730
S 5 – Mo 4	0.55	2.34310
S 65 – Mo 4	0.64	2.36593
S 138 – Mo 4	0.57	2.41099
S 132 – Mo 4	0.46	2.50006
S 102 – Mo 4	0.27	2.60575
S 98 – Mo 5	0.31	2.36449
S 103 – Mo 5	0.45	2.43617
S 136 – Mo 5	0.33	2.44773
S 61 – Mo 5	0.50	2.47537
S 139 – Mo 5	0.20	2.49435
S 66 – Mo 5	0.38	2.55565
As 18 – Mo 5	0.14	2.62478

Table A9. Bond populations for mq As_2S_3:V.

Bond	Population	Length (Å)
S 62 – V 1	0.44	2.35153
S 97 – V 1	0.64	2.36834
S 43 – V 1	0.51	2.45819
S 96 – V 1	0.48	2.55119
V 1 – As 2	0.20	2.62421
S 117 – V 1	0.21	2.66196
S 37 – V 2	0.49	2.39509
S 71 – V 2	0.52	2.44170
S 104 – V 2	0.32	2.51202
S 51 – V 2	0.27	2.51837
S 48 – V 2	0.29	2.54396
V 2 – As 69	0.02	2.69034
S 50 – V 3	0.63	2.28939
S 136 – V 3	0.41	2.40071
S 18 – V 3	0.43	2.41271
S 103 – V 3	0.44	2.47104
S 66 – V 3	0.37	2.48602
S 100 – V 3	0.46	2.54992
S 21 – V 4	0.73	2.27300
S 5 – V 4	0.51	2.35213
S 65 – V 4	0.63	2.36482
S 132 – V 4	0.46	2.40702
S 138 – V 4	0.52	2.44573
S 102 – V 4	0.25	2.67791
S 61 – V 5	0.57	2.31356
S 98 – V 5	0.33	2.38004
S 103 – V 5	0.46	2.44358
S 136 – V 5	0.32	2.47183
S 139 – V 5	0.19	2.51564
S 66 – V 5	0.37	2.59185

Table A10. Bond populations for mq-GOpt As_2S_3:V.

Bond	Population	Length (Å)
S 97 – V 1	0.65	2.29484
S 96 – V 1	0.57	2.32880
S 62 – V 1	0.38	2.35927
S 43 – V 1	0.49	2.37497
S 117 – V 1	0.18	2.60801
V 1 – As 2	0.10	2.65447
S 37 – V 2	0.51	2.28793
S 104 – V 2	0.49	2.32588
S 71 – V 2	0.44	2.42616
S 51 – V 2	0.27	2.48525
S 48 – V 2	0.25	2.55042
V 2 – As 69	0.05	2.87340
S 50 – V 3	0.63	2.25890
S 136 – V 3	0.41	2.34574
S 103 – V 3	0.43	2.35365
S 18 – V 3	0.46	2.36615
S 66 – V 3	0.35	2.42930
S 100 – V 3	0.42	2.51927
S 21 – V 4	0.73	2.19812
S 5 – V 4	0.49	2.30330
S 65 – V 4	0.60	2.32754
S 138 – V 4	0.53	2.37150
S 132 – V 4	0.41	2.49042
S 102 – V 4	0.23	2.64811
S 98 – V 5	0.34	2.29003
S 139 – V 5	0.23	2.38030
S 61 – V 5	0.54	2.38320
S 103 – V 5	0.39	2.42524
S 136 – V 5	0.29	2.48977
S 66 – V 5	0.39	2.52080
V 5 – As 18	0.08	2.66532

Table A11. Bond populations for mq As_2S_3:W.

Bond	Population	Length (Å)
S 97 – W 1	0.76	2.36834
S 43 – W 1	0.61	2.45819
S 96 – W 1	0.53	2.55119
As 2 – W 1	0.29	2.62421
S 117 – W 1	0.24	2.66196
S 37 – W 2	0.58	2.39509
S 71 – W 2	0.60	2.44170
S 104 – W 2	0.41	2.51202
S 51 – W 2	0.31	2.51837
S 48 – W 2	0.34	2.54396
As 69 – W 2	0.10	2.69034
S 50 – W 3	0.74	2.28939
S 136 – W 3	0.44	2.40071
S 18 – W 3	0.53	2.41271
S 103 – W 3	0.50	2.47104
S 66 – W 3	0.42	2.48602
S 100 – W 3	0.55	2.54992

Table A11. *Cont.*

Bond	Population	Length (Å)
S 21 – W 4	0.92	2.27300
S 5 – W 4	0.65	2.35213
S 65 – W 4	0.75	2.36482
S 132 – W 4	0.58	2.40702
S 138 – W 4	0.63	2.44573
S 102 – W 4	0.31	2.67791
S 61 – W 5	0.65	2.31372
S 98 – W 5	0.38	2.37996
S 103 – W 5	0.55	2.44368
S 136 – W 5	0.37	2.47186
S 139 – W 5	0.21	2.51564
S 66 – W 5	0.39	2.59173
As 18 – W 5	0.14	2.65071

Table A12. Bond populations for mq-GOpt As_2S_3:W.

Bond	Population	Length (Å)
S 62 – W 1	0.55	2.34022
S 97 – W 1	0.74	2.34139
S 43 – W 1	0.62	2.41789
S 96 – W 1	0.58	2.46070
S 117 – W 1	0.24	2.58213
As 2 – W 1	0.17	2.68840
S 37 – W 2	0.58	2.35692
S 71 – W 2	0.58	2.43016
S 104 – W 2	0.46	2.47766
S 51 – W 2	0.35	2.48555
S 48 – W 2	0.28	2.56779
As 69 – W 2	0.20	2.67965
S 50 – W 3	0.69	2.35406
S 103 – W 3	0.53	2.40099
S 18 – W 3	0.57	2.41828
S 100 – W 3	0.60	2.43290
S 136 – W 3	0.43	2.43514
S 66 – W 3	0.39	2.46608
S 21 – W 4	0.93	2.24250
S 5 – W 4	0.61	2.35681
S 65 – W 4	0.71	2.36997
S 138 – W 4	0.63	2.40442
S 132 – W 4	0.54	2.47843
S 102 – W 4	0.32	2.60427
S 98 – W 5	0.40	2.38134
S 103 – W 5	0.50	2.43729
S 61 – W 5	0.59	2.45037
S 136 – W 5	0.37	2.46057
S 139 – W 5	0.26	2.49339
S 66 – W 5	0.41	2.54839
As 18 – W 5	0.31	2.60879

References

1. Goryunova, N.; Kolomiets, B. Electrical properties and structure in system of Selenide of Tl, Sb, and As. *Zhurnal Tekhnicheskoi Fiz.* **1955**, *25*, 2669.
2. Kastner, M. Bonding Bands, Lone-Pair Bands and Impurity States in Chalcogenide Semiconductors. *Phys. Rev. Lett.* **1972**, *28*, 355–357. [CrossRef]
3. Kolobov, A.V. *Photo-Induced Metastability in Amorphous Semiconductors*; Wiley VCH: Weinheim, Germany, 2003; p. 436.

4. Tanaka, K.; Shimakawa, K. *Amorphous Chalcogenide Semiconductors and Related Materials*; Springer: New York, NY, USA, 2011. [CrossRef]
5. Kolobov, A.V.; Oyanagi, H.; Tanaka, K.; Tanaka, K. Structural study of amorphous selenium by in situ EXAFS: Observation of photoinduced bond alternation. *Phys. Rev. B* **1997**, *55*, 726–734. [CrossRef]
6. Kolobov, A.V.; Tominaga, J. *Two-Dimensional Transition-Metal Dichalcogenides*; Springer International Publishing: Cham, Switzerland, 2016. [CrossRef]
7. Goudsmit, S.A.; Richards, P.I. The Order of Electron Shells in Ionized Atoms. *Proc. Natl. Acad. Sci. USA* **1964**, *51*, 664–671. Erratum in *Proc. Natl. Acad. Sci. USA* **1964**, *51*, 906. [CrossRef] [PubMed]
8. Pyykkö, P. Is the Periodic Table all right ("PT OK")? *EPJ Web Conf.* **2016**, *131*, 01001. [CrossRef]
9. Klechkovskii, V. On the First Appearance of Atomic Electrons With l, n, nr and n + l Given. *J. Exp. Theor. Phys.* **1956**, *3*, 125.
10. Klechkovskii, V. Justification of the Rule for Successive Filling of (n + l) Groups. *J. Exp. Theor. Phys.* **1962**, *14*, 334.
11. Wilson, J.A.; Salvo, F.J.D.; Mahajan, S. Charge-Density Waves in Metallic, Layered, Transition-Metal Dichalcogenides. *Phys. Rev. Lett.* **1974**, *32*, 882–885. [CrossRef]
12. Wilson, J.; Salvo, F.D.; Mahajan, S. Charge-density waves and superlattices in the metallic layered transition metal dichalcogenides. *Adv. Phys.* **1975**, *24*, 117–201. [CrossRef]
13. McMillan, W.L. Landau theory of charge-density waves in transition-metal dichalcogenides. *Phys. Rev. B* **1975**, *12*, 1187–1196. [CrossRef]
14. Gor'kov, L.P. Strong electron-lattice coupling as the mechanism behind charge density wave transformations in transition-metal dichalcogenides. *Phys. Rev. B* **2012**, *85*, 165142. [CrossRef]
15. Shen, D.W.; Xie, B.P.; Zhao, J.F.; Yang, L.X.; Fang, L.; Shi, J.; He, R.H.; Lu, D.H.; Wen, H.H.; Feng, D.L. Novel Mechanism of a Charge Density Wave in a Transition Metal Dichalcogenide. *Phys. Rev. Lett.* **2007**, *99*, 216404. [CrossRef] [PubMed]
16. Dai, J.; Calleja, E.; Alldredge, J.; Zhu, X.; Li, L.; Lu, W.; Sun, Y.; Wolf, T.; Berger, H.; McElroy, K. Microscopic evidence for strong periodic lattice distortion in two-dimensional charge-density wave systems. *Phys. Rev. B* **2014**, *89*, 165140. [CrossRef]
17. Neto, A.H.C. Charge Density Wave, Superconductivity and Anomalous Metallic Behavior in 2D Transition Metal Dichalcogenides. *Phys. Rev. Lett.* **2001**, *86*, 4382–4385. [CrossRef]
18. Hellmann, S.; Rohwer, T.; Kalläne, M.; Hanff, K.; Sohrt, C.; Stange, A.; Carr, A.; Murnane, M.; Kapteyn, H.; Kipp, L.; et al. Time-domain classification of charge-density-wave insulators. *Nat. Commun.* **2012**, *3*, 1069. [CrossRef]
19. Rossnagel, K. On the origin of charge-density waves in select layered transition-metal dichalcogenides. *J. Phys. Condens. Matter* **2011**, *23*, 213001. [CrossRef]
20. Hooper, H.O.; de Graaf, A.M. (Eds.) *Amorphous Magnetism*; Springer: New York, NY, USA, 1973. [CrossRef]
21. Levy, R.A.; Hasegawa, R. (Eds.) *Amorphous Magnetism II*; Springer: New York, NY, USA, 1977. [CrossRef]
22. Kaneyoshi, T. *Amorphous Magnetism*; CRC Press: Boca Raton, FL, USA, 1984; p. 190.
23. Fairman, R.; Ushkov, B. (Eds.) *Semiconducting Chalcogenide Glass II*; Academic Press: Cambridge, MA, USA, 2004; Volume 79, p. 307.
24. Konstantinou, K.; Mavračić, J.; Mocanu, F.C.; Elliott, S.R. Simulation of Phase-Change-Memory and Thermoelectric Materials using Machine-Learned Interatomic Potentials: Sb_2Te_3. *Phys. Status Solidi (b)* **2020**, *258*, 2000416. [CrossRef]
25. Mocanu, F.C.; Konstantinou, K.; Mavračić, J.; Elliott, S.R. On the Chemical Bonding of Amorphous Sb_2Te_3. *Phys. Status Solidi Rapid Res. Lett.* **2020**, *15*, 2000485. [CrossRef]
26. Mullen, D.J.E.; Nowacki, W. Refinement of the crystal structures of realgar, AsS and orpiment, As_2S_3. *Z. Krist.* **1972**, *136*, 48–65. [CrossRef]
27. Caravati, S.; Bernasconi, M.; Kühne, T.D.; Krack, M.; Parrinello, M. Coexistence of tetrahedral- and octahedral-like sites in amorphous phase change materials. *Appl. Phys. Lett.* **2007**, *91*, 171906. [CrossRef]
28. Perdew, J.P.; Burke, K.; Ernzerhof, M. Generalized Gradient Approximation Made Simple. *Phys. Rev. Lett.* **1996**, *77*, 3865–3868. [CrossRef] [PubMed]
29. Blöchl, P.E. Projector augmented-wave method. *Phys. Rev. B* **1994**, *50*, 17953–17979. [CrossRef] [PubMed]
30. Kresse, G.; Joubert, D. From ultrasoft pseudopotentials to the projector augmented-wave method. *Phys. Rev. B* **1999**, *59*, 1758–1775. [CrossRef]
31. Kresse, G.; Hafner, J. Ab initio molecular dynamics for liquid metals. *Phys. Rev. B* **1993**, *47*, 558–561. [CrossRef]
32. Kresse, G.; Furthmüller, J. Efficiency of ab-initio total energy calculations for metals and semiconductors using a plane-wave basis set. *Comput. Mater. Sci.* **1996**, *6*, 15–50. [CrossRef]
33. Kresse, G.; Furthmüller, J. Efficient iterative schemes for ab initio total-energy calculations using a plane-wave basis set. *Phys. Rev. B* **1996**, *54*, 11169–11186. [CrossRef]
34. Pfrommer, B.G.; Côté, M.; Louie, S.G.; Cohen, M.L. Relaxation of Crystals with the Quasi-Newton Method. *J. Comput. Phys.* **1997**, *131*, 233–240. [CrossRef]
35. Press, W.H.; Flannery, B.P.; Teukolsky, S.A.; Vetterling, W.T. *Numerical Recipes in C*; Cambridge University Press: Cambridge, UK, 1992; p. 994.
36. Shanno, D.F. Conjugate Gradient Methods with Inexact Searches. *Math. Oper. Res.* **1978**, *3*, 244–256. [CrossRef]
37. Eyert, V. A Comparative Study on Methods for Convergence Acceleration of Iterative Vector Sequences. *J. Comput. Phys.* **1996**, *124*, 271–285. [CrossRef]

38. Perdew, J.P.; Wang, Y. Accurate and simple analytic representation of the electron-gas correlation energy. *Phys. Rev. B* **1992**, *45*, 13244–13249. [CrossRef]
39. Tkatchenko, A.; Scheffler, M. Accurate Molecular Van Der Waals Interactions from Ground-State Electron Density and Free-Atom Reference Data. *Phys. Rev. Lett.* **2009**, *102*, 073005. [CrossRef] [PubMed]
40. Segall, M.D.; Lindan, P.J.D.; Probert, M.J.; Pickard, C.J.; Hasnip, P.J.; Clark, S.J.; Payne, M.C. First-principles simulation: Ideas, illustrations and the CASTEP code. *J. Phys. Condens. Matter* **2002**, *14*, 2717–2744. [CrossRef]
41. Clark, S.J.; Segall, M.D.; Pickard, C.J.; Hasnip, P.J.; Probert, M.I.J.; Refson, K.; Payne, M.C. First principles methods using CASTEP. *Z. Krist. Cryst. Mater.* **2005**, *220*, 567–570. [CrossRef]
42. Vanderbilt, D. Soft self-consistent pseudopotentials in a generalized eigenvalue formalism. *Phys. Rev. B* **1990**, *41*, 7892–7895. [CrossRef] [PubMed]
43. Mulliken, R.S. Electronic Population Analysis on LCAO–MO Molecular Wave Functions. II. Overlap Populations, Bond Orders and Covalent Bond Energies. *J. Chem. Phys.* **1955**, *23*, 1841–1846. [CrossRef]
44. Mulliken, R.S. Electronic Population Analysis on LCAO–MO Molecular Wave Functions. I. *J. Chem. Phys.* **1955**, *23*, 1833–1840. [CrossRef]
45. Monkhorst, H.J.; Pack, J.D. Special points for Brillouin-zone integrations. *Phys. Rev. B* **1976**, *13*, 5188–5192. [CrossRef]
46. Knotek, P.; Kutálek, P.; Černošková, E.; Vlček, M.; Tichý, L. The density, nanohardness and some optical properties of As–S and As–Se chalcogenide bulk glasses and thin films. *RSC Adv.* **2020**, *10*, 42744–42753. [CrossRef]
47. Perdew, J.P. Density functional theory and the band gap problem. *Int. J. Quantum Chem.* **2009**, *28*, 497–523. [CrossRef]
48. Perdew, J.P.; Zunger, A. Self-interaction correction to density-functional approximations for many-electron systems. *Phys. Rev. B* **1981**, *23*, 5048–5079. [CrossRef]
49. Krbal, M.; Prokop, V.; Kononov, A.A.; Pereira, J.R.; Mistrik, J.; Kolobov, A.V.; Fons, P.J.; Saito, Y.; Hatayama, S.; Shuang, Y.; et al. Amorphous-to-Crystal Transition in Quasi-Two-Dimensional MoS_2: Implications for 2D Electronic Devices. *ACS Appl. Nano Mater.* **2021**, *4*, 8834–8844. [CrossRef]
50. Huo, N.; Li, Y.; Kang, J.; Li, R.; Xia, Q.; Li, J. Edge-states ferromagnetism of WS_2 nanosheets. *Appl. Phys. Lett.* **2014**, *104*, 202406. [CrossRef]
51. Kolomiets, B. Electrical and optical properties of vitreous chalcogenide semiconductor films. *Thin Solid Films* **1976**, *34*, 1–7. [CrossRef]
52. Ignatiev, F.; Karpov, V.; Klinger, M. Atomic critical potentials and structure of non-single-well potentials in glasses. *J. Non-Cryst. Solids* **1983**, *55*, 307–323. [CrossRef]

Disclaimer/Publisher's Note: The statements, opinions and data contained in all publications are solely those of the individual author(s) and contributor(s) and not of MDPI and/or the editor(s). MDPI and/or the editor(s) disclaim responsibility for any injury to people or property resulting from any ideas, methods, instructions or products referred to in the content.

Communication

Artificial HfO₂/TiOₓ Synapses with Controllable Memory Window and High Uniformity for Brain-Inspired Computing

Yang Yang [1,2,3], Xu Zhu [1,2,3], Zhongyuan Ma [1,2,3,*], Hongsheng Hu [1,2,3], Tong Chen [1,2,3], Wei Li [1,2,3], Jun Xu [1,2,3], Ling Xu [1,2,3] and Kunji Chen [1,2,3]

1. School of Electronic Science and Engineering, Nanjing University, Nanjing 210093, China
2. Collaborative Innovation Center of Advanced Microstructures, Nanjing University, Nanjing 210093, China
3. Jiangsu Provincial Key Laboratory of Photonic and Electronic Materials Sciences and Technology, Nanjing University, Nanjing 210093, China
* Correspondence: zyma@nju.edu.cn

Citation: Yang, Y.; Zhu, X.; Ma, Z.; Hu, H.; Chen, T.; Li, W.; Xu, J.; Xu, L.; Chen, K. Artificial HfO₂/TiOₓ Synapses with Controllable Memory Window and High Uniformity for Brain-Inspired Computing. *Nanomaterials* **2023**, *13*, 605. https://doi.org/10.3390/nano13030605

Academic Editor: Dong-Wook Kim

Received: 11 December 2022
Revised: 25 January 2023
Accepted: 30 January 2023
Published: 2 February 2023

Copyright: © 2023 by the authors. Licensee MDPI, Basel, Switzerland. This article is an open access article distributed under the terms and conditions of the Creative Commons Attribution (CC BY) license (https://creativecommons.org/licenses/by/4.0/).

Abstract: Artificial neural networks, as a game-changer to break up the bottleneck of classical von Neumann architectures, have attracted great interest recently. As a unit of artificial neural networks, memristive devices play a key role due to their similarity to biological synapses in structure, dynamics, and electrical behaviors. To achieve highly accurate neuromorphic computing, memristive devices with a controllable memory window and high uniformity are vitally important. Here, we first report that the controllable memory window of an HfO₂/TiOₓ memristive device can be obtained by tuning the thickness ratio of the sublayer. It was found the memory window increased with decreases in the thickness ratio of HfO₂ and TiOₓ. Notably, the coefficients of variation of the high-resistance state and the low-resistance state of the nanocrystalline HfO₂/TiOₓ memristor were reduced by 74% and 86% compared with the as-deposited HfO₂/TiOₓ memristor. The position of the conductive pathway could be localized by the nanocrystalline HfO₂ and TiO₂ dot, leading to a substantial improvement in the switching uniformity. The nanocrystalline HfO₂/TiOₓ memristive device showed stable, controllable biological functions, including long-term potentiation, long-term depression, and spike-time-dependent plasticity, as well as the visual learning capability, displaying the great potential application for neuromorphic computing in brain-inspired intelligent systems.

Keywords: artificial synapse; memristor; brain-inspired computing

1. Introduction

With the big data era coming, artificial neural networks, as a game-changer to break up the bottleneck of classical von Neumann architectures, have attracted great interest due to their high massive information processing speed [1–3]. As a unit of artificial neural networks, memristive devices play a key role in achieving a biological function because of their similarity to biological synapses in structure, dynamics, and electrical behaviors [4–9]. To ensure that artificial synapses can be used in neuromorphic computing, memristive devices with a tunable memory window and high uniformity are vitally important. Among the candidates for artificial synapses, HfO₂/TiOₓ memristors are preferred for their compatibility with CMOS and remarkable characteristics of resistance switching [10–21]. Compared with memristors based on the low-dimensional structure of TiOₓ from nanoparticles to nanorods [22–24], the advantage of HfO_y/TiOₓ bilayers lies in two main aspects. First, the barrier of the HfO₂ and TiOₓ sublayer can be modulated by tuning the dielectric permittivity of TiOₓ (~80) and HfO₂ (~22) [25]. Second, the conductive pathway's morphology can be controlled by tuning the thicknesses of the HfO₂ and TiOₓ sublayers due to the various concentrations of oxygen vacancies in the two kinds of sublayers. To obtain an HfO₂/TiOₓ bilayer memristor with a tunable memory window and high uniformity, we fabricated HfO₂/TiOₓ bilayers with different thickness ratios by the atomic layer deposition (ALD) method. Compared with the as-deposited HfO_2/TiO_x bilayer, the coefficients of variation in the high-resistance state and the low-resistance state of nanocrystalline HfO_2/TiO_x memristors could be reduced by 74% and 86%. It was found that the improved uniformity of the HfO_2/TiO_x memristive device was related to the permanent nanocrystalline HfO_2 and TiO_x dots, which could localize the position of the conductive pathway in the HfO_2/TiO_x bilayer. The disconnection and formation of conductive pathways occurred mainly in the thin TiO_x layer, which led to a substantial improvement in the switching uniformity. When the thickness ratio of the HfO_2/TiO_x sublayer increased from 1:4 to 4:1, the memory window increased from 1.0 to 1.8 V. The nanocrystalline HfO_2/TiO_x bilayer memristor also exhibited multiple conductance states similar to the weight of biological synapses updating under pulses with different voltage widths and numbers. The stable, controllable operation and multilevel characteristics enabled the successful implementation of biological functions, including long-term potentiation, long-term depression, and spike-time-dependent plasticity, as well as visual capability, demonstrating the great potential application for neuromorphic systems in the AI period.

2. Experimental Details

Ahead of preparing the thin films, the Si substrate was cleaned following a standard Radio Corporation of American cleaning method. Then, the Ti/Pt bilayers were deposited on the P–Si substrate via electron beam evaporation as the bottom electrode. The thicknesses of the Ti and Pt were 5 nm and 40 nm, respectively. Subsequently, the HfO_2/TiO_x bilayers were grown on the surface of the Ti/Pt bilayer in an atomic layer deposition (ALD) system with $Hf[N(C_2H_5)CH_3]_4$ (TEMAH) and $Ti[N(CH_3)_2]_4$ (TDMAT) as the main precursors. The different thickness ratios of TiO_x and HfO_2 were achieved by tuning the deposition time. Three kinds of HfO_2/TiO_x bilayers with different thickness ratios, including 4:1, 1:1, and 1:4, were obtained, which were named H12T3, H7.5T7.5, and H3T12. The thicknesses of HfO_x in H12T3, H7.5T7.5, and H3T12 were 12, 7.5, and 3 nm. The thicknesses of TiO_2 in H12T3, H7.5T7.5, and H3T12 were 3 nm, 7.5 nm, and 12 nm. The total thickness of the HfO_2/TiO_x bilayers remained at 15 nm. To obtain nanocrystalline HfO_2/TiO_x, the H12T3 was annealed at 700 °C for 30 min under a pure N_2 atmosphere. After thermal annealing, the 40 nm-thick Ti top electrodes were deposited by using electron beam evaporation through a shadow mask. The atomic concentration ratios of the HfO_2/TiO_x bilayer were determined through an XPS test using a PHI 5000 Versa Probe (Ulvac-Phi Inc., Chigasaki, Japan). The microstructure of the annealed H12T3 was analyzed by high-resolution cross-section transmission electron microscopy (HRXTEM) with a JEOL 2100F electron microscope (JEOL Inc., Tokyo, Japan) operated at 200 kV. An Agilent B1500A semiconductor (Agilent Inc., Santa Clara, CA, USA) analyzer was used to explore the electrical behaviors of the HfO_2/TiO_x memristor under the atmosphere.

3. Results and Discussion

Figure 1a shows a schematic diagram and the measurement setup of the HfO_2/TiO_x-bilayer memristor device. An HRXTEM micrograph (JEOL Inc., Tokyo, Japan) of an annealed H12T3 bilayer is displayed in Figure 1b. The HfO_2/TiO_x bilayer with a clear interface can be observed between the Ti bottom electrode (BE) and the Pt top electrode (TE). The thicknesses of HfO_2 and TiO_x sublayers were 12 nm and 3 nm, respectively. It is obvious that nanocrystalline HfO_2 and TiO_x dots were formed in both the HfO_2 and TiO_x sublayers. The crystalline lattice parameters of the nanocrystalline TiO_x and HfO_2 were 0.35 nm and 0.32 nm, respectively. It is interesting that the nanocrystalline TiO_x and HfO_2 dots were connected at the interface of TiO_x and HfO_2, which supplied a channel for resistive switching.

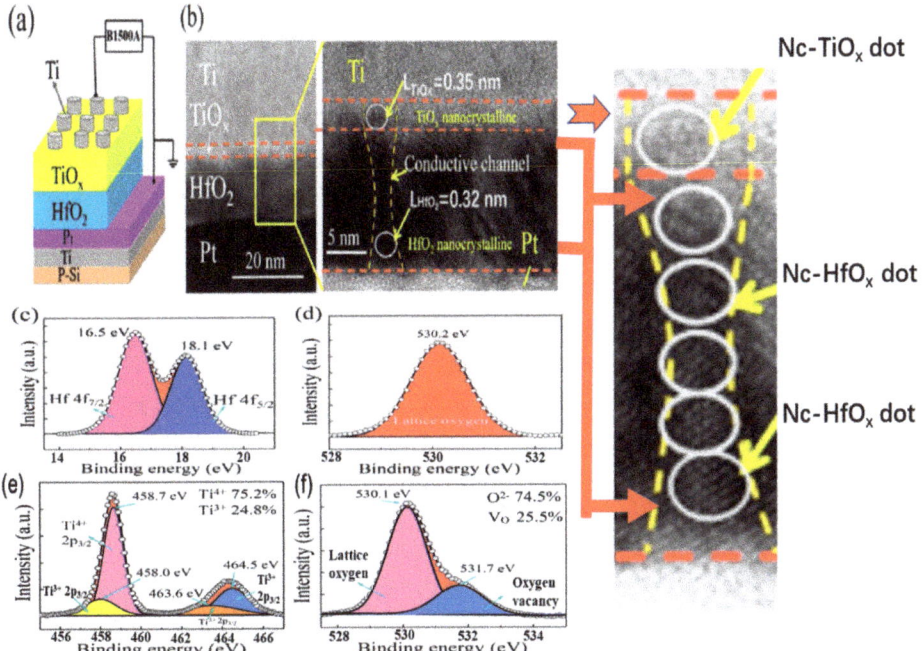

Figure 1. (a) Schematic diagram of Ti/TiO$_x$/HfO$_2$/Pt memristor device; (b) cross-sectional TEM image of H12T3 bilayers after annealing; (c,d) XPS spectra of Hf 4f and O1s in the HfO$_2$ sublayer; (e,f) XPS spectra of O1s and Ti 2p in the TiO$_x$ sublayer.

The atomic configuration of the annealed H12T3 bilayers was analyzed through the XPS spectra in Figure 1. The XPS spectrum of the HfO$_2$ sublayer can be fit by two binding energy peaks located at 16.5 eV and 18.1 eV, as shown in Figure 1c, which correspond to Hf^{4+} 4f$_{7/2}$ and Hf^{4+} 4f$_{5/2}$, respectively. There was only one peak located at 530.2 eV, as presented in Figure 1d, which originated from O 1s. The existence of Hf^{4+} and O^{2-} in the XPS spectra indicates that the stoichiometry of the HfO$_2$ sublayer met the standard chemical ratio. Regarding the TiO$_x$ sublayer, the corresponding XPS spectra are displayed in Figure 1e,f. Two peaks at 458.7 eV and 464.5 eV were detected, which were related to Ti^{4+} 2p$_{3/2}$ and Ti^{3+} 2p$_{3/2}$, respectively [26–30]. The atomic percentages of Ti^{4+} and Ti^{3+} were 75.2% and 24.8%. As shown in Figure 1f, the peak corresponding to O 1s could be fitted with two peaks, including lattice oxygen (at 530.1 eV) and non-lattice oxygen (at 531.7) [31,32]. The atomic percentages of lattice oxygen and non-lattice oxygen were 74.5% and 25.5%. They correspond to the oxide component and defective oxides, respectively. The defect-induced oxidation state of Ti^{3+} verified the existence of oxygen vacancies. It is worth noting that the atomic percentage of Ti^{3+} in the Ti 2p spectra was 24.8% and the atomic percentage of the oxygen vacancies in the O 1s spectra was 25.5%. The high concentrations of Ti^{3+} and oxygen vacancies reveal that a large number of oxygen vacancies coexisted with the nanocrystalline TiO$_x$ and HfO$_2$ dots in the TiO$_x$ sublayer.

Figure 2a displays the electroforming characteristics of the as-deposited HfO$_2$/TiO$_x$ memristor with different thickness ratios of HfO$_2$ and TiO$_x$ during the positive DC sweeping with a compliance current (Icc) of 10 mA. It can be observed that the forming voltage increased from 2.3 V to 4.4 V as the thickness of TiO$_x$ decreased from 12 nm to 3 nm. This was because the number of oxygen vacancies in the TiO$_x$ sublayer was higher than that of the HfO$_2$ sublayer, which made the main contribution to the formation of the conductive pathway. As the thickness of the TiO$_x$ sublayer decreased, the diameter of the oxygen conductive pathway became thinner. Thus, the total resistance of the HfO_2/TiO_x memristor increased, which is the reason why a higher voltage was needed to complete the forming process. The typical bipolar resistive switching characteristic was detected from the three devices with different HfO_2/TiO_x thickness ratios, as shown in Figure 2b. The set process was observed under a positive bias voltage, while the reset process occurred under a negative bias voltage. The resistances of the low-resistance state (LRS) for all three devices were basically consistent, while the resistances of the high-resistance state (HRS) increased gradually as the thickness of the HfO_2 sublayer increased from 3 to 12 nm. The cell-to-cell distribution of the memory window for the three kinds of devices is shown in Figure 2c. It is obvious that the memory window increased to 100 as the thickness of TiO_x decreased from 12 nm to 3 nm. The H12T3 device showed the largest memory window compared with the other two devices, which was related to the largest resistance of the HRS.

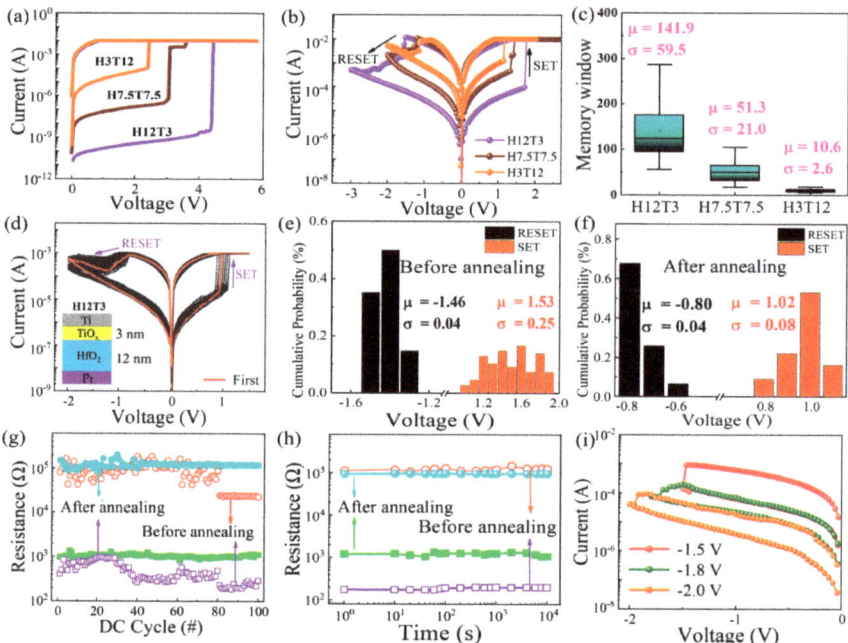

Figure 2. (**a**,**b**) Formation characteristics and bipolar resistive switching of the HfO_2/TiO_x memristor with different sublayer thickness ratios from 4/1 to 1/4, including the devices named H12T3, H7.5T7.5, and H3T12; (**c**) memory window distribution of the HfO_2/TiO_x memristors with different sublayer thickness ratios from 4/1 to 1/4, including the devices named H12T3, H7.5T7.5, and H3T12; (**d**) I–V characteristics of the H12T3 memristor after annealing; (**e**,**f**) statistical distribution of the set/reset voltage for the H12T3 memristor before and after annealing; (**g**,**h**) endurance and retention characteristics of the H12T3 memristor before and after annealing; (**i**) multilevel I–V characteristics of the RRAM device under different reset voltages.

Considering that the H12T3 device had the largest memory window, it was chosen to be annealed to form the nanocrystalline HfO_2/TiO_x device. The resistive switching characteristics of the nanocrystalline HfO_2/TiO_x device are displayed in Figure 2d. Good uniformity and repeatability were observed for the nanocrystalline HfO_2/TiO_x device after 100 switching cycles compared with those of the as-deposited device. The statistical distribution of the set and reset voltages is displayed in Figure 2e,f. It is worth noting that the average set voltages were reduced from 1.53 V to 1.02 V after annealing. Meanwhile, the average reset voltage decreased from −1.46 V to −0.8 V after annealing. The distribution consistency of the set voltage decreased from 0.25 to 0.08 after annealing. To analyze the variability in the set voltage and reset voltage after 100 cycles, we calculated the coefficient of variation, which was equal to the ratio of the standard deviation (σ) and the mean value (u) of V_{SET} or V_{RESET} in absolute value. Compared with the as-deposited HfO_2/TiO_x bilayer, the coefficient of variation in the high-resistance state and the low-resistance state of the nanocrystalline HfO_2/TiO_x memristor were reduced 74% and 86%. The endurance and retention characteristics of the nanocrystalline HfO_2/TiO_x device are presented in Figure 2g,h. It is evident that the resistances of the HRS and LRS for the nanocrystalline HfO_2/TiO_x device were retained well after 100 cycles, exhibiting good stability compared with the as-deposited HfO_2/TiO_x device. The memory window of 100 could be retained after 10^4 s, indicating good retention characteristics. As the set voltage increased from -1.5 to -2.0 V, multiple resistance states were obtained, as shown in Figure 2i. The tunable resistance state is beneficial for constructing artificial synapses for neuromorphic computing.

To investigate the resistive switching mechanism of the HfO_2/TiO_x RRAM device, the resistive switching characteristics of the H12T3 device before and after annealing are plotted in Figure 3a,b. The set process characteristics of the H12T3 device before and after annealing are replotted in double logarithmic scales, as shown in Figure 3c,d. In the low-voltage area of the HRS, the slopes of the H12T3 device before and after annealing were 1.09 and 1.15, respectively. The carrier transportation was in agreement with the ohmic conduction model. The thermal excitation in the conduction band was the origin of the mobile electrons. As the voltage increased, the initial region of the HRS showed slopes of 1.78 and 1.84, corresponding to the devices before and after annealing, respectively. The carrier transportation obeyed the SCLC model. After switching to LRS, the slope of the H12T3 device before and after annealing decreased from 1.78 and 1.84 to 1.15 and 1.04, respectively. This means that the ohmic conduction mode plays a dominant role in carrier transportation.

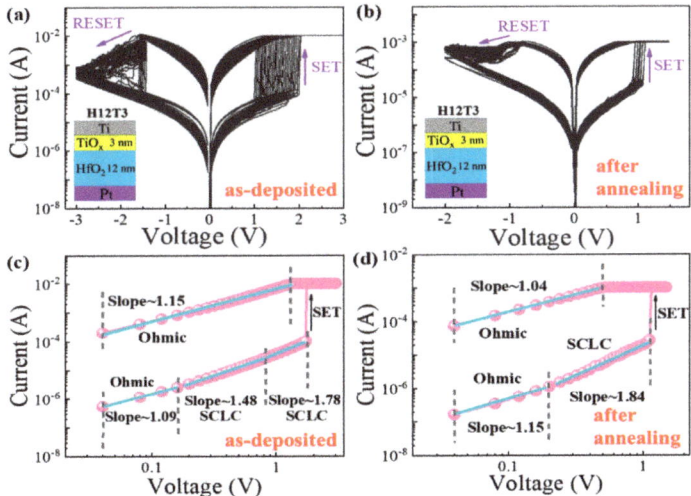

Figure 3. (**a**,**b**) Resistive switching characteristics of HfO_2/TiO_x RRAM (H12T3) before and after annealing; (**c**,**d**) set process characteristics of the H12T3 device before and after annealing replotted in double logarithmic scales.

To reveal the relationship between the TiO_x thickness and the diameter of the conductive filament in more detail, we produced a schematic resistive pathway model related to different thickness ratios of TiO_x/HfO_x in an RRAM device. As shown in Figure 4a, there were many oxygen vacancies in the TiO_x sublayer of the TiO_x/HfO_x memristor with a thickness ratio of 4/1 in its initial state. When positive bias was applied to the Ti top electrode, more oxygen vacancies could be produced as the oxygen ions moved toward the Pt bottom electrode. Thus, a thicker conductive pathway could be formed in the TiO_x sublayer, which combined with the oxygen vacancies in the HfO_x sublayer to form a gradual conductive pathway. In contrast, the number of oxygen vacancies in the thinner TiO_x sublayer decreased as the thickness ratio of the TiO_x and HfO_x decreased from 4/1 to 1/4. Therefore, a thinner conductive pathway could be constructed in the TiO_x sublayer, which connected to the oxygen vacancies in the HfO_x sublayer to form the thinnest conductive pathway, as shown in Figure 4b. When negative bias was applied on the Ti top electrode, the oxygen ions near the bottom electrode could migrate toward the top electrode to neutralize the oxygen vacancies, causing the conductive pathway to disconnect. Therefore, the TiO_x/HfO_x RRAM device could be switched from the LRS to the HRS. The gradual change in conductance is similar to the synaptic weight update in a biological synaptic device.

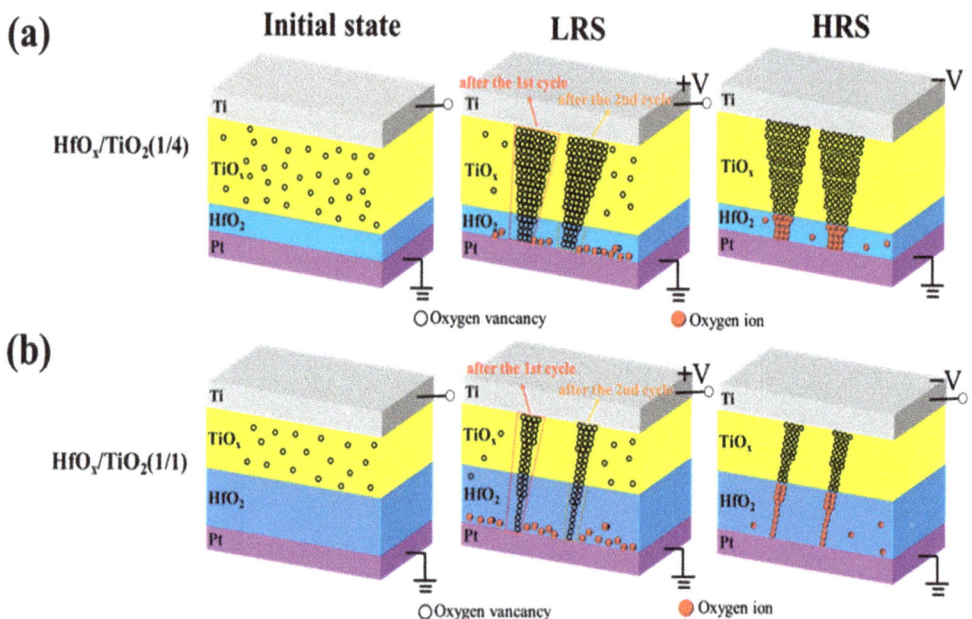

Figure 4. (**a**,**b**) Schematic resistive switching model of as-grown HfO_2/TiO_x memristor with thickness ratio of 1/4 and 1/1 corresponding to HfO_x and TiO_x sublayer, respectively.

Figure 5a,b show the resistive switching pathway model of the RS behaviors in the H12T3 device before and after annealing. For the as-grown HfO_2/TiO_x memristor, oxygen vacancies existed in the TiO_x sublayer at the initial state. Under a positive voltage, the electric field could drive the O^{2-} to move to the Ti electrode and leave a great number of oxygen vacancies in the top electrode; thus, Ti is an oxygen affinity electrode. Once the oxygen vacancies in the thin film reached a high level, the thin, tapered conductive filament composed of oxygen vacancies in the HfO_2 film could connect to the thick conductive filament in the TiO_x film, switching the device to the LRS. Under the negative bias, O^{2-} could move back to the Pt electrode from the boundary of the Ti electrode and neutralize the oxygen vacancies, which would break up the oxygen vacancy conductive filament. As the distribution of oxygen vacancies was uneven in the TiO_x sublayer, the position of the conduction pathway varied as the cycles increased. In contrast to the TiO_x/HfO_x device before annealing, nanocrystalline TiO_x and HfO_x were formed in the TiO_x and HfO_x

sublayers after annealing, as revealed by the HRTEM in Figure 1. Nanocrystalline TiO_x and HfO_x acted as a presetting conductive channel, which could be connected to oxygen vacancies to form the whole conductive pathway under positive bias. The introduction of nanocrystalline TiO_x and HfO_2 supplied a "fixed position" to confine the growth and rupture of oxygen vacancies during the RS behaviors. In contrast to the device after annealing, the distribution of the oxygen vacancy conductive filament under a positive bias was random in the as-deposited device without the confinement provided by nanocrystalline TiO_x and HfO_2, as shown in Figure 3c. Under a negative bias, not all oxygen vacancies in the nanopathway were accurately neutralized by O^{2-}, which resulted in fluctuations in the switching parameters during the successive cycles. It is evident that nanocrystalline TiO_x and HfO_2 are beneficial for improving the uniformity of conductive nanopathways, while the formation and neutralization of oxygen vacancies were responible for the bridging and rupture of conductive nanopathways. We also noticed that the set voltage of the device after annealing was reduced as the conductivity of nanocrystalline TiO_x and HfO_2 was higher than that of amorphous TiO_x and HfO_2. It was further revealed that nanocrystalline TiO_x and HfO_2 were the main contributors to the conductive pathway.

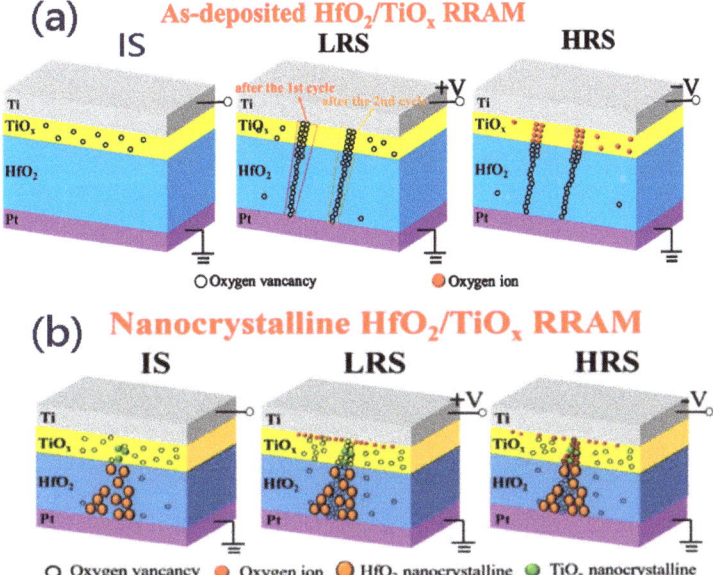

Figure 5. (a,b) Schematic description of the resistive switching pathway model of H12T3 before and after annealing.

A schematic illustration of an artificial synapse based on the nc-HfO_2/TiO_x device is shown in Figure 6a. The changeable conductance of the nc-HfO_2/TiO_x memristor was analogous to the connection strength between biological synapses [33,34]. The long-term potentiation (LTP) and long-term depression (LTD) characteristics of nc-HfO_2/TiO_x were detected by adjusting the pulse amplitude and numbers, as shown in Figure 6b,c. Regarding the potentiation process, the conductance continued to increase as the pulse amplitude increased from 0.7 to 1.25 V under the same pulse width of 100 us. This indicates that the strength of the synaptic connection between two neurons was strengthened. It is noteworthy that the conductance tended to be saturated when the pulse number increased to 20. Regarding the depression process, the conductance continued to decrease as the negative pulse amplitude increased from −1 to −1.32 V under the same pulse width of 100 us. This meant that the strength of the synaptic connection between two neurons was weakened. The rule of the depression process was consistent with the potentiation process

in general. Figure 6d shows the spike-timing-dependent plasticity (STDP) of nc-HfO$_2$/TiO$_x$ synapses, which is an important rule for learning and memory. The synaptic weight could be modulated by the relative timing of the pre-spike and the post-spike, which was determined by the gradual change in the conductance. As revealed in Figure 6d, when the pre-spike was ahead of/dropped behind the post-spike, the connection strength of the synapse between the two neurons was strengthened/weakened. It can be seen that a larger conductance change was caused by a closer spike timing, which was consistent with the Hebbian learning rule. The total conductance change in nc-HfO$_2$/TiO$_x$ synapses is defined as ΔG, which can be expressed as the following equation [35]:
(|G_after-G_before|)/(Min(G_after,G_before))

$$\Delta G = \frac{|G_{\text{after}} - G_{\text{before}}|}{\text{Min}(G_{\text{after}}, G_{\text{before}})} \quad (1)$$

where G$_{\text{before}}$ and G$_{\text{after}}$ are the conductance before and after the application of the spike. ΔG exhibited an increasing tendency when the pre-spike was ahead of the post-spike (Δt > 0) during the potentiation process, while it displayed a decreasing tendency when the post-spike was ahead of the pre-spike (Δt < 0) during the depression process. The pulse response of nc-HfO$_2$/TiO$_x$ synapses showed that the maximum value of ΔG could reach 650% for potentiation and 1400% for depression. The successful implementation of biosynaptic function ensured further application in neuromorphic computing. In order to visualize the information transferred to the HfO$_2$/TiO$_x$ neural network, 5 × 5 synaptic arrays based on the Ti/TiO$_x$/HfO$_2$/Pt memristor were applied to image recognition, as shown in Figure 6e. The conductance is represented by the color level. In the initial state, the conductance of all synapses ranged randomly. A unit device distributed in the shape of a "T" in the synapse array was selected to input 10 consecutive stimulus pulses with a duration of 1 us and amplitude of 1.15 V. The conductance of the HfO$_2$/TiO$_x$ unit device increased obviously after the application of stimulus pulses. Then, the number of consecutive stimulus pulses increased from 10 to 15, 20, and 25. After training by the 25 consecutive stimulus pulses with the same duration and amplitude, the image clearness of "T" reached a high level, as shown in the lower panel of Figure 6e. The visualized evolution of weight values showed that the HfO$_2$/TiO$_x$ memristor crossbar arrays had the potential to mimic real neurotransmission in a biological system.

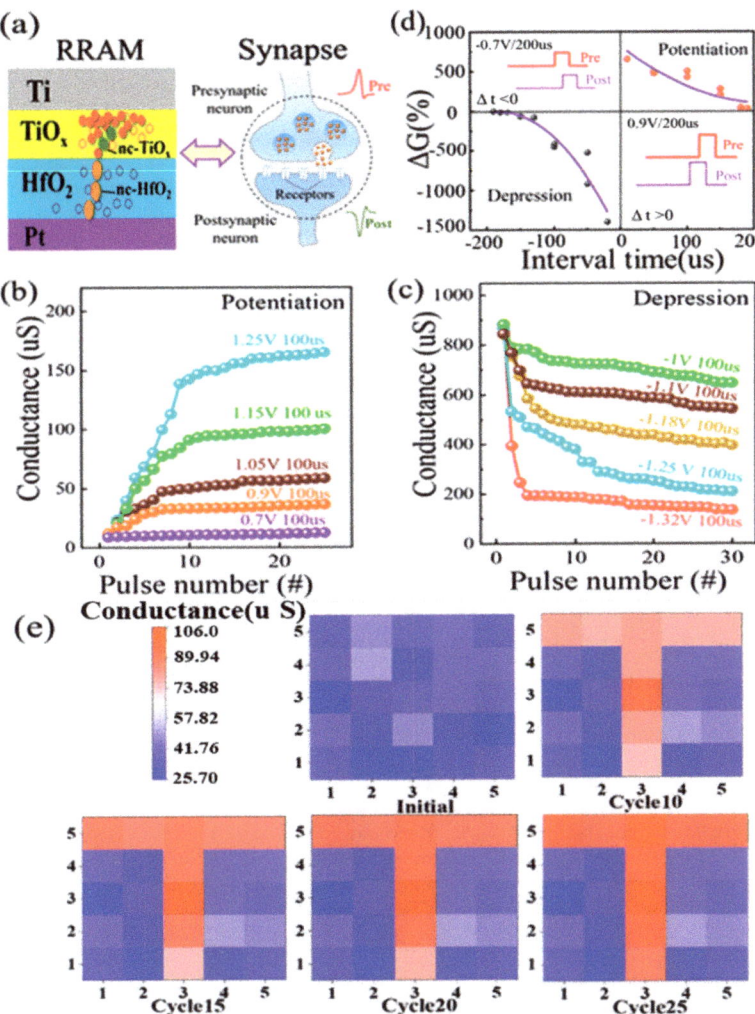

Figure 6. (a) Schematic illustration of the similarity between a biological synapse and an HfO_2/TiO_x memristor, displaying the diffusion of neurotransmitters among synaptic neurons and an electrical synapse based on the top electrode, switching matrix, and bottom electrode. (b,c) Long-term potentiation and depression of HfO_2/TiO_x synapse by changing the pulse amplitude under the same pulse duration; (d) spike-timing-dependent plasticity of HfO_2/TiO_x synapse as a function of interval time between pre-spike and post-spike; (e) simulated image memorization of HfO_2/TiO_x synapse under consecutive pulses on the selected synapse.

4. Conclusions

In summary, we successfully obtained an artificial synapse based on an HfO_2/TiO_x memristive crossbar with a controllable memory window and high uniformity. The controllable memory window could be obtained by tuning the thickness ratio of the sublayers. The position of the conductive pathway could be localized by the nanocrystalline HfO_2 and TiO_2 dot, leading to a substantial improvement in the switching uniformity. Notably, the coefficient of variation in the high-resistance state and the low-resistance state of the nanocrystalline HfO$_2$/TiO$_x$ memristor could be reduced by 74% and 86% compared with those of the as-deposited HfO$_2$/TiO$_x$ memristor. The nanocrystalline HfO$_2$/TiO$_x$ memristive device showed stable, controllable biological functions, including long-term potentiation, long-term depression, and spike-time-dependent plasticity, as well as visual learning capability, which provides a hardware base for their integration into the next generation of brain-inspired chips.

Author Contributions: Conceptualization, Z.M., X.Z. and Y.Y.; methodology, Y.Y., T.C. and W.L.; software, Y.Y.; validation, X.Z., H.H., J.X. and K.C.; formal analysis, Y.Y. and L.X.; investigation, X.Z.; resources, Y.Y.; data curation, X.Z.; writing—original draft preparation, Y.Y. and X.Z.; writing—review and editing, Z.M.; visualization, Z.M.; supervision, Z.M.; project administration, Z.M.; funding acquisition, Z.M. All authors have read and agreed to the published version of the manuscript.

Funding: This study was supported by the National Nature Science Foundation of China (Grant Nos. 61634003, 61571221, 11774155, 61921005 and 61735008), Six Talent Peaks Project in Jiangsu Province (DZXX-001) and the National Key R&D program of China (2018YFB2200101).

Data Availability Statement: The data that support the findings of this study are available from the corresponding authors upon reasonable request.

Conflicts of Interest: The authors declare that they have no competing interest.

References

1. Yao, P.; Wu, H.Q.; Gao, B.; Tang, J.S.; Zhang, Q.T.; Zhang, W.Q.; Yang, J.J.; Qian, H. Fully hardware-implemented memristor convolutional neural network. *Nature* **2020**, *577*, 641–646. [CrossRef]
2. Tong, L.; Peng, Z.R.; Lin, R.F.; Li, Z.; Wang, Y.L.; Huang, X.Y.; Xue, K.H.; Xu, H.Y.; Liu, F.; Xia, H.; et al. 2D materials-based homogeneous transistor-memory architecture for neuromorphic hardware. *Science* **2021**, *373*, 1353–1358. [CrossRef]
3. Li, L.H.; Xue, K.H.; Zou, L.Q.; Yuan, J.H.; Sun, H.J.; Miao, X.S. Multilevel switching in Mg-doped HfO$_x$ memristor through the mutual-ion effect. *Appl. Phys. Lett.* **2021**, *119*, 7. [CrossRef]
4. Rehman, M.M.; Mutee ur Rehman, H.M.; Kim, W.Y.; Hassan Sherazi, S.S.; Rao, M.W.; Khan, M.; Muhammad, Z. Biomaterial-based nonvolatile resistive memory devices toward ecofriendliness and biocompatibility. *ACS Appl. Electron. Mater.* **2021**, *3*, 2832–2861. [CrossRef]
5. Prezioso, M.; Merrikh-Bayat, F.; Hoskins, B.D.; Adam, G.C.; Likharev, K.K.; Strukov, D.B. Training and operation of an integrated neuromorphic network based on metal-oxide memristors. *Nature* **2015**, *521*, 61–64. [CrossRef]
6. Li, C.; Hu, M.; Li, Y.N.; Jiang, H.; Ge, N.; Montgomery, E.; Zhang, J.M.; Song, W.H.; Davila, N.; Graves, C.E.; et al. Analogue signal and image processing with large memristor crossbars. *Nat. Electron.* **2014**, *1*, 52–59. [CrossRef]
7. Yao, P.; Wu, H.; Gao, B.; Eryilmaz, S.B.; Huang, X.; Zhang, W.; Zhang, Q.; Deng, N.; Shi, L.; Wong, H.-S.P. Face classification using electronic synapses. *Nat. Commun.* **2016**, *8*, 15199. [CrossRef]
8. Zhuang, P.P.; Ma, W.Z.; Liu, J.; Cai, W.W.; Lin, W.Y. Progressive RESET induced by Joule heating in hBN RRAMs. *Appl. Phys. Lett.* **2021**, *118*, 6. [CrossRef]
9. Dawson, J.A.; Robertson, J. Nature of Cu Interstitials in Al$_2$O$_3$ and the Implications for Filament Formation in Conductive Bridge Random Access Memory Devices. *J. Phys. Chem. C* **2016**, *120*, 14474–14483. [CrossRef]
10. Ismail, M.; Chand, U.; Mahata, C.; Nebhen, J.; Kim, S. Demonstration of synaptic and resistive switching characteristics in W/TiO$_2$/HfO$_2$/TaN memristor crossbar array for bioinspired neuromorphic computing. *J. Mater. Sci. Technol.* **2022**, *96*, 94–102. [CrossRef]
11. Ye, C.; Deng, T.; Zhang, J.; Shen, L.; He, P.; Wei, W.; Wang, H. Enhanced resistive switching performance for bilayer HfO$_2$/TiO$_2$ resistive random access memory. *Semicond. Sci. Technol.* **2016**, *31*, 105005. [CrossRef]
12. Liu, J.; Yang, H.F.; Ji, Y.; Ma, Z.Y.; Chen, K.J.; Zhang, X.X.; Zhang, H.; Sun, Y.; Huang, X.F.; Oda, S. An electronic synaptic device based on HfO$_2$/TiO$_x$ bilayer structure memristor with self-compliance and deep-RESET characteristics. *Nanotechnology* **2018**, *29*, 10. [CrossRef]
13. Kim, Y.G.; Lv, D.X.; Huang, J.S.; Bukke, R.N.; Chen, H.P.; Jang, J. Artificial Indium-Tin-Oxide Synaptic Transistor by Inkjet Printing Using Solution-Processed ZrO(x)Gate Dielectric. *Phys. Status Solidi A-Appl. Mat.* **2020**, *217*, 8. [CrossRef]
14. Abbas, Y.; Han, I.S.; Sokolov, A.S.; Jeon, Y.R.; Choi, C. Rapid thermal annealing on the atomic layer-deposited zirconia thin film to enhance resistive switching characteristics. *J. Mater. Sci.-Mater. Electron.* **2019**, *31*, 903–909. [CrossRef]
15. Lubben, M.; Karakolis, P.; Ioannou-Sougleridis, V.; Normand, P.; Dimitrakis, P.; Valov, I. Graphene-Modified Interface Controls Transition from VCM to ECM Switching Modes in Ta/TaOx Based Memristive Devices. *Adv. Mater.* **2015**, *27*, 6202–6207. [CrossRef] [PubMed]
16. Yang, J.J.; Zhang, M.X.; Strachan, J.P.; Miao, F.; Pickett, M.D.; Kelley, R.D.; Medeiros-Ribeiro, G.; Williams, R.S. High switching endurance in TaOx memristive devices. *Appl. Phys. Lett.* **2010**, *97*, 3. [CrossRef]

17. Zhang, R.L.; Huang, H.; Xia, Q.; Ye, C.; Wei, X.D.; Wang, J.Z.; Zhang, L.; Zhu, L.Q. Role of Oxygen Vacancies at the TiO_2/HfO_2 Interface in Flexible Oxide-Based Resistive Switching Memory. *Adv. Electron. Mater.* **2019**, *5*, 7. [CrossRef]
18. Batko, I.; Batkova, M. Memristive behavior of $Nb/NbO_x/Nb$ structures prepared by local anodic oxidation. *Mater. Today Proc.* **2016**, *3*, 803–809. [CrossRef]
19. Chang, Y.F.; Fowler, B.; Zhou, F.; Chen, Y.C.; Lee, J.C. Study of self-compliance behaviors and internal filament characteristics in intrinsic SiOx-based resistive switching memory. *Appl. Phys. Lett.* **2016**, *108*, 5. [CrossRef]
20. Jaafar, A.H.; Al Chawa, M.M.; Cheng, F.; Kelly, S.M.; Picos, R.; Tetzlaff, R.; Kemp, N.T. Polymer/TiO_2 Nanorod Nanocomposite Optical Memristor Device. *J. Phys. Chem. C* **2021**, *125*, 14965–14973. [CrossRef]
21. El Mesoudy, A.; Lamri, G.; Dawant, R.; AriasZapata, J.; Gliech, P.; Beilliard, Y.; Ecoffey, S.; Ruediger, A.; Alibart, F.; Drouin, D. Fully CMOS-compatible passive TiO_2-based memristor crossbars for in-memory computing. *Microelectron. Eng.* **2022**, *255*, 111706. [CrossRef]
22. Khan, M.; Mutee Ur Rehman, H.M.; Tehreem, R.; Saqib, M.; Rehman, M.M.; Kim, W.Y. All-Printed Flexible Memristor with Metal-Non-Metal-Doped TiO_2 Nanoparticle Thin Films. *Nanomaterials* **2022**, *12*, 2289. [CrossRef] [PubMed]
23. Covi, E.; Brivio, S.; Serb, A.; Prodromakis, T.; Fanciulli, M.; Spiga, S. IEEE HfO_2-based Memristors for Neuromorphic Applications based memristors for neuromorphic applications. In Proceedings of the 2016 IEEE International Symposium on Circuits and Systems (ISCAS), Montréal, QC, Canada, 22–25 May 2016.
24. Covi, E.; Brivio, S.; Fanciulli, M.; Spiga, S. Synaptic potentiation and depression in Al:HfO_2-based memristor. *Microelectron. Eng.* **2015**, *147*, 41–44. [CrossRef]
25. Yu, S.M.; Li, Z.W.; Chen, P.Y.; Wu, H.Q.; Gao, B.; Wang, D.L.; Wu, W.; Qian, H. IEEE Binary Neural Network with 16 Mb RRAM Macro Chip for Classification and Online Training. In Proceedings of the 2016 IEEE International Electron Devices Meeting (IEDM), San Francisco, CA, USA, 3–7 December 2016; IEEE: New York, NY, USA, 2016.
26. Deng, T.F.; Ye, C.; Wu, J.J.; He, P.; Wang, H. Improved performance of $ITO/TiO_2/HfO_2/Pt$ random resistive accessory memory by nitrogen annealing treatment. *Microelectron. Reliab.* **2016**, *57*, 34–38. [CrossRef]
27. Sanjines, R.; Tang, H.; Berger, H.; Gozzo, F.; Margaritondo, G.; Levy, F. Electronicstructure of anatase TiO_2 oxide. *J. Appl. Phys.* **1994**, *75*, 2945–2951. [CrossRef]
28. Bharti, B.; Kumar, S.; Lee, H.N.; Kumar, R. Formation of oxygen vacancies and Ti^{3+} state in TiO_2 thin film and enhanced optical properties by air plasma treatment. *Sci. Rep.* **2016**, *6*, 32355. [CrossRef]
29. Woo, J.; Padovani, A.; Moon, K.; Kwak, M.; Larcher, L.; Hwang, H. Linking Conductive Filament Properties and Evolution to Synaptic Behavior of RRAM Devices for Neuromorphic Applications. *IEEE Electron. Dev. Lett.* **2017**, *38*, 1220–1223. [CrossRef]
30. Su, C.Y.; Liu, L.; Zhang, M.Y.; Zhang, Y.; Shao, C.L. Fabrication of Ag/TiO_2 nanoheterostructures with visible light photocatalytic function via a solvothermal approach. *Crystengcomm* **2012**, *14*, 3989–3999. [CrossRef]
31. Huang, C.Y.; Huang, C.Y.; Tsai, T.L.; Lin, C.A.; Tseng, T.Y. Switching mechanism of double forming process phenomenon in ZrO_x/HfO_y bilayer resistive switching memory structure with large endurance. *Appl. Phys.* **2014**, *104*, 4. [CrossRef]
32. Tsai, T.L.; Chang, H.Y.; Lou, J.J.C.; Tseng, T.Y. A high performance transparent resistive switching memory made from $ZrO_2/AlON$ bilayer structure. *Appl. Phys. Lett.* **2016**, *108*, 4. [CrossRef]
33. Zuo, F.; Panda, P.; Kotiuga, M.; Li, J.R.; Kang, M.G.; Mazzoli, C.; Zhou, H.; Barbour, A.; Wilkins, S.; Narayanan, B.; et al. Habituation based synaptic plasticity and organismic learning in a quantum perovskite. *Nat. Commun.* **2017**, *8*, 7. [CrossRef] [PubMed]
34. Efros, A.L.; Delehanty, J.B.; Huston, A.L.; Medintz, I.L.; Barbic, M.; Harris, T.D. Evaluating the potential of using quantum dots for monitoring electrical signals in neurons. *Nat. Nanotechnol.* **2018**, *13*, 278–288. [CrossRef] [PubMed]
35. Ryu, J.H.; Kim, S. Artificial synaptic characteristics of TiO_2/HfO_2 memristor with self-rectifying switching for brain-inspired computing. *Chaos Solitons Fractals* **2020**, *140*, 1220–1223. [CrossRef]

Disclaimer/Publisher's Note: The statements, opinions and data contained in all publications are solely those of the individual author(s) and contributor(s) and not of MDPI and/or the editor(s). MDPI and/or the editor(s) disclaim responsibility for any injury to people or property resulting from any ideas, methods, instructions or products referred to in the content.

Article

Tracing the Si Dangling Bond Nanopathway Evolution in a-SiN$_x$:H Resistive Switching Memory by the Transient Current

Tong Chen [1,2,3], Kangmin Leng [1,2,3], Zhongyuan Ma [1,2,3,*], Xiaofan Jiang [1,2,3], Kunji Chen [1,2,3], Wei Li [1,2,3], Jun Xu [1,2,3] and Ling Xu [1,2,3]

[1] The School of Electronic Science and Engineering, Nanjing University, Nanjing 210093, China
[2] Collaborative Innovation Center of Advanced Microstructures, Nanjing University, Nanjing 210093, China
[3] Jiangsu Provincial Key Laboratory of Photonic and Electronic Materials Sciences and Technology, Nanjing University, Nanjing 210093, China
* Correspondence: zyma@nju.edu.cn

Abstract: With the big data and artificial intelligence era coming, SiN$_x$-based resistive random-access memories (RRAM) with controllable conductive nanopathways have a significant application in neuromorphic computing, which is similar to the tunable weight of biological synapses. However, an effective way to detect the components of conductive tunable nanopathways in a-SiN$_x$:H RRAM has been a challenge with the thickness down-scaling to nanoscale during resistive switching. For the first time, we report the evolution of a Si dangling bond nanopathway in a-SiN$_x$:H resistive switching memory can be traced by the transient current at different resistance states. The number of Si dangling bonds in the conducting nanopathway for all resistive switching states can be estimated through the transient current based on the tunneling front model. Our discovery of transient current induced by the Si dangling bonds in the a-SiN$_x$:H resistive switching device provides a new way to gain insight into the resistive switching mechanism of the a-SiN$_x$:H RRAM in nanoscale.

Keywords: resistive switching memory; transient current; trap state

Citation: Chen, T.; Leng, K.; Ma, Z.; Jiang, X.; Chen, K.; Li, W.; Xu, J.; Xu, L. Tracing the Si Dangling Bond Nanopathway Evolution in a-SiN$_x$:H Resistive Switching Memory by the Transient Current. *Nanomaterials* **2023**, *13*, 85. https://doi.org/10.3390/nano13010085

Academic Editor: Patrick Fiorenza

Received: 7 December 2022
Revised: 20 December 2022
Accepted: 22 December 2022
Published: 24 December 2022

Copyright: © 2022 by the authors. Licensee MDPI, Basel, Switzerland. This article is an open access article distributed under the terms and conditions of the Creative Commons Attribution (CC BY) license (https://creativecommons.org/licenses/by/4.0/).

1. Introduction

As the key hardware unit of neuromorphic computing chips, resistive random access memory (RRAM) is considered the most promising candidate because of its excellent scalability, fast speed, and good endurance [1–6]. Among the next generation of RRAMs, silicon nitride-based-RRAM devices have attracted great interest in recent years because of their low operating current, stable switching behavior, and full compatibility with Si-based CMOS integration technology [7–14]. In particular, a controllable conductive nanopathway could be achieved by tuning the Si dangling bond conducting paths in hydrogenated silicon nitride (a-SiN$_x$:H) films with different N/Si ratios, which is similar to the tunable weight of biological synapse [12]. The programming current has been successfully reduced to the lowest record in Si-based RRAM. However, with the down-scaling of a-SiN$_x$:H films to the nanometer scale, an effective way to detect trap states related to the dangling bonds in resistive switching is a challenge. Here we first report that the transient current is more favorable for observing the trap states in ultra-thin a-SiN$_x$:H RRAM with tunable N/Si ratios. The dynamic evolution of the Si dangling bonds was revealed when the device was switched to different resistance states. We analyze the internal formation mechanism of the transient current. In contrast with other techniques [15], the transient current provides a more effective way to analyze the different distributions of dangling bonds at the programming and erasing state, which is crucial to illuminate the dynamic evolution of the conducting paths in trap-dominated RRAM devices.

2. Materials and Methods

To fabricate the RRAM device, a P^+-Si substrate with a low resistivity of 0.004–0.0075 Ωcm was prepared as the bottom electrode. The oxide on the back surface of the P^+-Si substrate was removed, and a thin Al layer was evaporated to the back side of the silicon substrate, which can reduce the contact resistance. A 7-nm-thick a-SiN$_x$:H film was deposited on the substrate in a plasma-enhanced chemical deposition system at 250 °C using silane and ammonia as the reaction gases. The N/Si ratio x was adjusted by varying the flow ratio of silane and ammonia. Subsequently, 100 nm-thick Al was thermally evaporated on the surface of a-SiN$_x$:H as the top electrode using a shadow mask with a diameter of 200 μm to form the final Al/a-SiN$_x$:H/P^+-Si device structure. The atomic concentration ratios of N/Si were determined through XPS measurement at a depth of 5 nm from the film surface. The ESR spectra were measured using the Bruker EMX-10/12 system. The microstructure of the sample was revealed by high-resolution transmission electron microscopy (HRTEM) using a Tecnai G2 F20 electron microscope operating at 200 kV. To detect the transient current related to the traps in all the resistance states, an Agilent B1500A semiconductor analyzer was used to generate the bias signal and record the current intensity. The signals were applied to the top Al electrode with the substrate grounded.

3. Results

Figure 1a–e shows the schematic illustration of the transient current measurement of an Al/a-SiN$_x$:H/P^+-Si device at different resistive switching states. Following each resistive switching operation, as shown in Figure 1a-1c, the charging process and transient current measurement were carried out, as displayed in Figure 1d–e. All the electrical measurements were carried out in the original system without position and state changes, so the transient current can be used to trace the resistance state evolution. During the charging process, a constant voltage was imposed on the device for 100 s. Then, the bias voltage was removed, and we measured the current flowing through the device immediately. The oscillogram of the applied voltage for the transient current measurement is shown in Figure 1f. A cross-sectional HRTEM photo of Al/a-SiN$_{1.17}$:H/P^+-Si device is presented in Figure 1g.

Figure 2a–c shows the time-dependent transient current of the Al/a-SiN$_x$:H/p^+-Si devices at the initial state with tunable N/Si ratios from 1.17 to 0.62. The amplitude of the charging voltage is smaller than that of the forming voltage, which ranges from 0.25 V to 1.75 V, to ensure that the devices remain in their initial states. All the charging currents were measured after the charging time of 100 s. It is found that the transient current intensity is enhanced with the charging voltage increasing when the N/Si ratio is fixed. Under the same applied voltage, the transient current increases as the N/Si ratio decreases. The time-dependent transient current is plotted as a log coordinate, and their slopes are −1. According to the demonstration of Dumin [13], if there are a large number of traps generated in oxides, the discharging current of traps will be the main contributor to the transient current, which obeys 1/t time dependence, and the transient current intensity is proportional to the trap density. Here a slope of −1 in the I–t curves for all devices means the transient current exactly fits the 1/t time dependence. As shown in Figure 2d, the transient current intensity increases as the N/Si ratio decreases from 1.17 to 0.62.

To reveal the role that H played in our devices, we analyzed the FTIR spectra of as-deposited SiN$_x$:H films and the one annealed at 600 °C in a vacuum, as seen in Figure 2e,f. The intensity of the Si-H peak from the device with the SiN$_x$:H films annealed at 600 °C decreases obviously compared with that of the pristine a-SiN$_x$:H films. This result means the number of Si-H bonds is reduced after annealing at 600 °C. Because the Si-H bond energy (318 kJ/mol) is lower than that of Si-N (355 kJ/mol), Si-H bonds are easier to be broken and form the Si dangling bonds. The corresponding transient current intensity of the device with pristine a-SiN$_x$:H films and the one annealed at 600 °C is displayed in Figure 2g. It is found that the transient current of the device with the SiN$_x$:H annealed at 600 °C is much larger than that of the pristine device. The above analysis confirms that Si dangling bonds induced by broken Si-H bonds make the main contribution to the transient current. The relation of the transient current and the Si dangling bond distribution in SiN$_x$:H films can be analyzed by the corresponding ESR spectra, which are presented in Figure 2h. A resonance peak with a g value of 2.0042 is detected from the SiN$_x$:H films, which is related to the paramagnetic center of the Si dangling bonds. The intensity of the resonance peak increases as the N/Si ratio decreases from 1.17 to 0.62, which is in agreement with the changing trend of the transient current intensity. As reported by Robertson and Mo et al., as-deposited silicon nitride films contain many silicon dangling bonds, and they are amphoteric deep trap centers located near the middle of the band gap [16,17]. Here, after the charging voltage was removed, electrons released by Si dangling bonds formed the transient current. The number of Si dangling bonds increases as the N/Si ratio decreases [12], leading to the enhancement of the transient current under the same charging voltage. In contrast to the other Si/N ratio, the device with x = 1.17 has the smallest number of Si dangling bonds. The transient current can be detected from the device with the smallest number of Si dangling bonds, which ensures that the same changing trend of transient current can also be detected from the other ones with the larger number of Si dangling bonds. Because the transient current intensity reflects the number of Si dangling bonds in an a-SiN$_{1.17}$:H device, we applied it to trace the trap state evolution during the resistive switching process.

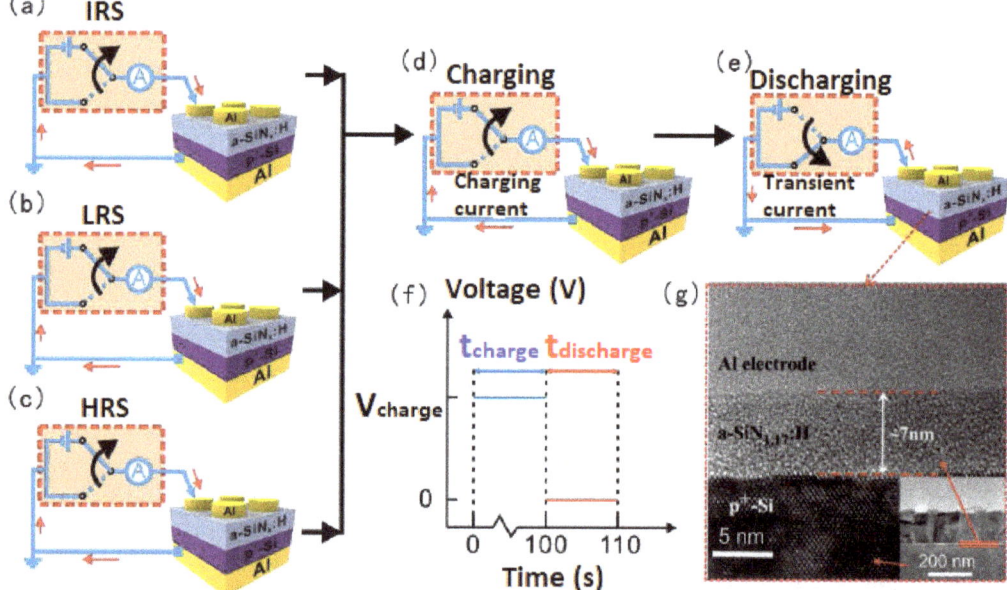

Figure 1. (a–e) Schematic illustration of the transient current measurement of an Al/a-SiN$_x$:H/P+-Si at different resistive switching states. (f) Oscillogram of the applied pulse for the transient current measurement. (g) Cross-sectional HRTEM photo of the Al/a-SiN$_{1.17}$:H/p+-Si. The inset shows the low-resolution TEM photo.

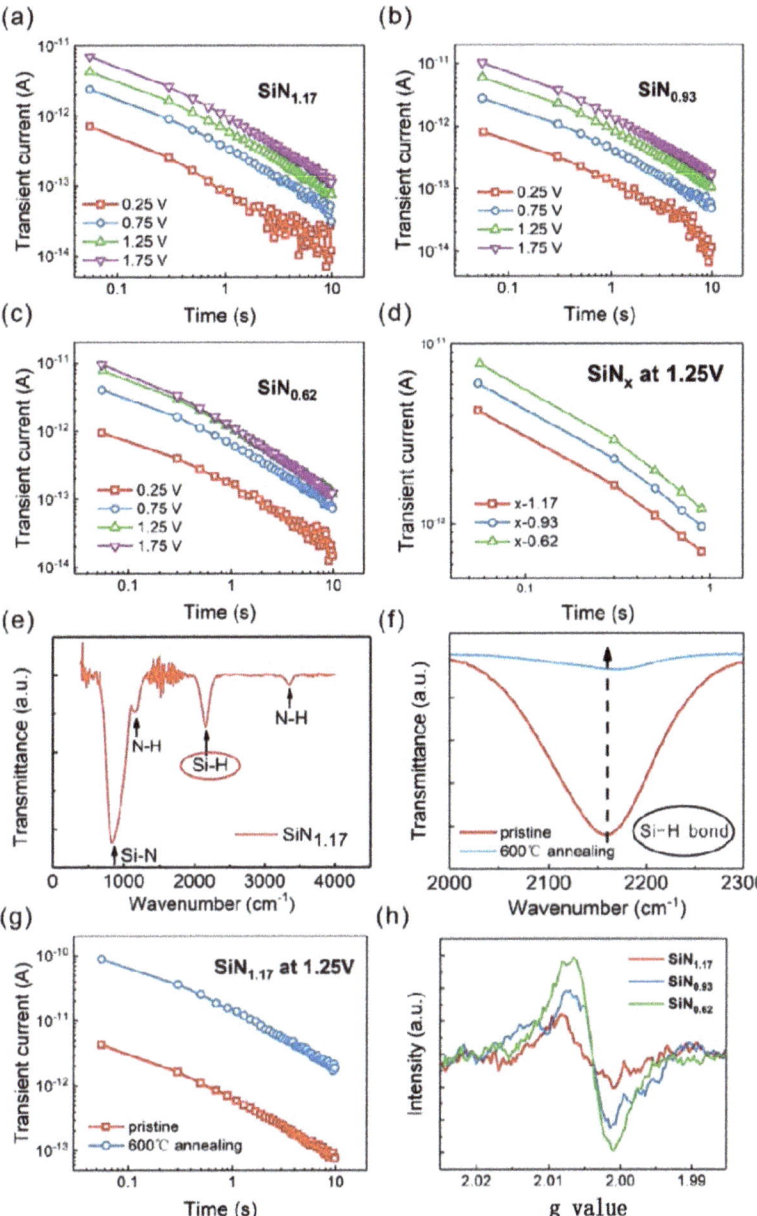

Figure 2. (**a**–**c**) Time-dependent transient current of the devices at the initial state with N/Si ratios of 1.17, 0.93, and 0.62. (**d**) Time-dependent transient current of the devices with N/Si ratios of 1.17, 0.93, and 0.62 at 1.25V. (**e**) The FTIR spectra of pristine a-SiN$_{1.17}$:H films. (**f**) The evolution of Si–H absorption peak of a-SiN$_{1.17}$:H in the magnified scale before and after 600 °C annealing. (**g**) Time-dependent transient current of a-SiN$_{1.17}$:H before and after 600 °C annealing. (**h**) ESR spectra of the devices at the initial state with N/Si ratios of 1.17, 0.93, and 0.62.

The DC sweep of the resistive switching process for an a-SiN$_{1.17}$:H device by continuously applying different reset voltages is shown in Figure 3a. The device switches from the initial resistance state (IRS) to the low resistance state (LRS) when the magnitude of forming voltage is increased to 4.1 V, with a compliance current of 10 µA. It is worth noting that the current of the Al/a-SiN$_x$:H/P$^+$-Si device decreases gradually with the reset voltage increasing, which is similar to the tunable weight of a biological synapse. Before the forming process, the transient current was measured after a charging voltage of 0.5 V for 100 s. The low voltage of 0.5 V ensures that the Si dangling bonds can be charged without damage to the Si-H bonds. Following the forming operation, the transient current of the LRS was measured, as shown in Figure 3b. The transient current obviously increases from 1.1 to 10.5 pA after discharging for 0.3 s, which indicates that a large number of Si dangling bonds are generated after the forming process. Subsequently, the reset1 operation made the device switch from the LRS to the middle resistance state (MRS1) under a voltage of −1.36 V. The transient current of the MRS1 is slightly reduced compared with that of the LRS. It means that a part of the generated traps was annihilated during the reset1 operation. After the reset1 operation, we carried out the reset2, reset3, and reset4 operations to switch the device from MRS1 to MRS2, MRS3, and the high resistance state (HRS), respectively. In this case, the transient current decreases gradually with the number of reset operations increasing. It reveals that the generated traps were annihilated gradually during the reset operation. Figure 3c shows the endurance characteristics of the Al/a-SiN$_x$:H/P$^+$-Si device. It can be seen that a stable memory window and good reliability can be maintained after 200 consecutive cycles. As displayed in Figure 3d, the current of HRS and LRS is still equal to that of the initial state after a retention time of 10^5 s, which shows good retention characteristics. No noticeable degradation is observed in either state. Figure 3e presents the statistical distributions of SET/RESET currents of 26 devices. It is obvious that the operating voltages of these devices show good uniformity with small deviations.

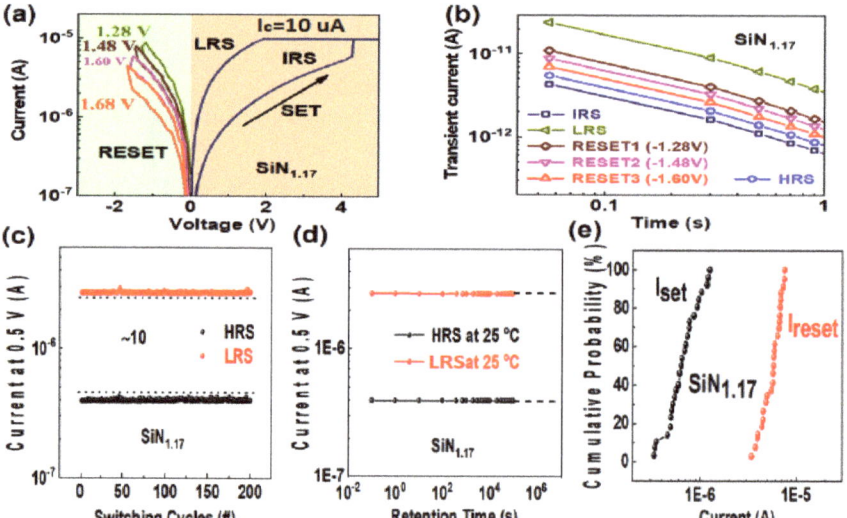

Figure 3. (**a**–**b**) Multilevel resistive switching I-V curves and the corresponding transient current of a-SiN$_{1.17}$:H-based RRAM device in different resistance states. (**c**) The endurance characteristics of the Al/a-SiN$_x$:H/P$^+$-Si device after 200 consecutive cycles. (**d**) The retention properties of the Al/a-SiN$_x$:H/P$^+$-Si device at room temperature. (**e**) The statistical distributions of SET/RESET currents of 26 devices.

The discharging of Si dangling bond centers can be described by the tunneling front model, in which at a given time t, the tunneling rate is sharply peaked spatially at a depth $x(t)$ from the Al/a-SiN$_x$:H interface [13,18–22]. The traps located closer than $x(t)$ are emptied, with those beyond $x(t)$ still occupied. The tunneling front depth $x(t)$ increases logarithmically with time, as

$$x(t) = \frac{1}{2\beta} \ln \frac{t}{t_0},\quad (1)$$

where t_0 is the characteristic tunneling time and β is a tunneling parameter that can be calculated as

$$\beta = \sqrt{\frac{2m_t^*}{\hbar^2} E_t},\quad (2)$$

where m_t^* denotes the whole effective mass and E_t denotes the trap energy level with respect to the top of the a-SiN$_x$:H valence band. Then the transient current can be determined by calculating the velocity of the tunneling front as

$$I(t) = qnvA = qN(x(t))\frac{dx(t)}{dt}A = \frac{qN(x(t))A}{2\beta t},\quad (3)$$

where q is the electronic charge, n is the carrier density, v is the tunneling front velocity, $N(x(t))$ is the spatial distribution of traps, and A is the area of conductive filamentary paths. In this way, we can estimate the lower limit of dangling bond density. In this study, t_0 is estimated to be 10^{-13} s [18,22], $m_t^* = 0.42m_0$ [23], and E_t = 3.1 eV [17]. According to this model, the trap distribution in the a-SiN$_x$:H films can be obtained from Equation (3) as

$$N(x(t)) = \frac{I(t)2\beta t}{qA},\quad (4)$$

In Figure 2a, the slope of the log I versus log t plot is -1, so the original Si dangling bond trap distribution is uniform in the IRS. Similarly shown in Figure 3b, the slopes of the transient current of the LRS and HRS are also close to -1, indicating that the generated Si dangling bond distribution is uniform. According to Equation (4) and the data in Figure 3b, the discharged original Si dangling bond density in the a-SiN$_{1.17}$:H device is calculated to be larger than 10^{17} cm^{-3}, while the Si dangling bond trap density generated in the LRS is calculated to be larger than 10^{18} cm^{-3}. Here the change in the transient current of the LRS and HRS is related to the generation and re-passivation of Si dangling bonds during the RS process [24,25]. The conducting filament evolution between the IRS, LRS, and HRS is illustrated in Figure 4a–c. As for the forming operation, the conducting paths are composed of Si dangling bonds because the weak Si-H bond can be broken by the forming voltage. The abundant newly generated Si dangling bonds result in a significant increase in the transient current following the set operation. After the reset process, the Si dangling bond conducting paths are partially broken because some Si dangling bonds are passivated by H. Thus, the transient current intensity decreases slightly. When the device switches back to the LRS, some Si dangling bonds are produced due to the broken Si-H bond. Thus, the transient current intensity increases slightly. As displayed in Figure 4d,e, electrons were trapped in Si dangling bonds during the charging process. After the bias voltage was removed, the electrons were released from the trap centers, which is the origin of the transient current.

Figure 5 shows the energy band diagrams of the charging and discharging process, respectively, for an a-SiN$_{1.17}$:H device in different resistance states. In the IRS, thermally-excited electrons are injected into the a-SiN$_{1.17}$:H film from the P$^+$ electrode and trapped in these original Si dangling bond trap centers during the charging process. After the removal of the bias, the electrons tunnel from the trap centers to the P$^+$ electrode to form the transient current. For the LRS, additional electrons can be captured by the newly generated Si dangling bonds. When the bias voltage is removed, the transient current from both the original and generated Si dangling bonds forms the transient current. As a result, we observe a clear increase in transient current after the forming process. In the HRS, a relatively small number of electrons are captured by the residual Si dangling bond traps due to the passivation of Si dangling bonds by H, and the transient current can be reduced accordingly.

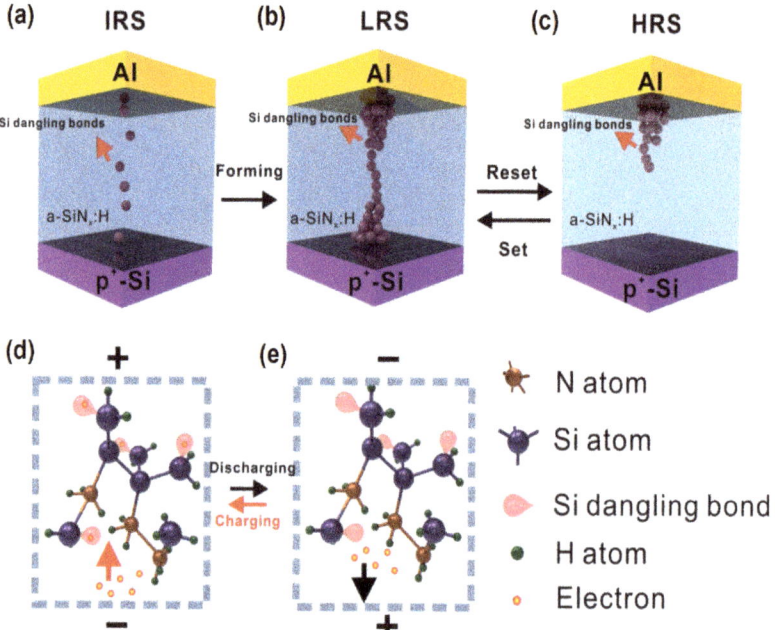

Figure 4. (**a–c**) The Si dangling bond distribution of the a-SiN$_{1.17}$:H device in the IRS, LRS, and HRS. (**d–e**) Schematic illustration of electrons trapped and released by Si dangling bond during the charging and the discharging process.

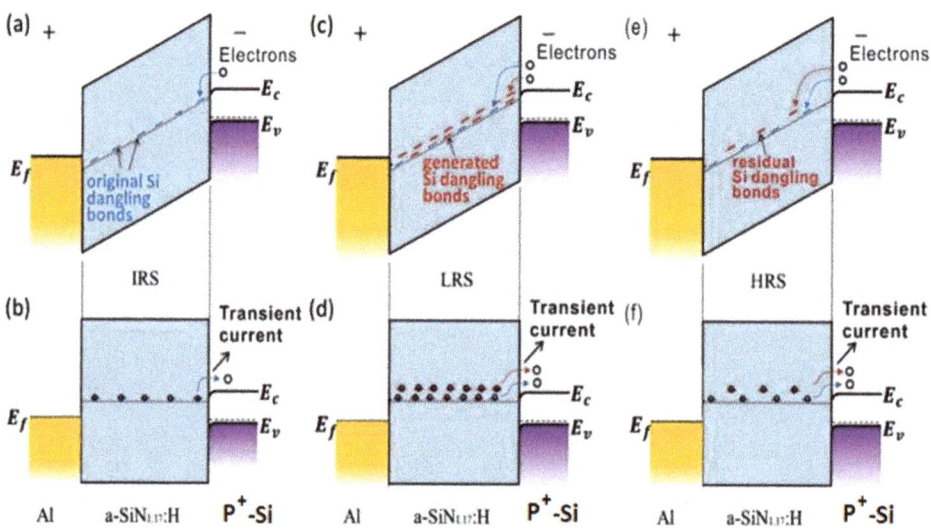

Figure 5. (**a–f**) The energy band diagrams of the Al/a-SiN$_{1.17}$:H/P$^+$-Si device in the IRS, LRS, and HRS during the charging and discharging process.

4. Conclusions

In summary, we successfully observed the evolution of Si dangling bond nanopathway in a-SiN$_x$:H RRAM devices at multiple resistance states by the transient current. The relationship of the transient current with Si dangling bonds for all resistive switching states is revealed in detail. Moreover, the density of the original and generated Si dangling bonds after the forming process can be qualitatively described based on the tunneling front model. The transient current is promising to trace the dynamic evolution of the conducting filament for trap-dominated resistive switching memory, opening a new avenue to insight into the resistive switching mechanism of the a-SiN$_x$:H RRAM in nanoscale for neuromorphic computing in the AI period.

Author Contributions: Conceptualization, Z.M. and T.C.; methodology, K.L. and W.L.; software, T.C.; validation, X.J., J.X. and K.C.; formal analysis, K.L. and L.X.; investigation, Z.M. and K.L.; resources, K.L.; data curation, T.C.; writing—original draft preparation, K.L.; writing—review and editing, Z.M.; visualization, Z.M.; supervision, Z.M.; project administration, Z.M.; funding acquisition, Z.M. All authors have read and agreed to the published version of the manuscript.

Funding: This study was supported by the National Nature Science Foundation of China (Grant Nos. 61634003, 61571221, 11774155, 61921005 and 61735008), Six Talent Peaks Project in Jiangsu Province (DZXX-001), and National Key R&D program of China (2018YFB2200101).

Institutional Review Board Statement: Not applicable.

Informed Consent Statement: Not applicable.

Data Availability Statement: The data that support the findings of this study are available from the corresponding authors upon reasonable request.

Conflicts of Interest: The authors declare no conflict of interest.

References

1. Yang, Y.; Huang, R. Probing Memristive Switching in Nanoionic Devices. *Nat. Electron.* **2018**, *5*, 274–287. [CrossRef]
2. Yang, Y.; Zhang, X.; Qin, L.; Zeng, Q.; Qiu, X.; Huang, R. Probing Nanoscale Oxygen Ion Motion in Memristive Systems. *Nat. Commun.* **2017**, *8*, 15173. [CrossRef] [PubMed]
3. Yan, X.; Zhao, J.; Liu, S.; Zhou, Z.; Liu, Q.; Chen, J.; Liu, X.Y. Memristor with Ag-Cluster-Doped TiO$_2$ Films as Artificial Synapse for Neuroinspired Computing. *Adv. Funct. Mater.* **2018**, *28*, 1705320. [CrossRef]
4. Zhang, Y.; Yang, T.; Yan, X.; Zhang, Z.; Bai, G.; Lu, C.; Jia, X.; Ding, B.; Zhao, J.; Zhou, Z. A Metal/Ba$_{0.6}$Sr$_{0.4}$TiO$_3$/SiO$_2$/Si Single Film Device for Charge Trapping Memory Towards a Large Memory Window. *Appl. Phys. Lett.* **2017**, *110*, 223501. [CrossRef]
5. Hong, S.M.; Kim, H.D.; An, H.M.; Kim, T.G. Effect of Work Function Difference Between Top and Bottom Electrodes on The Resistive Switching Properties of SiN Films. *IEEE Electron. Dev. Lett.* **2013**, *34*, 1181–1183. [CrossRef]
6. Kwon, J.Y.; Park, J.H.; Kim, T.G. Self-rectifying Resistive-switching Characteristics with Ultralow Operating Currents in SiO$_x$N$_y$/AlN Bilayer Devices. *Appl. Phys. Lett.* **2015**, *106*, 223506. [CrossRef]
7. Kim, H.D.; An, H.M.; Hong, S.M.; Kim, T.G. Forming-free SiN-based Resistive Switching Memory Prepared by RF Sputtering. *Phys. Status Solidi A* **2013**, *210*, 1822–1827. [CrossRef]
8. Jiang, X.; Ma, Z.; Yang, H.; Yu, J.; Wang, W.; Zhang, W.; Feng, D. Nanocrystalline Si Pathway Induced Unipolar Resistive Switching Behavior From Annealed Si-rich SiN$_x$/SiN$_y$ Multilayers. *J. Appl. Phys.* **2014**, *116*, 123705. [CrossRef]
9. Kim, H.D.; An, H.M.; Hong, S.M.; Kim, T.G. Unipolar Resistive Switching Phenomena in Fully Transparent SiN-based Memory Cells. *Semicond. Sci. Technol.* **2012**, *27*, 125020. [CrossRef]
10. Kim, S.; Chang, Y.F.; Kim, M.H.; Park, B.G. Improved Resistive Switching Characteristics in Ni/SiN$_x$/p^{++}-Si Devices by Tuning x. *Appl. Phys. Lett.* **2017**, *111*, 033509. [CrossRef]
11. Kim, S.; Park, B.G. Tuning Tunnel Barrier in Si$_3$N$_4$-based Resistive Memory Embedding SiO$_2$ for Low-power and High-density Cross-point Array Applications. *J. Alloy. Compd.* **2016**, *663*, 256–261. [CrossRef]
12. Jiang, X.; Ma, Z.; Xu, J.; Chen, K.; Xu, L.; Li, W.; Feng, D. a-SiNx: H-based Ultra-low Power Resistive Random Access Memory with Tunable Si Dangling Bond Conduction Paths. *Sci. Rep.* **2015**, *5*, 15762. [CrossRef] [PubMed]
13. Dumin, D.J.; Maddux, J.R. Correlation of Stress-induced Leakage Current in Thin Oxides with Trap Generation Inside the Oxides. *IEEE Trans. Electron. Dev.* **1993**, *40*, 986–993. [CrossRef]
14. Scott, R.S.; Dumin, D.J. The Charging and Discharging of High-voltage Stress-generated Traps in Thin Silicon Oxide. *IEEE Trans. Electron. Dev.* **1996**, *43*, 130–136. [CrossRef]
15. Vanheusden, K.; Seager, C.H.; Warren, W.T.; Tallant, D.R.; Voigt, J.A. Correlation Between Photoluminescence and Oxygen Vacancies in ZnO Phosphors. *Appl. Phys. Lett.* **1996**, *68*, 403–405. [CrossRef]

16. Robertson, J.; Powell, M.J. Gap States in Silicon Nitride. *Appl. Phys. Lett.* **1984**, *44*, 415–417. [CrossRef]
17. Mo, C.M.; Zhang, L.; Xie, C.; Wang, T. Luminescence of Nanometer-sized Amorphous Silicon Nitride Solids. *J. Appl. Phys.* **1993**, *73*, 5185–5188. [CrossRef]
18. Oldham, T.R.; Lelis, A.J.; McLean, F.B. Spatial Dependence of Trapped Holes Determined From Tunneling Analysis and Measured Annealing. *IEEE Trans. Nucl. Sci.* **1986**, *33*, 1203–1209. [CrossRef]
19. Benedetto, J.M.; Boesch, H.E.; McLean, F.B.; Mize, J.P. Hole Removal in Thin-gate MOSFETs by Tunneling. *IEEE Trans. Nucl. Sci.* **1985**, *32*, 3916–3920. [CrossRef]
20. Yamada, R.I.; Mori, Y.; Okuyama, Y.; Yugami, J.; Nishimoto, T.; Kume, H. Analysis of Detrap Current Due to Oxide Traps to Improve Flash Memory Retention. In Proceedings of the 2000 IEEE International Reliability Physics Symposium Proceedings. 38th Annual (Cat. No. 00CH37059), San Jose, CA, USA, 10–13 April 2000; Volume 9, pp. 200–204.
21. Xu, Z.; Pantisano, L.; Kerber, A.; Degraeve, R.; Cartier, E.; De Gendt, S.; Groeseneken, G. A Study of Relaxation Current in High-/Spl kappa/Dielectric stacks. *IEEE Trans. Electron. Dev.* **2004**, *51*, 402–408. [CrossRef]
22. Chen, H.M.; Lan, J.M.; Chen, J.L.; Yamin, L.J. Time-dependent and Trap-related Current Conduction Mechanism in Ferroelectric Pb(Zr_xTi_{1-x})O_3 Films. *Appl. Phys. Lett.* **1996**, *69*, 1713–1715. [CrossRef]
23. Yeo, Y.C.; Lu, Q.; Lee, W.C.; King, T.J.; Hu, C.; Wang, X.; Ma, T.P. Direct Tunneling Gate Leakage Current in Transistors With Ultrathin Silicon Nitride Gate Dielectric. *IEEE Electron Dev. Lett.* **2000**, *21*, 540–542.
24. Wang, Y.; Chen, K.; Qian, X.; Fang, Z.; Li, W.; Xu, J. The x Dependent Two Kinds of Resistive Switching Behaviors in SiOx Films with Different x Component. *Appl. Phys. Lett.* **2014**, *104*, 012112. [CrossRef]
25. Yu, J.; Ma, Z.; Wang, Y.; Ren, S.; Fang, Z.; Huang, X.; Wang, L. Improvement of Retention and Endurance Characteristics of Si Nanocrystal Nonvolatile Memory Device. In Proceedings of the 2014 12th IEEE International Conference on Solid-State and Integrated Circuit Technology (ICSICT), Guilin, China, 28–31 October 2014; Volume 14, pp. 1–3.

Disclaimer/Publisher's Note: The statements, opinions and data contained in all publications are solely those of the individual author(s) and contributor(s) and not of MDPI and/or the editor(s). MDPI and/or the editor(s) disclaim responsibility for any injury to people or property resulting from any ideas, methods, instructions or products referred to in the content.

Article

Controlling the Carrier Injection Efficiency in 3D Nanocrystalline Silicon Floating Gate Memory by Novel Design of Control Layer

Hongsheng Hu [1,2,3], Zhongyuan Ma [1,2,3,*], Xinyue Yu [1,2,3], Tong Chen [1,2,3], Chengfeng Zhou [1,2,3], Wei Li [1,2,3], Kunji Chen [1,2,3], Jun Xu [1,2,3] and Ling Xu [1,2,3]

1. School of Electronic Science and Engineering, Nanjing University, Nanjing 210093, China
2. Collaborative Innovation Center of Advanced Microstructures, Nanjing University, Nanjing 210093, China
3. Jiangsu Provincial Key Laboratory of Photonic and Electronic Materials Sciences and Technology, Nanjing University, Nanjing 210093, China
* Correspondence: zyma@nju.edu.cn

Abstract: Three-dimensional NAND flash memory with high carrier injection efficiency has been of great interest to computing in memory for its stronger capability to deal with big data than that of conventional von Neumann architecture. Here, we first report the carrier injection efficiency of 3D NAND flash memory based on a nanocrystalline silicon floating gate, which can be controlled by a novel design of the control layer. The carrier injection efficiency in nanocrystalline Si can be monitored by the capacitance–voltage (C–V) hysteresis direction of an nc-Si floating-gate MOS structure. When the control layer thickness of the nanocrystalline silicon floating gate is 25 nm, the C–V hysteresis always maintains the counterclockwise direction under different step sizes of scanning bias. In contrast, the direction of the C–V hysteresis can be changed from counterclockwise to clockwise when the thickness of the control barrier is reduced to 22 nm. The clockwise direction of the C–V curve is due to the carrier injection from the top electrode into the defect state of the SiN_x control layer. Our discovery illustrates that the thicker SiN_x control layer can block the transfer of carriers from the top electrode to the SiN_x, thereby improving the carrier injection efficiency from the Si substrate to the nc-Si layer. The relationship between the carrier injection and the C–V hysteresis direction is further revealed by using the energy band model, thus explaining the transition mechanism of the C–V hysteresis direction. Our report is conducive to optimizing the performance of 3D NAND flash memory based on an nc-Si floating gate, which will be better used in the field of in-memory computing.

Keywords: C–V memory window; nanocrystalline Si; floating gate memory

1. Introduction

With the development of the Internet in modern society, highly efficient processing of big data is faced with the challenge of breaking up the storage walls and power consumption walls induced by traditional von Neumann architecture. Recently, in-memory computing has attracted great interest, due to its ability to effectively improve the processing efficiency of big data. As a strong device candidate for in-memory computing, silicon-based 3D NAND flash memory with perfect compatibility with CMOS technology has attracted much attention [1–5]. However, the traditional 3D NAND flash, based on a polysilicon floating gate, is confronted with the current leakage problem, which is induced by random defects in the tunnel oxide layer after repeated erasing and writing [6–8]. As the charges stored in the floating-gate layer can be free to move laterally, it results in data loss. To solve the current leakage problems, the nanocrystal has been adapted to the floating-gate memory [8–11]. In contrast to the polysilicon floating gate, using discrete nanocrystals as the charge storage layer can avoid the free movement of charges laterally in the floating-gate layer, thus effectively preventing data loss caused by charge leakage [12–14]. In addition, the ultra-thin tunnel oxide layer in the nanocrystal floating-gate memory has the advantages of low power consumption and high erasing/programming speed [9–11]. The research on nanocrystal floating-gate memory is extensive, from traditional silicon germanium materials to third-generation semiconductor materials of SiC [15–17]. As reported by Jin et al. [15], floating-gate memory based on antimony-doped tin oxide nanoparticles has a maximum memory window of 85 V. However, there are 40 program/erase cycles, which has not been tried in 3D NAND devices. Lepadatu et al. carried out research on a new floating-gate MOS structure consisting of an HfO_2/floating gate of a single layer of Ge QDs in the HfO_2/tunnel HfO_2/p-Si wafers [16]. The memory window of 3.8 V shows a very slow capacitance decease, which is not adopted in 3D NAND devices. According to the research of Andrzej et al. [17], a nanocrystalline SiC floating-gate memory exhibits better charge retention characteristics than the conventional floating-gate memory. However, the program/erase speed is in the scale of S, which is not applied in 3D NAND memory. Compared with the above nanocrystalline floating-gate memory, nc-Si-based floating-gate memory not only has faster program and erase speeds, but also has high durability for 10^7 program/erase cycles. In particular, its high compatibility with modern microelectronics technology is beneficial to be integrated with computing in-memory chips. Our previous work focused on improving the density of 3D NAND flash memory by using double-layered nanocrystalline Si dots [18]. Up to now, there are few studies on how to improve the carrier injection efficiency of 3D NAND flash memory based on an nc-Si floating gate by novel design of the control gate thickness of the nano-silicon floating-gate memory.

In this paper, we first report that 3D NAND memories based on an nc-Si floating gate with high carrier injection efficiency could be obtained by novel design of the control barrier layer. The carrier injection efficiency can be monitored through the direction of capacitance–voltage (C–V) hysteresis of the nc-Si floating-gate MOS structure [19–26]. According to the principle of the floating-gate memory, the threshold voltage corresponding memory window can be expressed by the following equation.

$$\Delta V_{th} = \frac{q n_{nc}}{\varepsilon_{ox}} \left(t_{cntl} + \frac{1}{2} \frac{\varepsilon_{ox}}{\varepsilon_{Si}} t_{nc} \right) \qquad (1)$$

In the equation, ΔV_{th} is the threshold voltage shift, t_{cntl} is the thickness of the control nitride, t_{nc} is the size of a single nanocrystal, ε_{ox} and ε_{Si} are the dielectric constants of the SiO_2 tunneling layer and nc-Si, respectively, q is the magnitude of electronic charge, and n_{nc} is the density of the nanocrystals. According to the equation, the value of the memory window is related to the parameters of the floating gate, such as the density of the nc-Si, the thickness of the control layer, and the tunnel layer, etc. To improve the carrier injection efficiency, a novel design of the parameters is vitally important. Compared with the program/erase pulse cycle of an nc-Si floating-gate MOSFET, the C–V characteristic of an nc-Si floating-gate MOS structure can directly reflect the carrier injection efficiency without the influence of the source electrode and the drain electrode. Therefore, we focus on the C–V investigation of an nc-Si floating-gate MOS structure to improve the carrier injection in 3D nc-Si floating-gate memory. So far, the relationship between C–V hysteresis direction and carrier injection efficiency has been less reported in an nc-Si floating-gate memory [27,28]. Here, we first report that the C–V hysteresis direction of nc-Si floating-gate MOS structures could be shifted from counterclockwise to clockwise when the control layer thickness is reduced from 25 nm to 22 nm, which is influenced by the step size of the bias and the gate voltage [29]. When the thickness of the control layer reaches 25 nm, the direction of the C–V hysteresis can be maintained in the counterclockwise direction. However, when the thickness of the silicon nitride layer is reduced to 22 nm, a clockwise hysteresis curve will first appear in the low gate voltage scanning process. As the voltage increases, the window of the curve will gradually become smaller and then turn into a normal counterclockwise hysteresis curve. This is due to the carrier injection from the top electrode into the defect state of the SiN_x layer [30,31]. A thicker SiN_x control layer can block the transfer of carriers between the top electrode and the nc-Si dots, thereby improving the carrier injection efficiency from the Si substrate to the nc-Si dots. Combined with HRTEM and energy band theory, the relationship between the carrier injection efficiency and the C–V hysteresis direction is further revealed, which is beneficial to optimizing the performance of 3D NAND memory based on an nc-Si floating gate for future application in in-memory computing.

2. Materials and Methods

Figure 1 shows the schematic diagram of a 3D nc-Si floating-gate memory with three layers of a-Si:H channels. The substrate is a p-type silicon wafer with a crystal orientation of <100> and a resistivity of 6 to 9 Ω cm. To obtain the nc-Si floating-gate MOS structure, the Si substrate was first cleaned by standard RCA procedures [32]. The natural oxide layer grown on the surface was removed with dilute hydrofluoric acid. In order to fabricate the ultra-thin tunneling SiO_2 layer, the thermal dry oxidation method under a temperature of 850 °C was employed [33,34]. The thickness of the tunnel SiO_2 layer was 5 nm. Then, an a-Si layer was deposited on the tunnel SiO_2 layer by introducing SiH_4 into the PECVD chamber with an RF source frequency of 13.56 MHz [35,36]. The substrate temperature was 250 °C. Under the same circumstances, NH_3 and SiH_4 were decomposed to fabricate the SiN_x control layer on the surface of the a-Si:H layer. The thickness of the SiN_x control layer is 22 nm. To check the role of the SiN_x layer, a reference sample with an SiN_x thickness of 25 nm was fabricated by the same process. Finally, the samples were thermally post-treated at 1000 °C under ambient N_2 to form nc-Si dots in the a-Si:H layer. For the convenience of the electric measurements, aluminum (Al) top electrodes were thermally evaporated on the surfaces of the samples with a shadow mask to form a circular spot. Al was thermally evaporated at the back side of the Si substrate as the bottom electrode. After the annealing process, the nc-Si embedded floating-gate MOS cross-sectional structure can be directly revealed by high-resolution cross-section transmission electron microscopy (HRTEM) with a JEM2010 electron microscope working at 200 kV. The C–V, transfer, and output characteristic were measured by using an Agilent B1500A at room temperature.

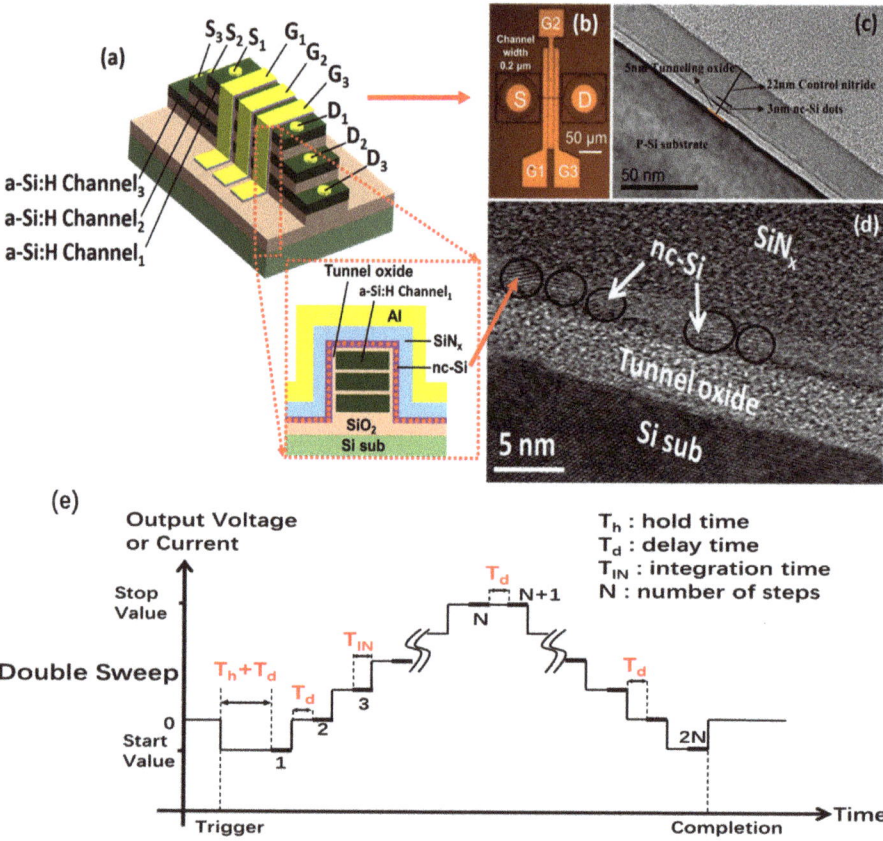

Figure 1. (a) Schematic diagram of 3D flash memory based on three-layered a-Si:H channels with a single-layered nc-Si dots floating-gate structure. (b) The optical image of 3D NAND flash memory based on the SiN_x/nc-Si(a-Si)/SiO_2 floating gate. (c,d) Ordinary and high resolution cross-section TEM photograph of the SiN_x/nc-Si(a-Si)/SiO_2 floating-gate MOS structure. (e) The time dependence of output voltage for double C–V measurement.

3. Results and Discussion

Figure 1a is a schematic diagram showing how the 3D nc-Si dots floating-gate flash memory based on three-layered a-Si:H channels is fabricated, according to the following steps. First, the surface of the Si substrate was covered by a thicker silicon oxide layer of 300 nm, which was grown by the wet oxidation method. Second, three layers of amorphous Si:H and SiO_2 were alternately deposited on the surface of the SiO_2 layer in the PECVD chamber at 300 °C. The thickness of the Si:H and SiO_2 layers was 70 nm and 100 nm, respectively. Then, the three layers of a-Si:H and SiO_2 were patterned by electron beam lithography and dry-etched to form the amorphous silicon channel. An nc-Si floating gate was grown on the surface of the three-layer a-Si channels, according to the preparation method of the MOS structure. In the following process, an a-Si:H terrace was obtained by etching to prepare for the drain electrode and the source electrode. The hole position of the drain electrode and the source electrode in each a-Si:H was different from each other, which was formed by etching of each a-Si:H terrace according to the pattern of the photolithography. Meanwhile, the holes of the source electrode and the drain electrode were obtained by etching the a-Si:H from the first to the third a-Si:H layers to expose the three a-Si:H layers. At last, the construction of the source electrodes and drain electrodes were completed by filling the holes of the source electrodes and drain electrodes with aluminum. Meanwhile, the metal of the gate electrodes was also deposited by thermal evaporation followed by lift-off technology to prepare for the electrical measurement. The C–V, erasing, and programming, as well as output characteristics, were measured by using an Agilent B1500A at room temperature. Under the bias of the gate, the nc-Si floating-gate unit on the two side walls of the 3D a-Si:H channel can be chosen to complete the writing and erasing functions separately. The site of the chosen nc-Si floating-gate cells decided by the position of the source electrode and the drain electrode, which range from the first layer to the third layer of the a-Si:H.

As revealed in Figure 1b, the nc-Si floating-gate MOS structure, including a tunnel oxide layer of 5 nm, an nc-Si layer of 3 nm, and a control layer of 22 nm, is clearly observed. It is evident that the nc-Si crystals of less than 3 nm are embedded in the a-Si sublayers, as presented in Figure 1c. A clear lattice image of nanocrystalline silicon can be seen, which corresponds to the crystal face index of (100). The C–V measurement diagram of the nc-Si floating-gate MOS device is shown in Figure 1d. During the C–V measurement, the top aluminum electrode is applied with the gate voltage, and the bottom aluminum electrode is grounded. During the double C–V measurement, the step size can be changed from 20 mV/s to 2500 mV/s. As illustrated in Figure 1e, the definition of the step size can be expressed by the Formula (2). It depends on the output voltage (V_{step}) and the time of each scanning step, which can be divided into delay time (T_d) and integration time (T_{IN}).

$$v = \frac{V_{step}}{T_d + T_{IN}} \quad (2)$$

We change the value of the T_{IN} to tune the step size. During the C–V measurement for the nc-Si floating-gate MOS structure device, we make the output voltage (V_{step}) remain the same.

Figure 2a shows the double C–V characteristic curves of the SiN_x/nc-Si/SiO_2 floating-gate MOS structure under different step sizes, from 25 to 500 mv/s. The thickness of the control layer is 22 nm. The frequency is 1 MHz. The inset of each figure shows the variation of the flat band voltages (V_{fb}s) with the bias increasing. As indicated in Figure 2a, under the step size of 20 mV/s, the direction of C–V hysteresis remains counterclockwise with the scanning bias increase from (−2 V, 2 V) to (−8 V, 8 V), and the maximum window is 1.14 V. It is interesting to find that a clockwise C–V hysteresis window appears under the bias from −2 V to 2 V when the step size increases to 250 mV/s, as shown in Figure 2b. When the bias changes from −3 V to 3 V, the clockwise window expands to a maximum of about 0.35 V. With the bias changing from −4 V to 4 V, the clockwise window begins to reduce. It is worth noting that the direction of the hysteresis curve changes to counterclockwise under the bias voltage from −5 V to 5 V. The counterclockwise window is continuously enhanced with the bias increasing from (−6 V, 6 V) to (−7 V, 7 V). The memory window reaches a saturation value of 0.93 V with the bias increasing from −8 to 8 V. After the step size is enhanced to 500 mV/s, the clockwise direction can be detected under the lower bias range from (−2 V, 2 V) to (−5 V, 5 V). In contrast with the step size of 250 mV/s, the critical bias corresponding to the direction transition from clockwise to counterclockwise is increased from −6.3 V to 6.3 V, as shown in Figure 2c.

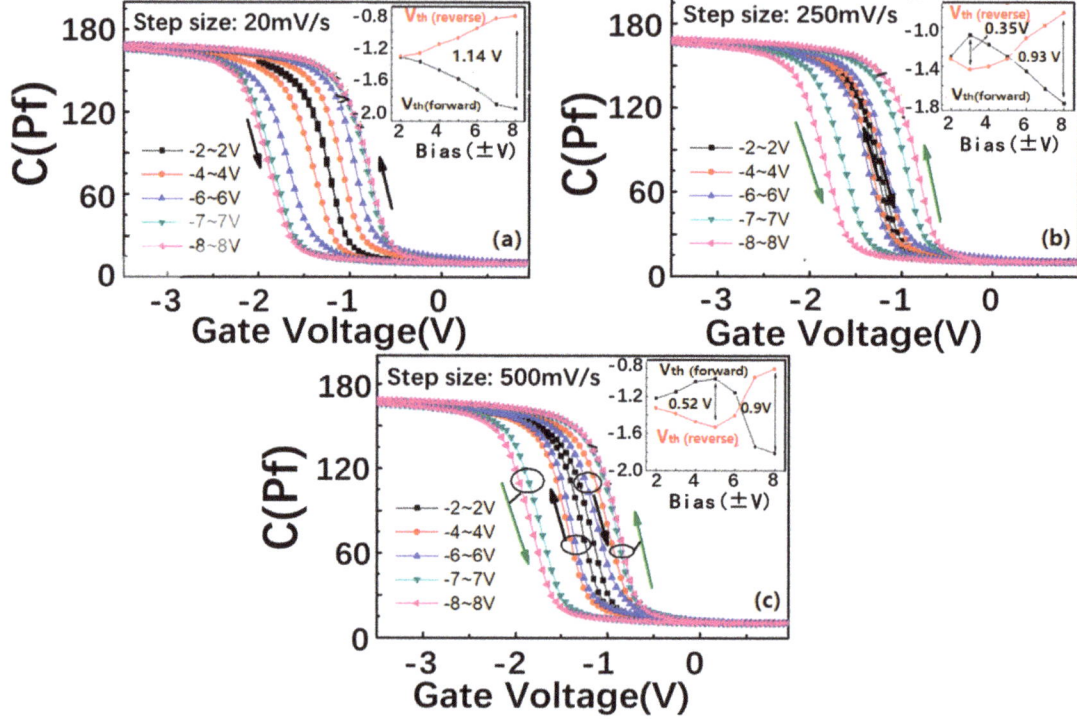

Figure 2. Double C–V sweeps of the SiN_x/nc-Si(a-Si)/SiO_2 floating-gate MOS structure in different scanning bias ranges. The step sizes are (**a**) 20 mV/s, (**b**) 250 mV/s, and (**c**) 500 mV/s, respectively. The insets show the scanning bias-dependent V_{fb} under different step sizes.

Figure 3 shows the schematic diagram of the counterclockwise C–V hysteresis formation from the SiN_x/nc-Si(a-Si)/SiO_2 floating-gate MOS structure. As displayed in Figure 3a,b, the free electrons accumulated on the surface of the Si substrate under a positive bias. With the positive bias reaching a critical value, the electrons can tunnel through the SiO_2 layer into the nc-Si layer, as shown in Figure 3c. The transfer of electrons contributes to a positive movement of the flat band voltage. The control SiN_x layer can prevent electrons moving from the nc-Si to the upper electrode just like a blockade. As a result, the electrons can be stored in the nc-Si quantum dots in the nc-Si layer. Similarly, when the gate voltage changes from positive to negative, the free holes tunnel from the substrate into the nc-Si layer, causing the flat band voltage to move negatively, as shown in Figure 3d–f. Therefore, a counterclockwise C–V hysteresis window is formed.

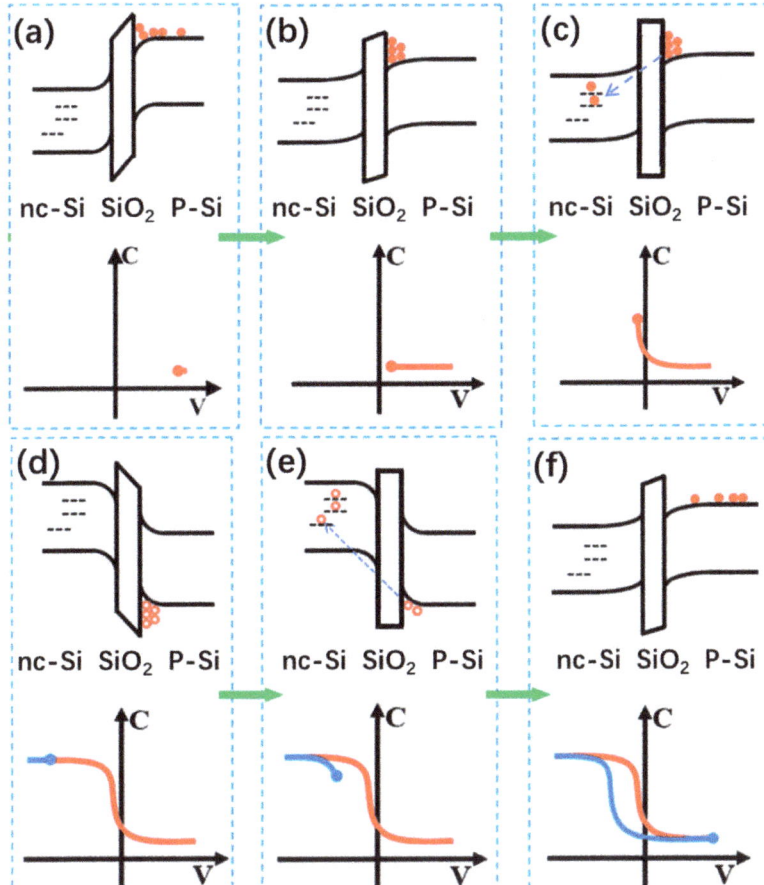

Figure 3. Formation mechanism of the counterclockwise C–V hysteresis of the SiN_x/nc-Si(a-Si)/SiO_2 floating-gate MOS structure. (**a–c**) Under the positive gate voltage, the electrons tunnel from the substrate into the nc-Si layer, leading to the positive movement of the flat band voltage as shown by the red lines. (**d–f**) Under the negative gate voltage, the holes tunnel from the substrate into the nc-Si layer, leading to the negative movement of the flat band voltage as shown by the blue lines. And a counterclockwise C–V window is formed. The hollow circles and the solid circles represent the holes and electrons, respectively. The blue arrow represents the direction of electron tunneling and hole tunneling.

In an attempt to provide insight into the formation mechanism of the clockwise C–V hysteresis, the C–V characteristics of the SiN_x/nc-Si(a-Si)/SiO_2 floating-gate MOS structure with a thicker SiN_x control layer of 25 nm were tested, as shown in Figure 4. It is interesting to find that the clockwise C–V window cannot be observed with the step size increasing from 250 mV/s to 2500 mV/s, which proves that the clockwise C–V hysteresis is related to the thickness of the SiN_x control layer. When the SiN_x control layer is thinner, the corresponding barrier is lower, which is easier for holes moving from the top electrode to the defects in the nitride SiN_x control. In particular, the defects in the SiN_x control have lower energy levels than the Fermi level of the aluminum, and the number of the holes injected into the defects is higher than that of the electrons tunneling from the Si substrate to the nc-Si layer under the same gate bias. Thus, the holes from the top electrode become the main contributor to the charge movement, which determines that the movement direction of the flat band voltage is reverse that induced by the electrons from the substrate. Therefore, we can detect the formation of the clockwise C–V hysteresis.

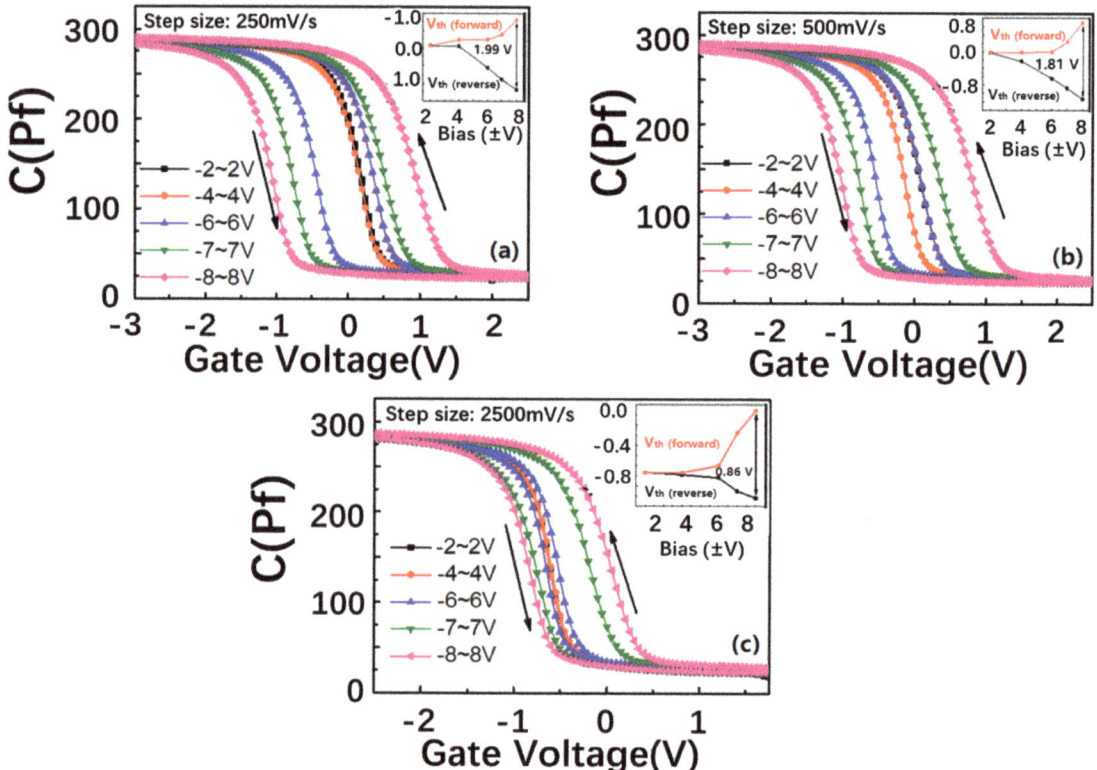

Figure 4. Double C–V sweeps of the fabricated SiN_x/nc-Si(a-Si)/SiO_2 floating-gate MOS structure with a thicker nitride SiN_x control layer in different scanning bias ranges, and the ramp rates are (**a**) 250 mV/s, (**b**) 500 mV/s, and (**c**) 2500 mV/s, respectively. The insets show the scanning bias range is dependent on the V_{fb}s at different ramp rates.

To further reveal the relationship between the carrier injection efficiency and the direction of the C–V memory window, we elaborate the energy band gap model of the SiN_x/nc-Si/SiO_2 floating-gate MOS structure with thicker and thinner SiN_x control layers under different step sizes of bias [28,37]. As presented in Figure 5a–c, when the SiN_x control layer of the nc-Si floating-gate MOS structure is thicker, a larger quantity of electrons tunnel from the Si substrate to the nc-Si layer under the positive bias. The higher barrier of the thicker SiN_x control layer can reduce the movement of holes from the top electrode to the SiN_x control layer. Therefore, the number of the electrons tunneling from the substrate to the nc-Si layer is larger than that of the holes stored in the defects of the SiN_x control layer from the Al top electrode. The electrons tunnelling from the substrate to the nc-Si layer results in the positive movement of the flat band voltage. Under the negative bias, the number of holes tunneling from the substrate to the nc-Si layer is larger than that of the electrons stored in the defects of the SiN_x control layer from the Al top electrode. The holes tunnelling from the substrate to the nc-Si layer result in the negative movement of the flat band voltage. Therefore, a counterclockwise hysteresis is formed. Even with the step size of the bias increasing from 250 to 2500 mV/s, the counterclockwise direction of the C–V memory window remains unchanged, as revealed in Figure 4a. When the SiN_x control layer is thinner, the lower barrier height of the SiN_x control layer is easier for the carriers moving from the top electrode to the SiN_x control layer, as displayed in Fig5.d-f. When the step size is 50 and 250 mV/s, the bias duration on the electrode is shorter. Under the lower bias, the number of the carriers tunneling from the Si substrate to the nc-Si layer is smaller than that of the carriers moving from the top electrode to the SiN_x control layer, which results in the reverse movement of the flat band. Thus, the clockwise memory window can be observed. It is worth noting the number of carriers tunnelling from the substrate to the nc-Si can be enhanced with the bias increasing. Under the lower bias, it is larger than the number of the carriers moving from the top electrode to the SiN_x control layer, so the direction of the C–V memory window will change to counterclockwise, as shown in Figure 5i,j. This is the reason why the clockwise memory window shows a changeable trend from the maximum to the minimum, as proved by Figure 4b,c. When the step size is reduced to 20 mV/s, the bias duration on the electrode gets longer. The number of the carriers tunneling from the substrate to the nc-Si layer is larger than that of the carriers stored in the defects of the SiN_x control layer from the Al top electrode. The counterclockwise memory window can be maintained from the low bias to the high bias, as shown by Figure 4a. The contribution of the SiN_x layer to the C–V hysteresis window can also be illustrated by comparing the two devices with different thickness of the SiN_x layer, which is shown in Figures 3b and 4a. Under the same step size of 250 mV/s, the memory window of the device with a thickness of 25 nm is 2.5 V, which is larger than the 1 V of the device with a thickness of 25 nm. Because the barrier height of the SiN_x layer can be enhanced with the thickness increasing, it is more difficult for carriers to move from the top electrode to the defect in the SiN_x. As a result, the carrier injection efficiency from the Si substrate into the nc-Si dots is enhanced, which leads to a larger memory window. Based on the above analysis, the thicker SiN_x control layer is beneficial to increasing the carrier injection efficiency.

The erasing and programming performance of the 3D NAND nc-Si floating-gate memory unit, based on a three layered a-Si:H channel with a thicker control SiN_x of 25 nm, is shown in Figure 6a,b. As displayed in Figure 6c, the corresponding output current is increased with the positive gate voltage changing from 1 to 5 V, displaying the N-type characteristic of the a-Si channel. A stable memory window of 1.21 V can be obtained under the programming and erasing voltages of +7 V and −7 V. The programming and erasing speed reaches 100 μs. It is worth noting that the stable memory window of 1.21 V can be maintained after 10^7 P/E cycles, as shown in Figure 6d. The novel performance of the 3D NAND nc-Si floating-gate memory unit is determined by the design of the nc-Si floating gate with a novel SiN_x control layer.

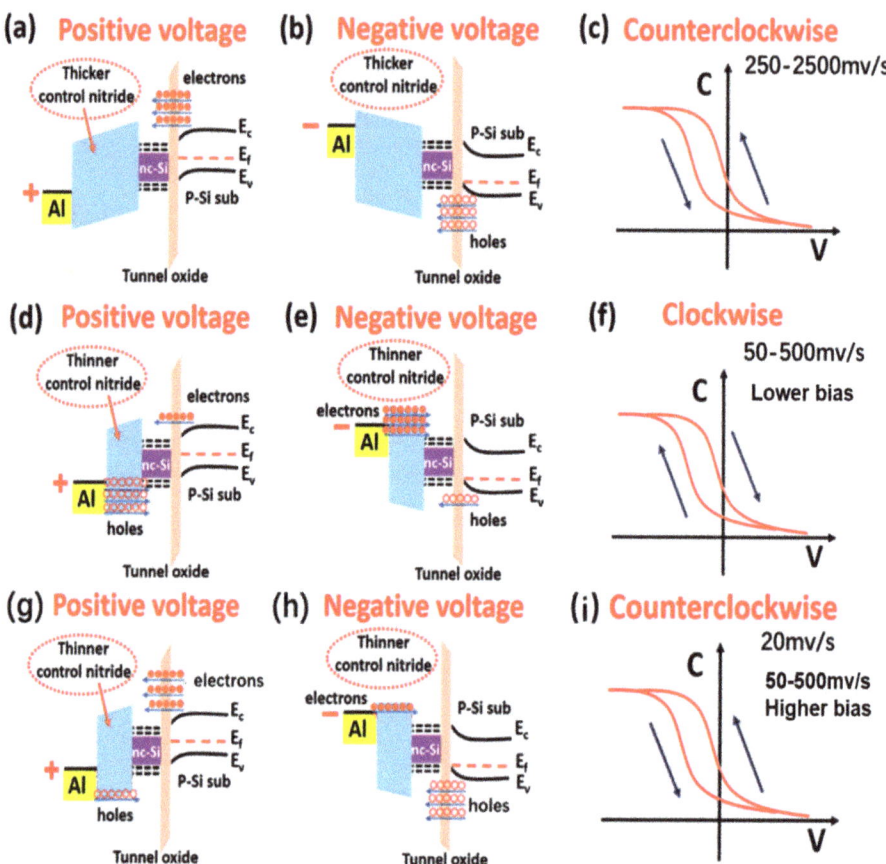

Figure 5. The energy band gap model of the SiN_x/nc-Si/SiO_2 floating-gate MOS structure with thicker SiN_x control layer under (**a**) positive and (**b**) negative voltages. (**c**) The counterclockwise hysteresis memory window of the SiN_x/nc-Si/SiO_2 floating-gate MOS structure with step size from 250 to 2500 mv/s. The energy band gap model of the SiN_x/nc-Si(a-Si)/SiO_2 floating-gate MOS structure with the thinner control SiN_x layer under (**d**) positive and (**e**) negative voltages. (**f**) The clockwise C–V hysteresis memory window of the SiN_x/nc-Si/SiO_2 floating-gate MOS structure with the step size from 50 to 500 mv/s. The energy band gap model of the SiN_x/nc-Si(a-Si)/SiO_2 floating-gate MOS structure with the thinner SiN_x control layer under (**g**) positive and (**h**) negative voltages. (**i**) The counterclockwise C–V hysteresis memory window of the SiN_x/nc-Si/SiO_2 floating-gate MOS structure with the step size of 20 mv/s. The hollow circles and solid circles represent the holes and electrons, respectively.

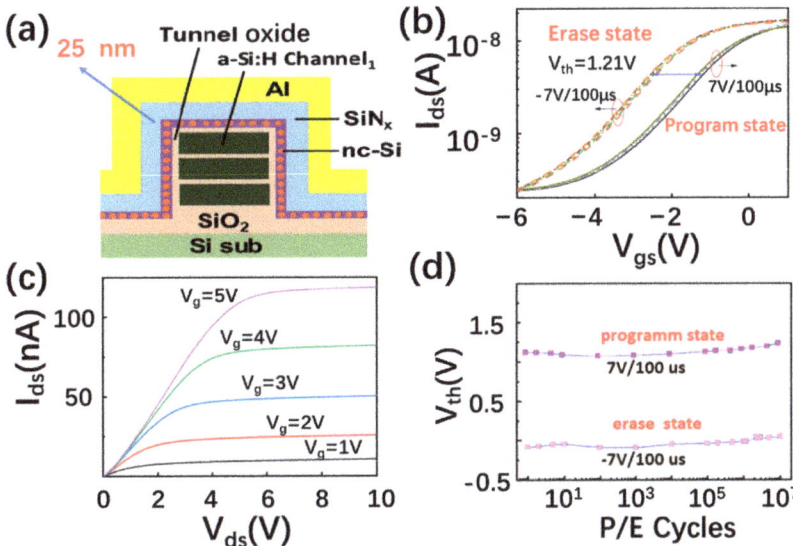

Figure 6. (a) The schematic diagram of the nc-Si floating-gate memory with a SiN_x thickness of 25 nm. (b) The erasing and programming characteristics of the 3D nc-Si floating-gate memory with a SiN_x thickness of 25 nm. The programming and erasing speeds are +7 V/100 us and −7 V/−100 us (c) The output characteristics of the 3D nc-Si floating-gate memory with a SiN_x thickness of 25 nm. (d) The endurance characteristics of the 3D nc-Si floating-gate memory with a SiN_x thickness of 25 nm after the 10^7 P/E cycle operations at +7 V/100 us and −7 V/100 μs.

4. Conclusions

In summary, carrier injection efficiency can be controlled in 3D NAND memory based on a nanocrystalline silicon floating gate by novel design of the control barrier layer. The C–V hysteresis direction of the nc-Si floating-gate MOS structure can change from counterclockwise to clockwise when the thickness of the control layer decreases from 25 to 22 nm, which is easily effected by the step size of the bias and gate voltage. As the thickness of the control layer reaches 25 nm, the direction of the C–V hysteresis remains counterclockwise. The thicker SiN_x control layer can block the transfer of carriers between the top electrode and the nc-Si dots, leading to enhanced carrier injection efficiency from the Si substrate to the nc-Si dots. HRTEM and the energy band theory model further reveal the relationship between the carrier injection efficiency and the direction of the C-V hysteresis, which is beneficial to optimizing the performance of 3D NAND memory based on an nc-Si floating-gate memory for the application of in-memory computation.

Author Contributions: Conceptualization, Z.M. and H.H.; formal analysis, Z.M., H.H., T.C. and C.Z.; writing—original draft preparation, X.Y. and H.H.; writing—review and editing, Z.M., K.C. and W.L.; supervision, Z.M., J.X. and L.X.; project administration, Z.M. and J.X.; funding acquisition, Z.M. All authors have read and agreed to the published version of the manuscript.

Funding: This study was supported by the National Nature Science Foundation of China (Grant Nos. 61634003, 61571221, 11774155, 61921005 and 61735008), and the National Key R&D program of China (2018YFB2200101). Research Fund for the Doctoral Program of the Higher Education of China (Grant No. 20130091110024), Six Talent Peaks Project in Jiangsu Province (DZXX-001).

Data Availability Statement: Data can be available upon request from the authors.

Conflicts of Interest: The authors declare no conflict of interest.

References

1. Bez, R.; Camerlenghi, E.; Modelli, A.; Visconti, A. Introduction to flash memory. *Proc. IEEE* **2003**, *91*, 489–502. [CrossRef]
2. Pavan, P.; Bez, R.; Olivo, P.; Zanoni, E. Flash memory cells-an overview. *Proc. IEEE* **1997**, *85*, 1248–1271. [CrossRef]
3. Wang, P.; Xu, F.; Wang, B.; Gao, B.; Wu, H.; Qian, H.; Yu, S. Three-Dimensional nand Flash for Vector–Matrix Multiplication. *IEEE Trans. Large Scale Integr. (VLSI) Syst.* **2019**, *27*, 988–991. [CrossRef]
4. Lee, G.H.; Hwang, S.; Yu, J.; Kim, H. Architecture and process integration overview of 3D NAND flash technologies. *Appl. Sci.* **2021**, *11*, 6703. [CrossRef]
5. Park, J.K.; Kim, S.E. A Review of Cell Operation Algorithm for 3D NAND Flash Memory. *Appl. Sci.* **2022**, *12*, 10697. [CrossRef]
6. Govoreanu, B.; Brunco, D.; Van Houdt, J. Scaling down the interpoly dielectric for next generation flash memory: Challenges and opportunities. *Solid-State Electron.* **2005**, *49*, 1841–1848. [CrossRef]
7. Tan, Y.-N.; Chim, W.-K.; Cho, B.J.; Choi, W.-K. Over-erase phenomenon in SONOS-type flash memory and its minimization using a hafnium oxide charge storage layer. *IEEE Trans. Electron. Devices* **2004**, *51*, 1143–1147. [CrossRef]
8. Jeon, Y.; Lee, M.; Moon, T.; Kim, S. Flexible nano-floating-gate memory with channels of enhancement-mode Si nanowires. *IEEE Trans. Electron. Devices* **2012**, *59*, 2939–2942. [CrossRef]
9. Tiwari, S.; Rana, F.; Hanafi, H.; Hartstein, A.; Crabbé, E.F.; Chan, K. A silicon nanocrystals based memory. *Appl. Phys. Lett.* **1996**, *68*, 1377–1379. [CrossRef]
10. Naito, S.; Ueyama, T.; Kondo, H.; Sakashita, M.; Sakai, A.; Ogawa, M.; Zaima, S. Fabrication and evaluation of floating gate memories with surface-nitrided Si nanocrystals. *Jpn. J. Appl. Phys.* **2005**, *44*, 5687. [CrossRef]
11. Qian, X.-Y.; Chen, K.-J.; Ma, Z.-Y.; Zhang, X.-G.; Fang, Z.-H.; Liu, G.-Y.; Jiang, X.-F.; Huang, X.-F. Performance improvement of nc-Si nonvolatile memory by novel design of tunnel and control layer. In Proceedings of the 2010 10th IEEE International Conference on Solid-State and Integrated Circuit Technology, Shanghai, China, 1–4 November 2010; pp. 944–946.
12. Velampati, R.S.R.; Hasaneen, E.-S.; Heller, E.; Jain, F.C. Floating gate nonvolatile memory using individually cladded monodispersed quantum dots. *IEEE Trans. Large Scale Integr. (VLSI) Syst.* **2017**, *25*, 1774–1781. [CrossRef]
13. Muraguchi, M.; Sakurai, Y.; Takada, Y.; Shigeta, Y.; Ikeda, M.; Makihara, K.; Miyazaki, S.; Nomura, S.; Shiraishi, K.; Endoh, T. Collective electron tunneling model in si-nano dot floating gate MOS structure. In *Key Engineering Materials*; Trans Tech Publications Ltd.: Wollerau, Switzerland, 2011; pp. 48–53.
14. Yang, J.S.; Kim, S.-I.; Kim, Y.T.; Cho, W.J.; Park, J.H. Electrical characteristics of nano-crystal Si particles for nano-floating gate memory. *Microelectron. J.* **2008**, *39*, 1553–1555. [CrossRef]
15. Jin, R.; Shi, K.; Qiu, B.; Huang, S. Photoinduced-reset and multilevel storage transistor memories based on antimony-doped tin oxide nanoparticles floating gate. *Nanotechnology* **2021**, *33*, 025201. [CrossRef]
16. Lepadatu, A.M.; Palade, C.; Slav, A.; Maraloiu, A.V.; Lazanu, S.; Stoica, T.; Logofatu, C.; Teodorescu, V.S.; Ciurea, M.L. Single layer of Ge quantum dots in HfO$_2$ for floating gate memory capacitors. *Nanotechnology* **2017**, *28*, 175707. [CrossRef]
17. Mazurak, A.; Mroczyński, R.; Beke, D.; Gali, A. Silicon-Carbide (SiC) Nanocrystal Technology and Characterization and Its Applications in Memory Structures. *Nanomaterials* **2020**, *10*, 2387. [CrossRef] [PubMed]
18. Yu, X.; Ma, Z.; Shen, Z.; Li, W.; Chen, K.; Xu, J.; Xu, L. 3D NAND Flash Memory Based on Double-Layer NC-Si Floating Gate with High Density of Multilevel Storage. *Nanomaterials* **2022**, *12*, 2459. [CrossRef]
19. Lu, F.; Gong, D.; Wang, J.; Wang, Q.; Sun, H.; Wang, X. Capacitance-voltage characteristics of a Schottky junction containing SiGe/Si quantum wells. *Phys. Rev. B* **1996**, *53*, 4623. [CrossRef]
20. Li, M.; Sah, C.-T. New techniques of capacitance-voltage measurements of semiconductor junctions. *Solid-State Electron.* **1982**, *25*, 95–99. [CrossRef]
21. Lee, Y.K. Study of hysteresis behavior of charges in fluorinated polyimide film by using capacitance-voltage method. *Mod. Phys. Lett. B* **2006**, *20*, 445–449. [CrossRef]
22. Choi, S.; Park, B.; Kim, H.; Cho, K.; Kim, S. Capacitance–voltage characterization of Ge-nanocrystal-embedded MOS capacitors with a capping Al$_2$O$_3$ layer. *Semicond. Sci. Technol.* **2006**, *21*, 378. [CrossRef]
23. Lee, H.S. A new approach for the floating-gate MOS nonvolatile memory. *Appl. Phys. Lett.* **1977**, *31*, 475–476. [CrossRef]
24. Albin, D.S.; del Cueto, J.A. Correlations of capacitance-voltage hysteresis with thin-film CdTe solar cell performance during accelerated lifetime testing. In Proceedings of the 2010 IEEE International Reliability Physics Symposium, Anaheim, CA, USA, 2–6 May 2010; pp. 318–322.
25. Rathkanthiwar, S.; Bagheri, P.; Khachariya, D.; Mita, S.; Pavlidis, S.; Reddy, P.; Kirste, R.; Tweedie, J.; Sitar, Z.; Collazo, R. Point-defect management in homoepitaxially grown Si-doped GaN by MOCVD for vertical power devices. *Appl. Phys. Express* **2022**, *15*, 051003. [CrossRef]
26. Garnett, E.C.; Tseng, Y.-C.; Khanal, D.R.; Wu, J.; Bokor, J.; Yang, P. Dopant profiling and surface analysis of silicon nanowires using capacitance–voltage measurements. *Nat. Nanotechnol.* **2009**, *4*, 311–314. [CrossRef] [PubMed]
27. Duan, T.; Ang, D.S. Capacitance hysteresis in the high-k/metal gate-stack from pulsed measurement. *IEEE Trans. Electron. Devices* **2013**, *60*, 1349–1354. [CrossRef]
28. Lu, Q.; Qi, Y.; Zhao, C.Z.; Zhao, C.; Taylor, S.; Chalker, P.R. Anomalous capacitance-voltage hysteresis in MOS devices with ZrO2 and HfO2 dielectrics. In Proceedings of the 2016 5th International Symposium on Next-Generation Electronics (ISNE), Hsinchu, Taiwan, 4–6 May 2016; pp. 1–2.

29. Kandpal, K.; Gupta, N.; Singh, J.; Shekhar, C. On the Threshold Voltage and Performance of ZnO-Based Thin-Film Transistors with a ZrO_2 Gate Dielectric. *J. Electron. Mater.* **2020**, *49*, 3156–3164. [CrossRef]
30. Agrawal, K.; Yoon, G.; Kim, J.; Chavan, G.; Kim, J.; Park, J.; Phong, P.D.; Cho, E.-C.; Yi, J. Improving Retention Properties of ALD-Al_xO_y Charge Trapping Layer for Non-Volatile Memory Application. *ECS J. Solid State Sci. Technol.* **2020**, *9*, 043002. [CrossRef]
31. Yang, C.; Gu, Z.; Yin, Z.; Qin, F.; Wang, D. Interfacial traps and mobile ions induced flatband voltage instability in 4H-SiC MOS capacitors under bias temperature stress. *J. Phys. D Appl. Phys.* **2019**, *52*, 405103. [CrossRef]
32. Bachman, M. RCA-1 Silicon Wafer Cleaning. Engineering of Microworld at the University of California. Irvine. 1999. Available online: https://phas.ubc.ca/~ampel/nanofab/sop/rca-clean-1.pdf (accessed on 16 January 2023).
33. Deal, B.E.; Grove, A. General relationship for the thermal oxidation of silicon. *J. Appl. Phys.* **1965**, *36*, 3770–3778. [CrossRef]
34. Massoud, H.Z.; Plummer, J.D.; Irene, E.A. Thermal oxidation of silicon in dry oxygen: Accurate determination of the kinetic rate constants. *J. Electrochem. Soc.* **1985**, *132*, 1745. [CrossRef]
35. Takagi, T.; Takechi, K.; Nakagawa, Y.; Watabe, Y.; Nishida, S. High rate deposition of a-Si: H and a-SiN_x: H by VHF PECVD. *Vacuum* **1998**, *51*, 751–755. [CrossRef]
36. Stannowski, B.; Schropp, R.; Wehrspohn, R.; Powell, M. Amorphous-silicon thin-film transistors deposited by VHF-PECVD and hot-wire CVD. *J. Non-Cryst. Solids* **2002**, *299*, 1340–1344. [CrossRef]
37. Mohapatra, P.; Dung, M.X.; Choi, J.-K.; Jeong, S.-H.; Jeong, H.-D. Effects of curing temperature on the optical and charge trap properties of InP quantum dot thin films. *Bull. Korean Chem. Soc.* **2011**, *32*, 263–272. [CrossRef]

Disclaimer/Publisher's Note: The statements, opinions and data contained in all publications are solely those of the individual author(s) and contributor(s) and not of MDPI and/or the editor(s). MDPI and/or the editor(s) disclaim responsibility for any injury to people or property resulting from any ideas, methods, instructions or products referred to in the content.

Article

Enhancement of the Electroluminescence from Amorphous Er-Doped Al₂O₃ Nanolaminate Films by Y₂O₃ Cladding Layers Using Atomic Layer Deposition

Yang Yang *, Haiyan Pei, Zejun Ye and Jiaming Sun *

Tianjin Key Lab for Rare Earth Materials and Applications, School of Materials Science and Engineering, Nankai University, Tianjin 300350, China
* Correspondence: mseyang@nankai.edu.cn (Y.Y.); jmsun@nankai.edu.cn (J.S.)

Abstract: Amorphous Al_2O_3-Y_2O_3:Er nanolaminate films are fabricated on silicon by atomic layer deposition, and ~1530 nm electroluminescence (EL) is obtained from the metal-oxide-semiconductor light-emitting devices based on these nanofilms. The introduction of Y_2O_3 into Al_2O_3 reduces the electric field for Er excitation and the EL performance is significantly enhanced, while the electron injection of devices and the radiative recombination of doped Er^{3+} ions are not impacted. The 0.2 nm Y_2O_3 cladding layers for Er^{3+} ions increase the external quantum efficiency from ~3% to 8.7% and the power efficiency is increased by nearly one order of magnitude to 0.12%. The EL is ascribed to the impact excitation of Er^{3+} ions by hot electrons, which stem from Poole-Frenkel conduction mechanism under sufficient voltage within the Al_2O_3-Y_2O_3 matrix.

Keywords: electroluminescence; erbium; Al_2O_3; Y_2O_3; atomic layer deposition

Citation: Yang, Y.; Pei, H.; Ye, Z.; Sun, J. Enhancement of the Electroluminescence from Amorphous Er-Doped Al₂O₃ Nanolaminate Films by Y₂O₃ Cladding Layers Using Atomic Layer Deposition. *Nanomaterials* 2023, 13, 849. https://doi.org/10.3390/nano13050849

Academic Editor: Mikhael Bechelany

Received: 14 February 2023
Revised: 22 February 2023
Accepted: 23 February 2023
Published: 24 February 2023

Copyright: © 2023 by the authors. Licensee MDPI, Basel, Switzerland. This article is an open access article distributed under the terms and conditions of the Creative Commons Attribution (CC BY) license (https://creativecommons.org/licenses/by/4.0/).

1. Introduction

Rare earth (RE) ions are generally efficient luminescence centers in various matrices. Nowadays diverse RE-doped insulating materials have been developed for the applications in solid state lasers and phosphors [1,2]. Erbium (Er) ions are one of the most researched luminescence centers due to their near-infrared (NIR) 1.53 μm emission which coincides with the window of optical telecommunication [3,4] Aiming for the realization of Si-integrated optoelectronics, the 1.53 μm electroluminescence (EL) from Er^{3+} ion has been researched extensively in many materials, including SiO_x, SiN_x, TiO_2 and ZnO [5–8]. However, the efficiencies of the devices based on these aforementioned materials are still far from practical application, due to the limitations in doping tolerance and excitation efficiency. Y_2O_3 is one of the attractive doping hosts for RE ions as the substitution of other RE^{3+} ions in Y_2O_3 is quite easy without charge compensation and severe lattice distortion. In addition, Y^{3+} ions are not luminescent and Y_2O_3 has a large bandgap (5.8 eV) and high stability [9,10]. In our previous study, Al_2O_3 has been proved to be a suitable matrix for the excitation of RE^{3+} ion to realize the EL emissions but the doping concentration is still limited [11–13]. Therefore, using Y_2O_3 as a cladding layer in Er-doped Al_2O_3 could utilize the merits of both oxides, the Er-clustering and resultant concentration quenching could be reduced while the optical-active Er^{3+} ions can be excited more effectively [14].

In this work, we fabricate the metal-oxide-semiconductor light-emitting devices (MOSLEDs) based on the amorphous Al_2O_3-Y_2O_3:Er nanolaminate films, which are deposited using atomic layer deposition (ALD). Due to the unique growth mechanism based on the successive self-limiting gas-surface reactions, ALD realizes the precise control of the thickness of different compositions with excellent homogeneity [15,16]. By alternating deposition sequence of Al_2O_3 and Y_2O_3, nanolaminate Al_2O_3-Y_2O_3:Er films with interlayers of different thicknesses are fabricated. Under sufficient forward bias, such devices exhibit ~1530 nm emissions originating from the infra-4f transitions of Er^{3+} ions. Inserting of the

Y_2O_3 cladding layers increases the external quantum efficiency (EQE) from 3% to 8.7% and almost upgrades the power efficiency (PE) by one order of magnitude, while the excitation and recombination of the Er^{3+} ions are not affected. We believe that this work contributes to the development of silicon-based light sources for integrated optoelectronic applications.

2. Experimental

The luminescent Al_2O_3-Y_2O_3:Er nanolaminates were grown on <100>-oriented n-type silicon (2–5 Ω·cm) using the thermal ALD system (NanoTech Savannah 100, Cambridge, MA, USA). The growth chamber was first evacuated to a base pressure of 0.3 Torr. Trimethylaluminum [TMA, $Al(CH_3)$], $Y(THD)_3$ and $Er(THD)_3$ (THD = 2,2,6,6-tetramethyl-3,5-heptanedionate) were used as the precursors for Al_2O_3, Y_2O_3 and Er_2O_3, respectively, with ozone acting as the oxidant. During the ALD process, the Al precursor was maintained at room temperature (RT), while Y and Er precursors were maintained at 180 °C and 190 °C, respectively. The precursor delivery lines were heated at 190 °C. N_2 was used as the carrier and purge gas with a flow rate of 20 sccm. The pulse time for Al and RE precursors are 0.015 s and 2 s, respectively. One growth cycle consists of one precursor pulse, the 5 s N_2 purge, a 1.8 s ozone pulse, and the 9 s N_2 purge. Based on the former research, the Er dopant cycles are fixed at 2, which are preferable concerning both the efficient doping and the absence of RE clustering [11,13,17,18]. The substrates were maintained at 350 °C, and the growth rates for the Al_2O_3, Y_2O_3 and Er_2O_3 films are calibrated to 0.79, 0.2 and 0.23 Å/cycle respectively, which agree well with the previous reports [19]. During the deposition, the dopant Er_2O_3 atomic layers were sandwiched in two cladding Y_2O_3 layers of designed thickness, and then the Al_2O_3 interlayers with certain thickness and these Y_2O_3-Er_2O_3-Y_2O_3 composite nanolaminates were deposited repeatedly to achieve the nanolaminates with the deposition sequence of Al_2O_3-Y_2O_3-Er_2O_3-Y_2O_3. In order to explore the Al_2O_3-Y_2O_3:Er nanofilms, firstly for the Al_2O_3-Y_2O_3:Er nanofilms of different Y_2O_3 cladding layers, the thickness of Al_2O_3 interlayers was fixed at 3 nm and the two Y_2O_3 cladding layers (x nm) in each supercycle were changed from 0 to 0.2, 0.5 and 1.0 nm (with their growth cycles varied from 10 to 50), the same repeat numbers of 16 for the supercycles resulted into the total thickness of 48.6, 55.0, 64.6, and 80.6 nm for the Al_2O_3-Y_2O_3(x nm):Er nanolaminates. The calculated nominal doping concentrations of Er are 0.51–0.34 at%. Secondly, for the Al_2O_3-Y_2O_3:Er nanofilms of different Al_2O_3 interlayers, the thickness of Y_2O_3 cladding layers were fixed at 0.2 nm and the Al_2O_3 interlayers (y nm) in each supercycle were changed from 0.5 to 1, 2, 3 and 5 nm. To achieve the Al_2O_3-Y_2O_3:Er nanofilms of the total thickness of ~65 nm, the repeat numbers of the supercycle were changed from 69 to 45, 27, 19 and 12 for the Al_2O_3(y nm)-Y_2O_3:Er nanolaminates. The calculated nominal doping concentrations of Er are 4.28–0.29 at%. Here the deposition velocities and the growth cycles in recipes, and the densities of oxides (Al_2O_3, Y_2O_3, Er_2O_3) are used to calculate the corresponding dopant amount of Er^{3+} ions. After the deposition, the films were annealed at 800 °C in N_2 atmosphere for 1 h to enable activation of the dopants. Subsequent device procedures were as previously mentioned [12,13,17,18], resulting in the multilayer-structured MOSLEDs of ZnO:Al/TiO_2-Al_2O_3/Al_2O_3-Y_2O_3:Er/Si/Al. The top ZnO:Al electrodes were lithographically patterned into 0.5 mm circular dots, while the TiO_2-Al_2O_3 nanolaminates were used to enhance the operation stability of the devices.

The film thickness was measured by an ellipsometer with a 632.8 nm He-Ne laser at an incident angle of 69.8°. The phase and the crystal structure of the films were identified by an X-ray diffractometer (XRD, D/max 2500/pc, Rigaku) using the Cu Kα radiation. To activate EL from the MOSLEDs, appropriate forward bias was applied with the negative voltage connecting to the n-Si substrates. EL and Current-Voltage (I–V) characteristics were recorded by a Keithley 2410 SourceMeter. The EL signal was collected by a 0.5 m monochromator and detected by an InGaAs detector connected to a Keithley 2010 multimeter. The absolute EL power from the device surface was measured using a calibrated Newport 1830-C optical power-meter with an 818-IR Sensor. All measurements were performed at RT.

3. Results and Discussion

The XRD patterns of all the Al_2O_3-Y_2O_3:Er films annealed at 800 °C confirm that the nanolaminates are amorphous, one representative XRD pattern from the nanolaminate using 3 nm Al_2O_3 interlayers and 0.2 nm Y_2O_3 cladding layers is shown in Figure 1. The Al_2O_3 layers are not crystalized at such a relatively low temperature of 800 °C that beneficial for the EL performance from RE-doped Al_2O_3 films, while the crystallization of the sub-nanometer Y_2O_3 layers is restricted [17,20]. The amorphous nanolaminate films are quite smooth under the observation of scanning electron microscope, with a root-square roughness of only 0.56–0.7 nm scanned by the atomic force microscopy (AFM, Dimension Icon, Bruker) [21].

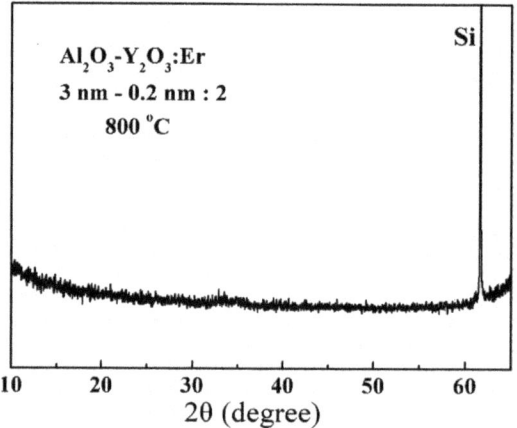

Figure 1. The XRD pattern for the representative Al_2O_3-Y_2O_3:Er nanolaminate film after annealing at 800 °C.

Figure 2a illustrates the schematic diagram for the MOSLEDs and the structure and deposition sequence of the luminescent nanolaminates. Figure 2b shows the NIR EL spectra of MOSLEDs based on the Al_2O_3-Y_2O_3:Er films of different Y_2O_3 cladding layers (with the thickness of x nm). The EL peaks centered at ~1530 nm correspond to the infra-4f $^4I_{13/2} \rightarrow {}^4I_{15/2}$ transitions of the Er^{3+} ions. The presence of other shoulder peaks is ascribed to the splitting levels associated with the Stark effect [22]. These EL peaks are similar in positions and sharps in the Al_2O_3-Y_2O_3:Er nanofilms with different Y_2O_3 cladding layers, thus the incorporation of Y_2O_3 cladding layers imposes no apparent effect on the Er^{3+} intra-4f transitions. In comparison with the Er-emissions from different matrices, the spectra also confirm that the Al_2O_3-Y_2O_3:Er films are amorphous due to the absence of companion peaks [20,23].

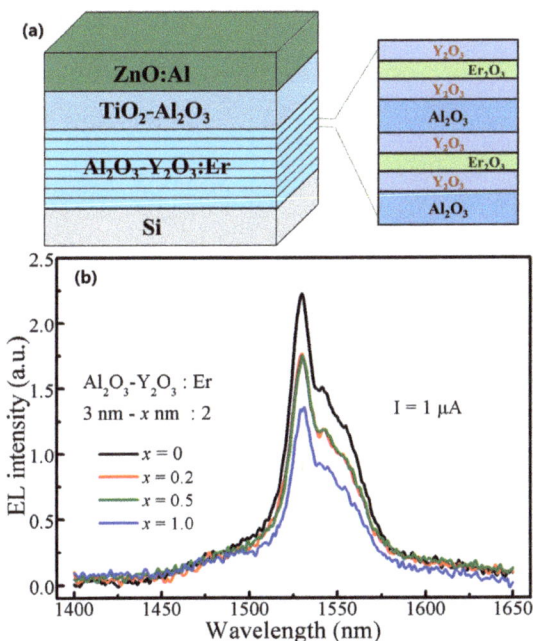

Figure 2. (a) The schematic diagram for the Al$_2$O$_3$-Y$_2$O$_3$:Er MOSLEDs and the luminescent nanolaminate films. (b) The NIR EL spectra for the Al$_2$O$_3$-Y$_2$O$_3$:Er MOSLEDs with different Y$_2$O$_3$ cladding layers (with the thickness of x nm) under the injection current of 1 µA.

Figure 3a presents the dependence of the 1530 nm EL intensities and the injection currents on the applied voltages for the Al$_2$O$_3$-Y$_2$O$_3$:Er MOSLEDs with different Y$_2$O$_3$ cladding layers (with the thickness of x nm). These EL–V and I–V curves are similar with our previous reports on the MOSLEDs based on RE-doped oxides, with the typical characteristic of MOS structures [13,18,24,25]. Beneath the threshold electric field, the defect states contribute to the low background currents. In the working voltage region, the currents increase exponentially until breakdown. The difference on the current injection will be discussed afterwards concerning the conduction mechanism. All the EL intensities also present an exponential relationship with the applied voltages until reaching saturation. The MOSLED with 0.2 nm Y$_2$O$_3$ cladding layers presents the highest EL intensity, with the lowest threshold voltage and the highest injection current. The devices with thicker Y$_2$O$_3$ layers underperform in EL intensities and the injection currents are restricted. Despite the uncertainty brought about by the device preparation, Y$_2$O$_3$ cladding layers with suitable thickness can effectively enhance the current injection and promote the EL emissions from these Al$_2$O$_3$-Y$_2$O$_3$:Er MOSLEDs. As previously reported, the incorporation of Y^{3+} ion makes the crystal field around Er^{3+} ions less symmetric and introduces distortion in the crystal field, moreover the Er^{3+} ions are dispersed to suppress the concentration quenching, resulting in the enhanced radiation probability [26]. Therefore, the 0.2 nm Y$_2$O$_3$ layers act as cladding layers that inhibit the Er-clustering, while the thicker Y$_2$O$_3$ layers inhibit the electron injection, which is ascribed to the higher dielectric index of Y$_2$O$_3$ and the disruptive interfaces among Al$_2$O$_3$ and Y$_2$O$_3$ interlayers.

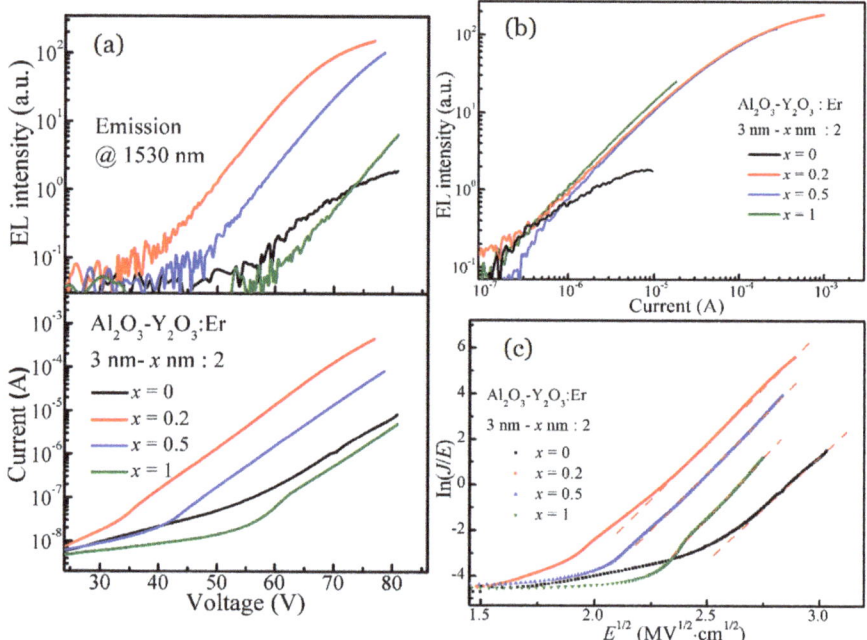

Figure 3. (a) The dependence of EL intensities and injection currents on the applied voltage for the Al$_2$O$_3$-Y$_2$O$_3$:Er MOSLEDs with different Y$_2$O$_3$ interlayers (with the thickness of x nm), and (b) the dependence of EL intensities on the injection currents for these devices. (c) The plot of ln(J/E) versus $E^{1/2}$ (P–F plots of the I–V characteristics) for the Al$_2$O$_3$-Y$_2$O$_3$:Er MOSLEDs with different Y$_2$O$_3$ interlayers.

Figure 3b shows the dependence of EL intensities on the injection currents for the Al$_2$O$_3$-Y$_2$O$_3$:Er MOSLEDs with different Y$_2$O$_3$ cladding layers. The threshold currents for all the devices are ~0.2 μA, the EL intensities and the injection currents present linear relationship. In comparison, the devices with different Y$_2$O$_3$ cladding layers exhibit similar EL, which increases more prominently than that based on the Al$_2$O$_3$:Er film. Y$_2$O$_3$ also lessens the saturation of EL intensities at higher injection currents. Considering the thick interlayers among RE layers (the Al$_2$O$_3$ interlayers with the thickness of at least 3 nm), the acceleration distance for hot electrons is sufficient; therefore the enhanced EL should result from more optical-active Er dopants as the Er^{3+} ions disperse into the Y$_2$O$_3$ layers and the Er-clustering is suppressed.

In our previously reported MOSLEDs based on RE-doped Al$_2$O$_3$, the RE-related EL is triggered by the direct impact excitation of the RE ions by the hot electrons accelerated under sufficient bias voltages [12,25]. As the I–V characterization are accordingly comparable, it is rational to ascribe the NIR EL from these Al$_2$O$_3$-Y$_2$O$_3$:Er MOSLEDs to the same mechanism. Considering the high bandgap of the matrix materials and the barrier for electrons to be injected from the Si substrates into the conduction band of the oxides, the current conduction of these MOSLEDs has been ascribe to the Poole-Frenkel (P-F) mechanism, in which the electrons hop via the defect-related trap states under sufficient electrical field [11,26,27]. In simplicity, the plot of the ln(J/E) versus $E^{1/2}$ presents linear relationship in P-F conduction mechanism, where J and E are the current density and the electric field, respectively [28,29]. Figure 3c shows the plots of the I–V characteristics derived from Figure 3a, the electrical fields across the luminescent films are roughly calculated in terms of electrostatics [30], and the well-defined linearity is established for all the MOSLEDs in the EL-enabling region. Thus the electron transport through the

Al$_2$O$_3$-Y$_2$O$_3$:Er nanofilms is governed by the P-F mechanism. These electrons tunnel into the conduction band of oxides and transport by hopping among trap states in the Al$_2$O$_3$-Y$_2$O$_3$ nanolaminates under sufficient electric field. Certain parts of the electrons are accelerated therein and become hot electrons that excite the Er^{3+} ions by inelastic impact, the subsequent recombination gives rise to the characteristic EL emissions. Apparently, the Y$_2$O$_3$ cladding layers decrease the working electric field prominently. As mentioned in the discussion on the I–V characteristics, the Y$_2$O$_3$ cladding layers increase both the injection currents and EL intensities, we conclude that ultrathin Y$_2$O$_3$ layers introduce defect sites within Al$_2$O$_3$, via which electrons transport by the P-F hopping mechanism; therefore the injection currents are enhanced. Since the accelerated electrons collide with the doped Er^{3+} ions and contribute to the NIR EL, adding the aforementioned crystal field distortion and cluster dispersion effects of the Y$_2$O$_3$ on Er^{3+} ions, the EL performance are greatly enhanced by the Y$_2$O$_3$ cladding layers in Al$_2$O$_3$ films. However, the Y$_2$O$_3$ cladding layers should be thin enough to not impact the carrier transport which could be ascribed to the formation of distinct Al$_2$O$_3$-Y$_2$O$_3$ interfaces when using thicker Y$_2$O$_3$ cladding layers.

In evaluation of the thickness of Al$_2$O$_3$ interlayers on the EL performance, the dependence of the EL intensities from each dopant cycle on the injection currents for the Al$_2$O$_3$-Y$_2$O$_3$:Er MOSLEDs with different Al$_2$O$_3$ interlayers (with the thickness of y nm) are shown in Figure 4a, the thickness of Y$_2$O$_3$ cladding layers is the optimal 0.2 nm. Again, the EL intensities increase almost linearly with the injection currents. The difference on the EL–I–V characteristics among these MOSLEDs with different Al$_2$O$_3$ layers are small (not shown herein), and the increase in EL intensity with the Al$_2$O$_3$ thickness could be ascribed to the less concentration quenching of doped Er^{3+} ions together with the longer acceleration distance. When the thickness of Al$_2$O$_3$ declines, the EL intensity decreases greatly due to the cross-relaxation of Er^{3+} between adjacent dopant layers when the inter-distance (the Al$_2$O$_3$ thickness) is smaller enough, and the limited acceleration length for the hot electrons to gain energy to excite the Er^{3+} ions [18,21,31–33]. Cross-relaxation is a common phenomenon that occurs among the same ions or different ions of similar energy intervals. One ion in the excited state ($^4I_{13/2}$ in the case of Er^{3+} ion) transfer the energy to another one (in the ground state of $^4I_{15/2}$ in this case of Er^{3+} ions), excite the latter to higher energy levels ($^4I_{13/2}$) while relaxing itself to lower energy levels ($^4I_{15/2}$) without radiation. The interaction of energy transfer by cross-relaxation could finally disperse the excitation energy through phonons instead of luminescent emissions.

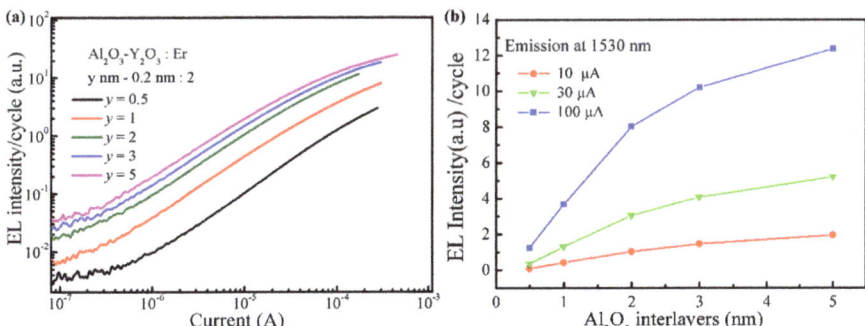

Figure 4. (**a**) The dependence of EL intensities on the injection currents for the Al$_2$O$_3$-Y$_2$O$_3$:Er MOSLEDs with different Al$_2$O$_3$ interlayers (with the thickness of y nm), herein the EL intensities are divided by the cycle numbers to manifest the emissions from each Er cycle. (**b**) The integrated EL intensity per cycle as a function of the thickness of the Al$_2$O$_3$ interlayers under different injection currents.

In the RE-doped Al$_2$O$_3$ MOSLEDs, the Al$_2$O$_3$ sublayer thickness affects the cross relaxation between excited RE ions, and the acceleration distance for injected electrons. Figure 4b shows the dependence of the integrated 1530 nm EL intensity per Er cycle on the thickness of Al_2O_3 interlayers under different injection currents. Under all these injection currents, with the increase in the thickness of Al_2O_3 interlayers, the contribution of single Er cycle to the EL intensity firstly increases and then saturates as the Al_2O_3 interlayer thickness reaches 3 nm. This is still in consistency with the common characteristic for the luminescent RE^{3+} ions in Al_2O_3 matrix that the distance for the presence of non-radiative interaction and adequate electron acceleration is around 3 nm [11,12,21,33].

Considering the total EL intensity from the MOSLEDs with Al_2O_3 interlayers of different thicknesses (marked as y nm here) shown in Figure 5a, the device using 3 nm Al_2O_3 interlayers presents the optimal emission intensity in the operation range, with the highest power density of 4.6 mW/cm^2. External efficiency is widely used to evaluate LED performance. Figure 5b shows the EQE and PE of these MOSLEDs based on different Al_2O_3-Y_2O_3:Er nanolaminate films. These EL efficiencies sustain a broad maximum, and fall down at higher currents. Generally, the EQE of the devices with 2–3 nm Al_2O_3 interlayers are the highest. As aforementioned, this phenomenon could be ascribed to the sufficient distance for electron acceleration and suppression of the cross-relaxation among adjacent Er_2O_3 dopant layers. The Y_2O_3 cladding layers somewhat decrease this critical distance which is beneficial for higher doping concentrations. The optimal device with 3/0.2 nm Al_2O_3/Y_2O_3 interlayers achieves the maximum EQE of 8.7% and a corresponding PE of 0.12%. These values are comparable to our Yb_2O_3:Er MOSLEDs but with lowered working voltages. In comparison, the control Al_2O_3:Er MOSLED presents only an EQE of 3% and a PE of 0.014%, much lower than the Al_2O_3-Y_2O_3:Er MOSLEDs. The Y_2O_3 cladding layers with suitable thickness enhance the efficiencies from the MOSLEDs to a great extent. We have found that by using a thicker luminescent layer, the efficiency of the Al_2O_3:RE MOSLED might be further increased to higher than 10% [25,34]. These efficiencies are superior to that from Si-based EL devises in literature, thus further optimization of the luminescent Al_2O_3-Y_2O_3:Er nanolaminates would supply potential light source for the applications in Si-based optoelectronics.

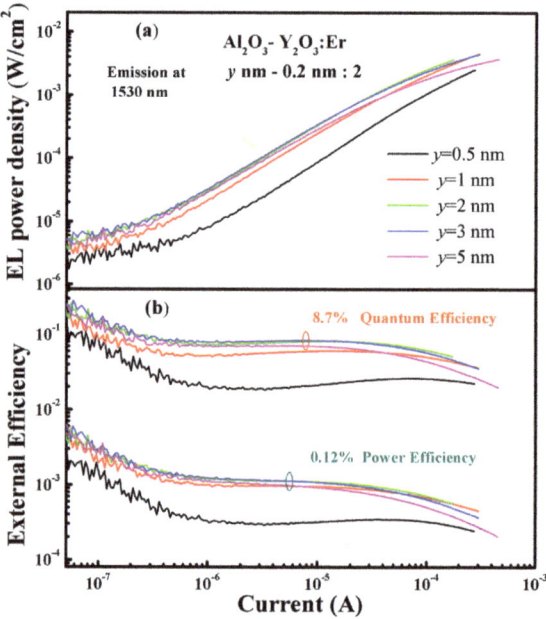

Figure 5. The dependence of (**a**) the EL power densities and (**b**) the external quantum efficiencies (the upper curves) and power efficiencies (the lower curves) on the injection currents for Al_2O_3-Y_2O_3:Er MOSLEDs using different Al_2O_3 interlayers (with the thickness of y nm).

4. Conclusions

In summary, significantly enhanced ~1530 nm NIR EL emissions are achieved from the MOSLEDs based on the amorphous Al_2O_3:Er nanolaminate films by the insertion of cladding Y_2O_3 sub-nanolayers, which are fabricated by ALD on Si substrates. The Y_2O_3 cladding layers reduce the threshold electric field for excitation and increase the radiative possibility of doped Er^{3+} ions, resulting in improved EL performance. The Al_2O_3-Y_2O_3:Er MOSLEDs with 0.2 nm Y_2O_3 and 3 nm Al_2O_3 interlayers present an EQE of 8.7% and a corresponding PE of 0.12%, which are much higher than that of the counterpart without Y_2O_3 cladding layers. The incorporation of Y_2O_3 does not change the electron injection mode under sufficient electric field that conforms to P-F mechanism, the resultant energetic electrons trigger the impact-excitation of Er^{3+} ions and subsequent EL emissions. The strategy of Y_2O_3-cladding by ALD can be employed to improve the EL performance from LEDs based on RE-doped oxides.

Author Contributions: Conceptualization, Y.Y. and J.S.; Methodology, H.P.; Validation, H.P. and Y.Y.; Formal analysis, H.P. and Y.Y.; Resources, Y.Y. and J.S.; Data curation, H.P. and Y.Y.; Writing-original draft preparation, Y.Y., H.P. and Z.Y.; Writing-review and editing, Y.Y.; Visualization, H.P. and Z.Y.; Supervision, Y.Y. and J.S.; Project administration, J.S.; Funding acquisition, Y.Y. and J.S. All authors have read and agreed to the published version of the manuscript.

Funding: This research was funded by the National Natural Science Foundation of China (No. 62275132 and 61705114).

Data Availability Statement: Data will be made available on request.

Conflicts of Interest: The authors declare no competing financial interest.

References

1. Bogaerts, W.; Chrostowski, L. Silicon Photonics Circuit Design: Methods, Tools and Challenges. *Laser Photonics Rev.* **2018**, *12*, 1700237. [CrossRef]
2. Crosnier, G.; Sanchez, D.; Bouchoule, S.; Monnier, P.; Beaudoin, G.; Sagnes, I.; Raj, R.; Raineri, F. Hybrid indium phosphide-on-silicon nanolaser diode. *Nat. Photonics* **2017**, *11*, 297–300. [CrossRef]
3. Sarkar, A.; Lee, Y.; Ahn, J.-H. Si nanomembranes: Material properties and applications. *Nano Res.* **2021**, *14*, 3010–3032. [CrossRef]
4. Makarova, M.; Sih, V.; Warga, J.; Li, R.; Dal Negro, L.; Vuckovic, J. Enhanced light emission in photonic crystal nanocavities with Erbium-doped silicon nanocrystals. *Appl. Phys. Lett.* **2008**, *92*, 161107. [CrossRef]
5. Yang, Y.; Jin, L.; Ma, X.; Yang, D. Low-voltage driven visible and infrared electroluminescence from light-emitting device based on Er-doped TiO_2/p^+-Si heterostructure. *Appl. Phys. Lett.* **2012**, *100*, 031103. [CrossRef]
6. Hernández Simón, Z.J.; Luna López, J.A.; de la Luz, A.D.; Pérez García, S.A.; Benítez Lara, A.; García Salgado, G.; Carrillo López, J.; Mendoza Conde, G.O.; Martínez Hernández, H.P. Spectroscopic Properties of Si-nc in SiO_x Films Using HFCVD. *Nanomaterials* **2020**, *10*, 1415. [CrossRef] [PubMed]
7. Yang, Y.; Li, Y.; Wang, C.; Zhu, C.; Lv, C.; Ma, X.; Yang, D. Rare-Earth Doped ZnO Films: A Material Platform to Realize Multicolor and Near-Infrared Electroluminescence. *Adv. Opt. Mater.* **2014**, *2*, 240–244. [CrossRef]
8. Guo, Y.; Lin, Z.X.; Huang, R.; Lin, Z.W.; Song, C.; Song, J.; Wang, X. Efficiency enhancement for SiN-based light emitting device through introduction of Si nanocones in emitting layer. *Opt. Mater. Express* **2015**, *5*, 969–976. [CrossRef]
9. Patra, A.; Friend, C.S.; Kapoor, R.; Prasad, P.N. Upconversion in Er^{3+}:ZrO_2 Nanocrystals. *J. Phys. Chem. B* **2002**, *106*, 1909–1912. [CrossRef]
10. Wang, X.J.; Yuan, G.; Isshiki, H.; Kimura, T.; Zhou, Z. Photoluminescence enhancement and high gain amplification of $Er_xY_{2-x}SiO_5$ waveguide. *J. Appl. Phys.* **2010**, *108*, 013506. [CrossRef]
11. Yang, Y.; Li, N.; Sun, J. Intense electroluminescence from Al_2O_3/Tb_2O_3 nanolaminate films fabricated by atomic layer deposition on silicon. *Opt. Express* **2018**, *26*, 9344–9352. [CrossRef]
12. Ouyang, Z.; Yang, Y.; Sun, J. Near-infrared electroluminescence from atomic layer doped Al_2O_3:Yb nanolaminate films on silicon. *Scr. Mater.* **2018**, *151*, 1–5. [CrossRef]
13. Wang, Y.; Yu, Z.; Yang, Y.; Sun, J. Bright red electroluminescence from Al_2O_3/Eu_2O_3 nanolaminate films fabricated by atomic layer deposition on silicon. *Scr. Mater.* **2021**, *196*, 113750. [CrossRef]
14. Rönn, J.; Karvonen, L.; Kauppinen, C.; Perros, A.P.; Peyghambarian, N.; Lipsanen, H.; Säynätjoki, A.; Sun, Z. Atomic Layer Engineering of Er-Ion Distribution in Highly Doped Er:Al_2O_3 for Photoluminescence Enhancement. *ACS Photonics* **2016**, *3*, 2040–2048. [CrossRef]
15. Leskelä, M.; Ritala, M. Atomic layer deposition (ALD): From precursors to thin film structures. *Thin Solid Film.* **2002**, *409*, 138–146. [CrossRef]

16. Lee, D.-J.; Kim, H.-M.; Kwon, J.-Y.; Choi, H.; Kim, S.-H.; Kim, K.-B. Structural and Electrical Properties of Atomic Layer Deposited Al-Doped ZnO Films. *Adv. Funct. Mater.* **2011**, *21*, 448–455. [CrossRef]
17. Ouyang, Z.; Yang, Y.; Sun, J. Electroluminescent Yb_2O_3:Er and $Yb_2Si_2O_7$:Er nanolaminate films fabricated by atomic layer deposition on silicon. *Opt. Mater.* **2018**, *80*, 209–215. [CrossRef]
18. Liu, Y.; Ouyang, Z.; Yang, L.; Yang, Y.; Sun, J. Blue Electroluminescent Al_2O_3/Tm_2O_3 Nanolaminate Films Fabricated by Atomic Layer Deposition on Silicon. *Nanomaterials* **2019**, *9*, 413. [CrossRef]
19. Tuomisto, M.; Giedraityte, Z.; Karppinen, M.; Lastusaari, M. Photon up-converting $(Yb,Er)_2O_3$ thin films by atomic layer deposition. *Phys. Status Solidi RRL* **2017**, *11*, 1700076. [CrossRef]
20. Xu, J.; Liu, J.; Yang, L.; Liu, J.; Yang, Y. Electroluminescent $Y_3Al_5O_{12}$ nanofilms fabricated by atomic layer deposition on silicon: Using Yb as the luminescent dopant and crystallization impetus. *Opt. Express* **2021**, *29*, 37–47. [CrossRef] [PubMed]
21. Liu, J.; Liu, Y.; Yang, Y.; Sun, J. Exploration of the green electroluminescence from Al_2O_3/Ho_2O_3 nanolaminate films fabricated by atomic layer deposition on silicon. *Opt. Mater.* **2020**, *107*, 110125. [CrossRef]
22. Gruber, J.B.; Burdick, G.W.; Chandra, S.; Sardar, D.K. Analyses of the ultraviolet spectra of Er^{3+} in Er_2O_3 and Er^{3+} in Y_2O_3. *J. Appl. Phys.* **2010**, *108*, 023109. [CrossRef]
23. Xu, J.; Yang, L.; Ma, Z.; Yang, Y.; Sun, J. Electroluminescent polycrystalline Er-doped $Lu_3Al_5O_{12}$ nanofilms fabricated by atomic layer deposition on silicon. *J. Alloys Compd.* **2021**, *885*, 160993. [CrossRef]
24. Sun, J.M.; Skorupa, W.; Dekorsy, T.; Helm, M.; Rebohle, L.; Gebel, T. Bright green electroluminescence from Tb^{3+} in silicon metal-oxide-semiconductor devices. *J. Appl. Phys.* **2005**, *97*, 123513. [CrossRef]
25. Sun, J.M.; Rebohle, L.; Prucnal, S.; Helm, M.; Skorupa, W. Giant stability enhancement of rare-earth implanted SiO_2 light emitting devices by an additional SiON protection layer. *Appl. Phys. Lett.* **2008**, *92*, 071103. [CrossRef]
26. Jiang, M.; Zhu, C.; Zhou, J.; Chen, J.; Gao, Y.; Ma, X.; Yang, D. Electroluminescence from light-emitting devices with erbium-doped TiO_2 films: Enhancement effect of yttrium codoping. *J. Appl. Phys.* **2016**, *120*, 163104. [CrossRef]
27. Rebohle, L.; Braun, M.; Wutzler, R.; Liu, B.; Sun, J.M.; Helm, M.; Skorupa, W. Strong electroluminescence from SiO_2-Tb_2O_3-Al_2O_3 mixed layers fabricated by atomic layer deposition. *Appl. Phys. Lett.* **2014**, *104*, 251113. [CrossRef]
28. Samanta, P.; Mandal, K.C. Hole injection and dielectric breakdown in 6H-SiC and 4H-SiC metal-oxide-semiconductor structures during substrate electron injection via Fowler–Nordheim tunneling. *Solid State Electron.* **2015**, *114*, 60–68. [CrossRef]
29. Zheng, C.Y.; He, G.; Chen, X.F.; Liu, M.; Lv, J.G.; Gao, J.; Zhang, J.W.; Xiao, D.Q.; Jin, P.; Jiang, S.S.; et al. Modification of band alignments and optimization of electrical properties of InGaZnO MOS capacitors with high-k HfO_xN_y gate dielectrics. *J. Alloys Compd.* **2016**, *679*, 115–121. [CrossRef]
30. Kim, W.; Park, S.I.; Zhang, Z.; Wong, S. Current Conduction Mechanism of Nitrogen-Doped AlO_x RRAM. *IEEE Trans. Electron Devices* **2014**, *61*, 2158–2163. [CrossRef]
31. Zhu, C.; Lv, C.; Gao, Z.; Wang, C.; Li, D.; Ma, X.; Yang, D. Multicolor and near-infrared electroluminescence from the light-emitting devices with rare-earth doped TiO_2 films. *Appl. Phys. Lett.* **2015**, *107*, 131103. [CrossRef]
32. Kim, Y.S.; Yun, S.J. Studies on polycrystalline ZnS thin films grown by atomic layer deposition for electroluminescent applications. *Appl. Surf. Sci.* **2004**, *229*, 105–111. [CrossRef]
33. Yuan, K.; Liu, Y.; Ou-Yang, Z.T.; Liu, J.; Yang, Y.; Sun, J. Resonant energy transfer between rare earth atomic layers in nanolaminate films. *Opt. Lett.* **2022**, *47*, 4897–4900. [CrossRef] [PubMed]
34. Yang, Y.; Ouyang, Z.; Liu, J.; Sun, J. Energy Transfer under Electrical Excitation and Enhanced Electroluminescence in the Nanolaminate Yb,Er Co-Doped Al_2O_3 Films. *Phys. Status Solidi RRL* **2019**, *13*, 1900137. [CrossRef]

Disclaimer/Publisher's Note: The statements, opinions and data contained in all publications are solely those of the individual author(s) and contributor(s) and not of MDPI and/or the editor(s). MDPI and/or the editor(s) disclaim responsibility for any injury to people or property resulting from any ideas, methods, instructions or products referred to in the content.

Article

Narrow UVB-Emitted YBO$_3$ Phosphor Activated by Bi^{3+} and Gd^{3+} Co-Doping

Zhimin Yu, Yang Yang * and Jiaming Sun *

School of Materials Science and Engineering, Tianjin Key Lab for Rare Earth Materials and Applications, Nankai University, Tianjin 300350, China
* Correspondence: mseyang@nankai.edu.cn (Y.Y.); jmsun@nankai.edu.cn (J.S.)

Abstract: Y$_{0.9}$(Gd$_x$Bi$_{1-x}$)$_{0.1}$BO$_3$ phosphors (x = 0, 0.2, 0.4, 0.6, 0.8, and 1.0, YGB) were obtained via high-temperature solid-state synthesis. Differentiated phases and micro-morphologies were determined by adjusting the synthesis temperature and the activator content of Gd^{3+} ions, verifying the hexagonal phase with an average size of ~200 nm. Strong photon emissions were revealed under both ultraviolet and visible radiation, and the effectiveness of energy transfer from Bi^{3+} to Gd^{3+} ions was confirmed to improve the narrow-band ultraviolet-B (UVB) ($^6P_J \rightarrow {}^8S_{7/2}$) emission of Gd^{3+} ions. The optimal emission was obtained from Y$_{0.9}$Gd$_{0.08}$Bi$_{0.02}$BO$_3$ phosphor annealed at 800 °C, for which maximum quantum yields (QYs) can reach 24.75% and 1.33% under 273 nm and 532 nm excitations, respectively. The optimal QY from the Gd^{3+}-Bi^{3+} co-doped YGB phosphor is 75 times the single Gd^{3+}-doped one, illustrating that these UVB luminescent phosphors based on co-doped YBO$_3$ orthoborates possess bright UVB emissions and good excitability under the excitation of different wavelengths. Efficient photon conversion and intense UVB emissions indicate that the multifunctional Gd^{3+}-Bi^{3+} co-doped YBO$_3$ orthoborate is a potential candidate for skin treatment.

Keywords: UVB emission; rare-earth orthoborate; gadolinium; phosphor; co-doping

Citation: Yu, Z.; Yang, Y.; Sun, J. Narrow UVB-Emitted YBO$_3$ Phosphor Activated by Bi^{3+} and Gd^{3+} Co-Doping. *Nanomaterials* **2023**, *13*, 1013. https://doi.org/10.3390/nano13061013

Academic Editors: Wojciech Pisarski and Antonios Kelarakis

Received: 9 February 2023
Revised: 24 February 2023
Accepted: 8 March 2023
Published: 11 March 2023

Copyright: © 2023 by the authors. Licensee MDPI, Basel, Switzerland. This article is an open access article distributed under the terms and conditions of the Creative Commons Attribution (CC BY) license (https://creativecommons.org/licenses/by/4.0/).

1. Introduction

Skin treatment using artificial sources of ultraviolet (UV) radiation in controlled conditions is well established, and narrow-band ultraviolet-B (UVB) therapy has been demonstrated to be effective against skin diseases and disorders such as psoriasis, vitiligo and hyperbilirubinemia (commonly known as infant jaundice) [1–4]. Phototherapy with narrow-band UVB (310–313 nm) as photosensitizers is believed to result from the direct interaction between the light of certain frequencies and tissues, causing a change in immune response [5–7]. Furthermore, during phototherapy investigations, it was observed that light belonging to longer wavelengths of the UVB region was more effective, while that of the shorter wavelengths was much less effective or even harmful [8,9]. Rare-earth (RE) orthoborates (RE-BO$_3$, RE = lanthanide, yttrium, and scandium) have aroused considerable interest due to their wide range of applications in plasma display panels and mercury-free fluorescent lamps [10,11]. In particular, YBO$_3$ is an excellent host for UV phosphors due to its high-vacuum UV transparency, exceptional optical damage thresholds, strong absorption in the UV range, and good chemical inertness [12–14]. Additionally, the YBO$_3$ phosphors exhibit a wide bandgap and high host-to-activator energy transfer efficiency at moderate RE^{3+} concentrations [15]. Therefore, it is of great interest to investigate RE-doped YBO$_3$ orthoborates for UVB treatments.

Among the RE ions, lanthanide gadolinium (Gd^{3+}) is of particular interest because of its ubiquitous nature (well known as U-spectrum) and the characteristic narrow-band UVB emission from $^6P_J \rightarrow {}^8S_{7/2}$ transitions [16,17]. The optical properties of Gd^{3+} ions have been widely studied, and many Gd-doped compounds can be used as efficient phosphors in the new generation of UV fluorescent lamps. Moreover, as a promising activator or sensitizer, the Bi^{3+} ion shows excellent emission and absorption ability in the UV region. Furthermore, the transitions of $^8S_{7/2} \rightarrow {}^6I_J$ (emission at ~270 nm) and $^8S_{7/2} \rightarrow {}^6P_J$ (at ~311 nm) of the Gd^{3+} ion overlap with the $^3P_1 \rightarrow {}^1S_0$ (at ~260 nm) transition of the Bi^{3+} ion in YBO_3 [18,19], which permits an efficient energy transfer from the Bi^{3+} to Gd^{3+} ions. Further research on improving the UVB emission has also been reported [18]. From a practical point of view, UV-emitting phosphors in well-defined regions are required for various applications. Keeping this in mind, we prepared a UVB-emitting Gd^{3+}-Bi^{3+} co-doped YBO_3 phosphor in this work, which can effectively achieve light conversion and UVB emissions. When UV fluorescence is irradiated on the surface of a skin wound, the activity of the mitochondrial catalase can increase in cells, which could promote the synthesis of proteins and the decomposition of adenosine triphosphate (ATP), ultimately healing the wound. A schematic representation of healing, which adopts phosphors as light conversion layers, is conceived in Figure 1.

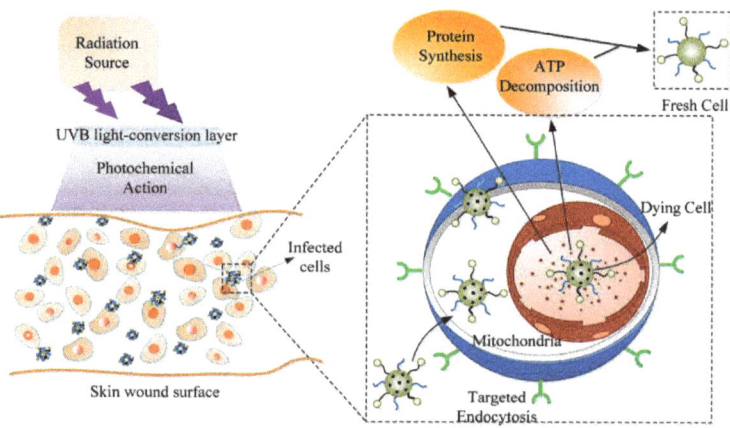

Figure 1. Scheme of light conversion layer for skin treatment using UVB emissions.

In this work, narrow-band UVB-emitting phosphors of Gd^{3+}-Bi^{3+} co-doped $Y_{0.9}(Gd_xBi_{1-x})_{0.1}BO_3$ (YGB, x = 0, 0.2, 0.4, 0.6, 0.8 and 1.0) were fabricated by high-temperature solid-state synthesis. The samples with the hexagonal phase and well-dispersed particles were characterized by XRD and SEM techniques, manifesting a micro-size of ~200 nm. The responses to UV and the visible (VIS) radiation of these YGB phosphors were compared, and sharp UVB luminescence was recorded with the adjustment of Gd^{3+} content. The sintering temperature indicated that co-doped Bi^{3+} ions enhanced the characteristic UVB luminescence from Gd^{3+} ions. The spectroscopic intensity parameters of YGB phosphors were derived from relative spectral power distributions, and the maximum quantum yields (QYs) at 313 nm were calculated at 24.75% and 1.33% under 273 nm and 532 nm excitations, respectively. YGB orthoborate phosphors with intense UVB emission could provide a viable approach for developing multifunctional composite materials for skin treatments.

2. Materials and Methods

The powders of $Y_{0.9}(Gd_xBi_{1-x})_{0.1}BO_3$ (x = 0, 0.2, 0.4, 0.6, 0.8, and 1.0, marked as YGB-0, YGB-0.2, YGB-0.4, YGB-0.6 YGB-0.8, and YGB-1.0, respectively) phosphors were prepared using high-purity reagents Y_2O_3 (99.9%), Gd_2O_3 (99.9%), Bi_2O_3 (A.R.), and H_3BO_3 (A.R.) as raw materials. The original chemicals for YGB with different Bi^{3+} contents as the designed sensitizer were mixed by grinding them in an agate mortar according to the stoichiometric ratio. The raw powders were transferred into alumina crucibles and pre-sintered at 500 °C for 1 h and then sintered at 700 °C, 800 °C, 900 °C and 1000 °C for 5 h. Afterwards, the samples were ground thoroughly after cooling.

The phase and the crystal structure of powders were identified by an X-ray diffractometer (XRD, MiniFlex 600, Rigaku, Tokyo, Japan) using Cu Kα radiation. Morphologies of the powders were analyzed by a field emission scanning electron microscope (SEM, JSM-7800F, JEOL, Tokyo, Japan) equipped with energy dispersive spectroscopy (EDS, X-MaxN 50, Oxford, Oxford, UK) using an accelerating voltage of 15 kV. Particle size distributions were measured in a Nanoparticle Analyzer (Zetasizer Nano-ZS, Malvern, UK). Photoluminescence (PL) spectra were recorded using a Keithley 2010 multimeter and the monochromator (λ500, Zolix, Beijing, China) equipped with a Si detector (DSi200, Zolix, Beijing, China). A commercial Xe lamp and two solid-state lasers emitting at 266 nm and 532 nm were used as the excitation sources for different excitation wavelengths. A standard PTFE diffuse reflective white plate (reflectivity greater than 99.9%) was used as a reference. The schematic diagram of the experimental setup is depicted in Figure 2. The relative spectral power distribution was obtained and calibrated by the Optical Power Meter (1830-C, Newport, Newport County, RI, USA). All measurements were performed at room temperature.

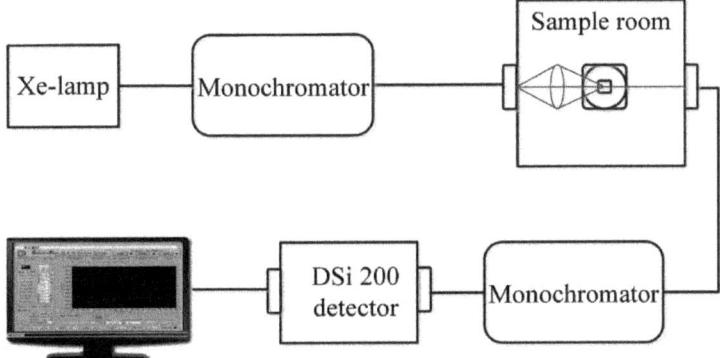

Figure 2. The schematic diagram of the experimental setup for the PL measurement.

3. Results

3.1. Structure and Morphology

Figure 3a shows the XRD patterns of YGB phosphors with different compositions annealed at 800 °C, the main peaks of which are in agreement with the JCPDS Card of YBO$_3$ (PDF#16-0277). The phosphors were confirmed as polycrystalline materials that possess a hexagonal crystal structure with space group P63/m and the cell parameters of a = b = 3.778 Å and c = 8.81 Å, similarly to what has been previously reported [20]. Moreover, the well-defined sharp diffraction peaks imply that these samples have high crystallinity, illustrating that Gd^{3+} and Bi^{3+} ions are substituted within the host. Some small impurity peaks are identified as Bi$_6$B$_{10}$O$_{24}$ (PDF#29-0228), which are attributed to the interaction of Bi$_2$O$_3$ and excess H$_3$BO$_3$ during the fabrication process [21]. The corresponding reaction equation is as follows: 2Bi$_2$O$_3$ + B$_2$O$_3$→Bi$_4$B$_2$O$_9$ and 3Bi$_4$B$_2$O$_9$ + 7B$_2$O$_3$→2Bi$_6$B$_{10}$O$_{24}$ [22]. Here, the XRD peaks (2θ = ~27.2°) of YGB samples with slightly smaller angles, in comparison with the standard YBO$_3$, should be attributed to the larger radius of Gd^{3+} (1.053 Å, 8-coordination) and Bi^{3+} (1.170 Å, 8-coordination) relative to that of Y^{3+} (1.019 Å, 8-coordination), according to Bragg's law [23–25], while the shift in diffraction peaks resulted from the different Gd^{3+}-Bi^{3+} contents in these phosphors. The slight change is also attributed to the surface charge redistribution of the crystal nucleus, induced by an inner-electron charge transfer between the doped ions and lattice cations [26,27]. Therefore, these results show that doped Gd^{3+} and Bi^{3+} ions do not affect the main crystal structure of YBO$_3$ and should be completely dissolved into the host lattice.

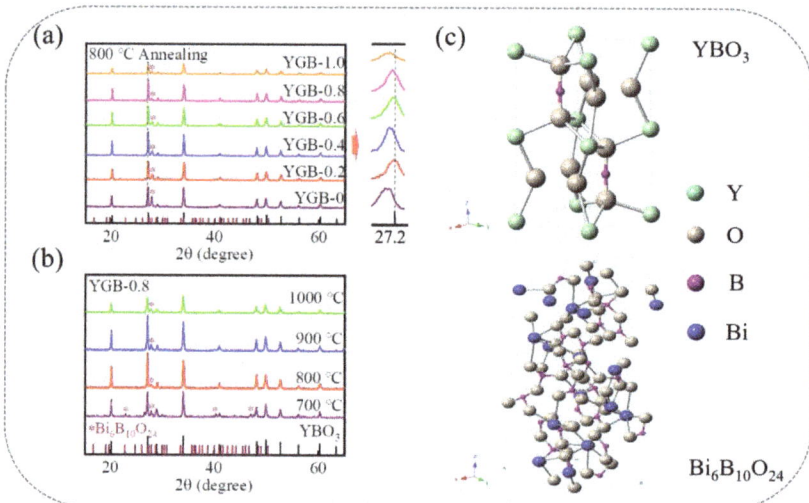

Figure 3. XRD patterns of (**a**) $Y_{0.9}(Gd_xBi_{1-x})_{0.1}BO_3$ (x = 0, 0.2, 0.4, 0.6, 0.8 and 1.0) phosphors annealed at 800 °C and (**b**) YGB-0.8 phosphors with different annealing temperatures ranging from 700 °C to 1000 °C. The * represents the diffraction peaks from $Bi_6B_{10}O_{24}$ phase. (**c**) The crystal structure of YGB systems, showing the coordination environment of YBO_3 and $Bi_6B_{10}O_{24}$.

As shown in Figure 3b, the impurities decrease, and the relative intensities of diffraction peaks initially increase and then decrease with the annealing temperature; this is attributed to the melting point of $Bi_6B_{10}O_{24}$ and the selectivity of the growth in the solid-state synthesis process [25,28]. In addition, the YGB crystal is composed of 8-coordinated Y^{3+} and 4- coordinated B^{3+} ions, which is illustrated in Figure 3c. Here, the Y^{3+} ions are 8- coordinated with two nonequivalent environments, while the B^{3+} ions and two interconnected BO_4 tetrahedral coordination form $(BO_3)^{3-}$ groups [29]. Moreover, due to the similar ionic radii of the 8-coordinated Y^{3+}, Bi^{3+}, and Gd^{3+} ions, Gd^{3+} and Bi^{3+} ions can easily substitute the Y^{3+} sites and form a solid solution of (Y,Gd,Bi)BO_3 crystals.

Take the $Y_{0.9}Gd_{0.08}Bi_{0.02}BO_3$ (YGB-0.8) phosphor as an example. The SEM images in Figure 4a–d show the typical morphologies of particles annealed at 700, 800, 900, and 1000 °C, revealing that the powders annealed at 800 °C and below possess an average size of ~200 nm and the regular morphology. To show this, the particle size distributions of the YGB-0.8 phosphor annealed at 800 °C are shown in Figure 4i. Here, the inset shows the macroscopic appearance of the sample exhibited under natural light irradiation. It can be observed that the particle size is mainly concentrated at ~200 nm, which is consistent with the SEM images. As shown in the SEM images, the powders with a narrow particle size distribution have been synthesized at lower sintering temperatures, and they possess a large effective surface area and weak atomic binding energy, resulting in the lower local symmetry of the YO_8 polyhedron and the surface defects of nanoparticles. When the annealing temperature exceeds a certain value, the crystal phase is gradually purified together with grain growth. Obviously, the morphology becomes more irregular in angularity, heterogeneity and compactness with the increase in sintering temperature, which is due to the changes in van der Waals attractions, while the small particle size may be caused by the distortion of anionic groups on the particle's surface [30–32]. The size of spherical particles is significantly larger after sintering at 1000 °C, while compositional particles lose their spherical shape and undergo significant aggregation, which is attributed to the higher activity of atoms on the particle's surface caused by the further decomposition of precursors. Under higher temperature annealing, the atoms could diffuse and combine with adjacent ones to form stable chemical bonds, leading to agglomeration [33,34]. No obvious changes in the morphology or particle size with various Gd^{3+} contents were observed at the same sintering temperature (not shown here), indicating that the doped Bi^{3+} and Gd^{3+} ions do not impact crystallization and grain growth. For the YGB-0.8 sample annealed at 800 °C, the homogeneous distributions of Gd, Bi, O, and Y elements are clearly observed by EDS, as shown in Figure 4e–h,j. The B element is undetected since its corresponding energy in the X-ray spectrum falls outside the scope. Moreover, high-packing densities, good slurry properties, and well-distributed particles in YGB systems are conducive to photon release.

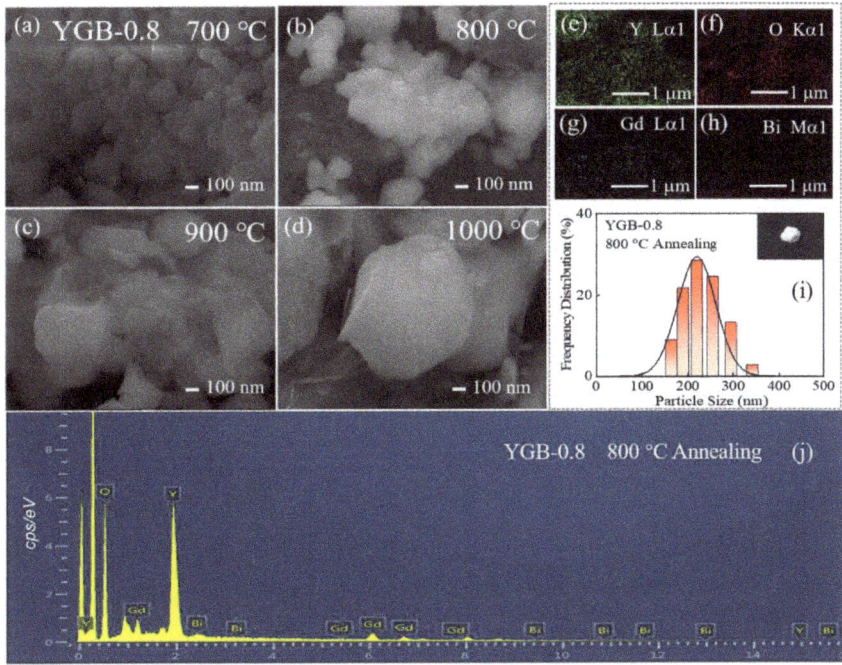

Figure 4. (a–d) SEM images of YGB-0.8 samples with different annealing temperatures ranging from 700 °C to 1000 °C. (e–h) Elemental mapping; (i) particle size distributions and (j) EDS spectrum of the YGB-0.8 sample annealed at 800 °C. Inset in (i): the macroscopic appearance of the sample exhibited under natural light irradiation.

3.2. Fluorescence Behaviors of YGB Phosphor

Figure 5a,b show the typical emission spectra of YGB phosphors with different Bi^{3+}/Gd^{3+} contents (x = 0, 0.2, 0.4, 0.6, 0.8 and 1.0) under UV (273 nm) and VIS (532 nm) excitations. Notably, strong UV emissions can also be obtained by up-conversion under the excitation of a 532 nm laser, and all samples show a narrow-band emission at 313 nm, which is attributed to the $^6P_{7/2} \rightarrow {}^8S_{7/2}$ transition of Gd^{3+} ions [6]. Emission intensity increases until the Gd^{3+} content exceeds x = 0.8, which results from the energy transfer between Bi^{3+} and Gd^{3+} ions; with the further increase in Gd^{3+} contents, concentration quenching occurs with the attenuation of emission intensity. In order to identify the change in spectral intensity more clearly, the dependence of the PL emission intensity at 313 nm on the Gd^{3+} content (x) in YGB phosphors under 273 nm and 532 nm excitations is illustrated in the inset of Figure 5b. Compared with the sample without Bi^{3+}, the weaker wide emission located around 440 nm resulted from the $6s^2 \rightarrow 6s6p$ transitions of Bi^{3+} ions. According to the photoluminescence excitation (PLE) spectra of these YGB phosphors in Figure 5c, monitored at 313 nm, the strongest excitation band centered at 273 nm should be attributed to the $^8S_{7/2} \rightarrow {}^6I_J$ transition of the Gd^{3+} ion, which well overlaps with the 253.7 nm line of mercury lamps [35], while the emission peak at 440 nm was derived from the $^3P_1 \rightarrow {^1S_0}$ transition of Bi^{3+} ions [36]. In particular, Figure 5d shows the PLE spectra of YGB samples with different Gd^{3+} contents in monitoring the 440 nm emission, the intensity of which decreases with the Gd^{3+} content, illustrating the energy transfer (ET) from Bi^{3+} to Gd^{3+} ions that consumes the excitation energy of Bi^{3+} ions.

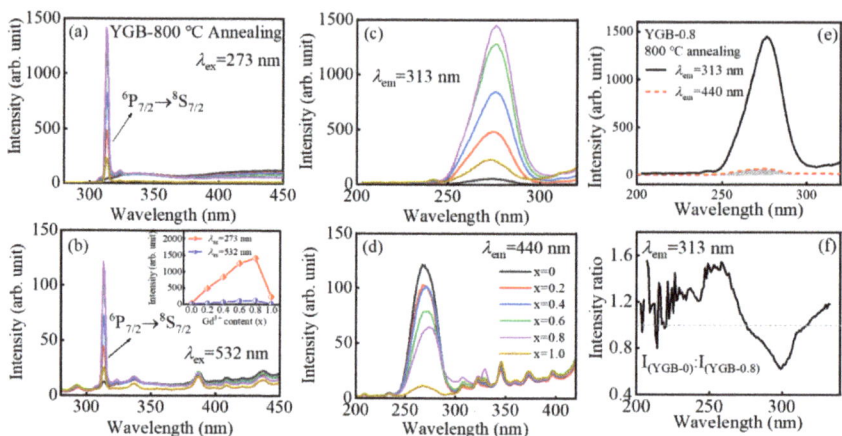

Figure 5. PL spectra of YGB phosphors (x = 0, 0.2, 0.4, 0.6, 0.8, and 1.0) under the (**a**) 273 nm and (**b**) 532 nm excitations, and their PLE spectra monitoring at (**c**) 313 nm and (**d**) 440 nm. Inset in (**b**): The dependence of PL emission intensities at 313 nm on the Gd^{3+} content (x) in YGB phosphors under 273 nm and 532 nm excitations. (**e**) Excitation spectra of YGB-0.8 phosphor with the spectral overlap presented by the shade and (**f**) the fluorescence intensity ratio of the 800 °C annealed YGB-0.8 and YGB-0 phosphors monitored at 313 nm.

The energy transfer depends on the overlap between the excitation band of the activator and the emission band of the sensitizer in the phosphors. Bi^{3+} ions have a $6s^2$ outer electronic configuration with a $^1S_0$ ground state, and the excited state has the configuration of 6s6p with $^3P_0$, $^3P_1$, $^3P_2$, and $^1P_1$ splitting levels. Due to the forbidden transitions of $^1S_0 \rightarrow {^3P_0}$ and $^1S_0 \rightarrow {^3P_2}$ by the electronic selection rules, the $^1S_0 \rightarrow {^3P_1}$ and $^1S_0 \rightarrow {^1P_1}$ transitions of Bi^{3+} ions are usually observed [37]. For the sample doped with only Bi^{3+} ions, it would first relax and transit into the lowest $^3P_1$ excited state and then return to the $^1S_0$ ground state via radiation. However, when Bi^{3+} and Gd^{3+} ions were co-doped into the host, energy transfer would occur, since the $^3P_1 \rightarrow {^1S_0}$ emission of Bi^{3+} effectively overlapped with the energy levels of Gd^{3+} ($^6P_{7/2}$, $^6P_{5/2}$, and $^6P_{3/2}$) [38]. For the $Y_{0.9}Gd_{0.08}Bi_{0.02}BO_3$ (YGB-0.8) sample annealed at 800 °C, the overlapped excitation spectra of Bi^{3+} and Gd^{3+} ions are shown in Figure 5e, confirming their efficient excitability in the short-wave UV region, which is advantageous for the resonance energy transfer from Bi^{3+} to Gd^{3+} ions. In order to further clarify the controversies over the ET from Bi^{3+} to Gd^{3+}, Figure 5f presents the Gd^{3+} fluorescence intensity ratio between 800 °C annealed YGB-0 and YGB-0.8 phosphors, monitored at 313 nm, which demonstrates the sensitizing effect of Bi^{3+} on Gd^{3+} ions. Compared with the sample without Gd^{3+} ions, the excitation energy of Bi^{3+} in Bi^{3+}-Gd^{3+} co-doped samples is transferred to the Gd^{3+} ion and leads to stronger fluorescence emissions from the Gd^{3+} ion in short-wave UVB radiation, resulting in increased excitability. In addition, the excitation peaks located at 258 nm are consistent with the characteristic excitation peaks of Bi^{3+} ($^1S_0 \rightarrow {^1P_1}$) [25,37], which further confirms the effectiveness of ET from Bi^{3+} to Gd^{3+} ions.

To further investigate the effect of sintering temperatures on luminous properties, the PL and PLE spectra measured at room temperature from the YGB-0.8 phosphors annealed at different temperatures are illustrated in Figure 6a–c. The intensity of the excitation peak at 313 nm increases with the annealing temperature until 800 °C. The grain size increases while the porosity decreases significantly with the increase in temperature, enhancing the luminous intensity. When the sintering temperature is higher than 800 °C, the decreased intensity is attributed to the accelerated volatilization of Bi^{3+} and the crystalline defects. Furthermore, the enhanced PL emission originates from the absorption of exciting UV light by co-doped Bi^{3+} ions, which transfer the energy to the Gd^{3+} ions [18,39]. This mechanism is schematically shown in Figure 6d. Firstly, phosphors absorb the UV light, which leads to the $^1S_0 \rightarrow ^3P_1$ transition of Bi^{3+} ions. The Bi^{3+} ions then transfer the energy non-radiatively to Gd^{3+} ions and ultimately realize UVB emissions from Gd^{3+} ions. Moreover, the smaller electronegativity of Gd^{3+} (1.20), compared to that of Y^{3+} (1.22) and Bi^{3+} ion (~2.02), allows an easier charge transfer, thus promoting PL emissions [19,40].

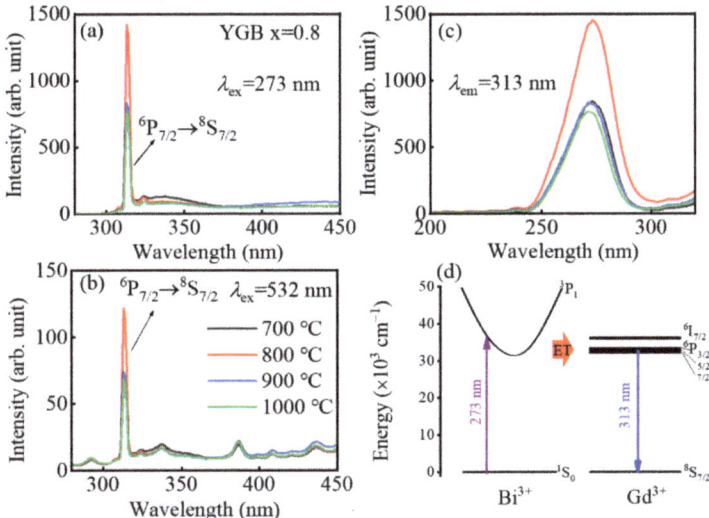

Figure 6. PL spectra of YGB-0.8 phosphors annealed from 700 °C to 1000 °C under (**a**) 273 nm and (**b**) 532 nm excitations. (**c**) Their PLE spectra monitored at 313 nm and (**d**) the schematic energy transfer mechanism from Bi^{3+} to Gd^{3+} ions.

In order to evaluate the optical property of Gd^{3+}-Bi^{3+} co-doped YGB phosphors with different Gd^{3+} contents and annealing temperatures, the relative spectral power distributions and relative photon distributions were determined and compared, as in Figure 7a–d. Under the 273 nm excitation, the sharp narrow-band UVB emission at 313 nm that originates from the $^6P_J \rightarrow ^8S_{7/2}$ transition increases significantly with the increase in Gd^{3+} content and annealing temperature, which reaches the maximum value while the Gd^{3+} content is x = 0.8 and is annealed at 800 °C. This should be attributed to the increased energy transfer caused by the reduced distance among Gd^{3+}-Gd^{3+} ion pairs [41]. The relative photon distribution provides fundamental information with respect to optical fields and relevant applications. Depending on the relative spectral power distribution $P(\lambda)$, photon distribution $N(\nu)$ can be deduced by $N(\nu) = \frac{\lambda^3}{hc}P(\lambda)$, where ν, λ, h, c, and $P(\lambda)$ represent wavenumber, wavelength, Planck constant, vacuum light velocity, and spectral power distribution, respectively [42]. Here, the abscissae of the distribution spectra were converted to a wavenumber (cm^{-1}) for accurate deconvolution. The net absorption and emission photon distribution curves of Gd^{3+}-Bi^{3+} co-doped YBO_3 phosphors were derived, as presented in Figure 7b,d, and their net emission and absorption intervals were selected at 30,300–33,300 cm^{-1} (corresponding to 313 nm) and 36,300–39,200 cm^{-1} (corresponding to 273 nm).

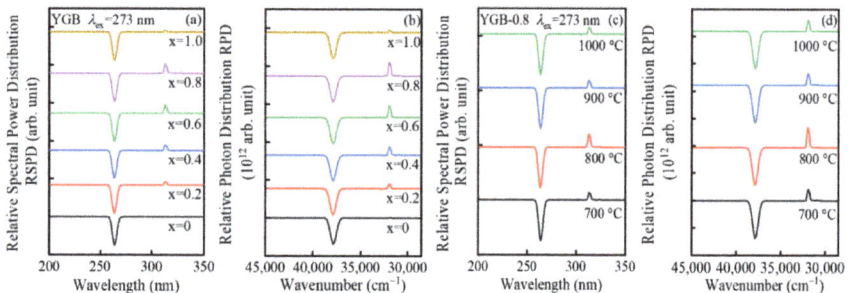

Figure 7. (**a**,**c**) Relative spectral power distributions and (**b**,**d**) relative photon distributions of $Y_{0.9}(Gd_xBi_{1-x})_{0.1}BO_3$ (x = 0, 0.2, 0.4, 0.6, 0.8 and 1.0) and YGB-0.8 samples with different annealing temperatures from 700 °C to 1000 °C under the 273 nm excitation.

In addition, upon the excitation under 532 nm VIS light, the samples still emit the up-conversion UVB emission at 313 nm. The spectral power distribution and the photon number distribution of all samples with different Gd^{3+} contents (x = 0, 0.2, 0.4, 0.6, 0.8, and 1.0) are displayed in Figure 8. With the increase in Gd^{3+} contents, the emission intensity increases until $Y_{0.9}Gd_{0.08}Bi_{0.02}BO_3$, because the transfer probability is proportional to the interaction between the sensitizer (Bi^{3+}) and the activator (Gd^{3+}) in both non-radiative and radiative resonance energy transfers. When x > 0.8, the intensity significantly decreases due to prominent concentration quenching caused by the reduced distance among Gd^{3+}-Gd^{3+} ions.

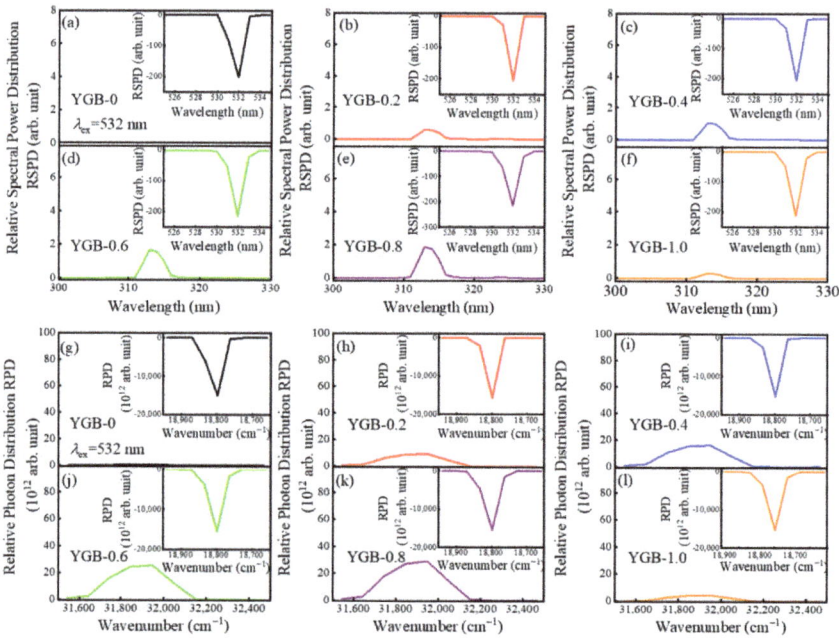

Figure 8. (**a**–**f**) Relative spectral power distributions and (**g**–**l**) relative photon distributions of $Y_{0.9}(Gd_xBi_{1-x})_{0.1}BO_3$ (x = 0, 0.2, 0.4, 0.6, 0.8 and 1.0) phosphors annealed at 800 °C under the 532 nm excitation.

According to Blasse [43,44], the energy would transfer from one activator to another until all energy is consumed. This phenomenon is regarded as concentration quenching in fluorescence, which is due to the non-radiative energy transfer among identical ions. Thus, the critical distance (R_C) is a parameter that is essential to understanding this phenomenon, which is calculated using the following equation: $R_c = 2\left[\frac{3V}{4\pi x_c N}\right]^{\frac{1}{3}}$, where V is the volume of the unit cell (in Å^3), x_c is critical concentration, and N is the number of Y^{3+}/Bi^{3+}/Gd^{3+} ions in the unit cell. Herein, the values are x_c = 0.08, N = 6, and V = 108.90 Å^3, and the critical distance R_C of the YGB phosphor is calculated to be about 7.57 Å. Meanwhile, the corresponding spectral power distribution and photon number distribution of YGB-0.8 phosphors annealed from 700 °C to 1000 °C under the 532 nm excitation were also derived, and they are shown in Figure 9 to demonstrate the up-conversion emission monitored at 313 nm and the optimal annealing temperature of 800 °C. These results verify the effectiveness of Gd^{3+}-Bi^{3+} co-doped phosphors in photon conversion and provide the theoretical basis for their application in skin treatments.

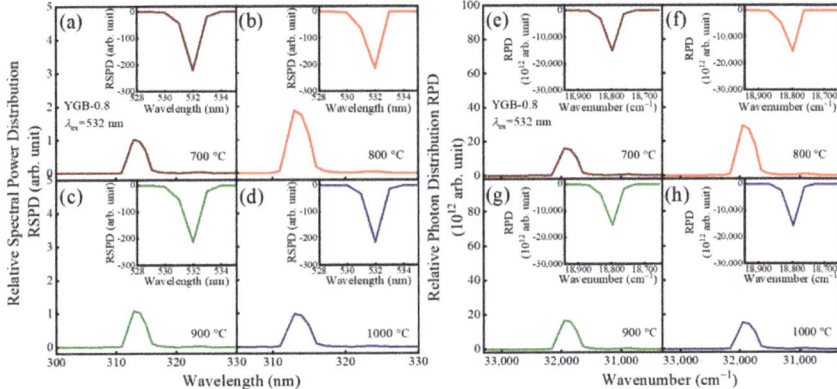

Figure 9. (a–f) Relative spectral power distributions and (e–h) relative photon distributions of YGB-0.8 samples with different annealing temperatures from 700 °C to 1000 °C under the 532 nm excitation.

The spectral parameters could also provide external quantum yields (QYs) to assess luminescence and laser materials, which are used to calculate the utilization efficiency of the absorbed photons for desired emissions, defined as the photon number ratio of emission and absorption. Namely, QY = emitted photons/absorbed photons = N_{em}/N_{abs}. Here, the maximum QY is derived to be 24.75% in a Y$_{0.9}$Gd$_{0.08}$Bi$_{0.02}$BO$_3$ sample annealed at 800 °C under the 273 nm excitation, which is larger than that of other Gd^{3+} ions doped phosphors [45,46], and it is 75 times the single Gd^{3+}-doped sample in this work. On the basis of these QYs, a higher photon release efficiency is achieved, which further exhibits the potential of Gd^{3+}-Bi^{3+} co-doped YBO$_3$ phosphors for UVB skin treatment and reflects the energy transfer effectiveness between Bi^{3+} and Gd^{3+} ions in these phosphors. Moreover, this phosphor maintains a unique up-conversion excitability in the VIS region with a QY of 1.33% under the excitation of 532 nm. The QY values for the different contents and annealing temperatures of these co-doped YGB phosphors, under the excitation of the 273 and 532 nm, are listed in Table 1. These results reveal that the Gd^{3+}-Bi^{3+} activated YBO$_3$ phosphors with up/down-conversion excitability exhibit excellent UVB emission performance.

Table 1. Quantum yields in Gd^{3+}-Bi^{3+} co-doped phosphors with different Gd^{3+} contents and sintering temperatures under 273 and 532 nm excitation.

Excitation Wavelength (nm)	External Quantum Yield QY (%)									
	Gd^{3+} Content (x) 800 °C Annealing						Sintering Temperature (°C)			
	0	0.2	0.4	0.6	0.8	1.0	700	800	900	1000
273	0.33	7.53	14.80	21.81	24.75	3.91	13.70	24.75	14.02	12.83
532	0.01	0.51	0.91	1.25	1.33	0.23	0.81	1.33	0.77	0.76

4. Conclusions

UVB-emitting $Y_{0.9}(Gd_xBi_{1-x})_{0.1}BO_3$ phosphors (x = 0, 0.2, 0.4, 0.6, 0.8, and 1.0) with the hexagonal phase and an average ~200 nm grain size were fabricated via the solid-state synthesis method. The enhanced PL emissions and the overlapped spectra verify the energy transfer between the Bi^{3+} and Gd^{3+} ions and a well-defined sharp and intense peak centered at 313 nm due to the $^6P_{7/2} \rightarrow {}^8S_{7/2}$ transitions of Gd^{3+} ions. The $Y_{0.9}Gd_{0.08}Bi_{0.02}BO_3$ phosphor annealed at 800 °C exhibits the highest QY values of 24.75% and 1.33% under the excitation of 273 nm and 532 nm, respectively, confirming that the system possesses excellent excitability in both UV and VIS regions. The optimal QY from the Gd^{3+}-Bi^{3+} co-doped YBO_3 phosphor is 75 times the single Gd^{3+}-doped sample. Bright and narrow UVB emissions resulting from efficient photon conversion demonstrate the multifunctional applications of Gd^{3+}-Bi^{3+}-activated YBO_3 phosphors and provide a new route for skin treatments.

Author Contributions: Conceptualization, Z.Y. and Y.Y.; methodology, Z.Y.; validation, Z.Y.; formal analysis, Z.Y. and Y.Y.; resources, Y.Y. and J.S.; data curation, Z.Y.; writing—original draft preparation, Z.Y.; writing—review and editing, Y.Y.; visualization, Z.Y. and Y.Y.; supervision, Y.Y. and J.S.; project administration, J.S.; funding acquisition, Y.Y. and J.S. All authors have read and agreed to the published version of the manuscript.

Funding: This research was funded by the National Natural Science Foundation of China (No. 62275132) and the Natural Science Foundation of Tianjin City (No. 21JCZDJC00020).

Institutional Review Board Statement: Not applicable.

Informed Consent Statement: Not applicable.

Data Availability Statement: Data will be made available upon request.

Conflicts of Interest: The authors declare no conflict of interest.

References

1. Chauhan, A.; Palan, C.; Sawala, N.; Omanwar, S. Synthesis and Photoluminescence study of LaF_3:Gd^{3+} phosphors for Phototherapy Application. *J. Emerg. Technol. Innov. Res.* **2022**, *9*, 382–387.
2. Hönigsmann, H.; Brenner, W.; Rauschmeier, W.; Konrad, K.; Wolff, K. Photochemotherapy for cutaneous T cell lymphoma: A follow-up study. *J. Am. Acad. Dermatol.* **1984**, *10*, 238–245. [CrossRef] [PubMed]
3. Scherschun, L.; Kim, J.J.; Lim, H.W. Narrow-band ultraviolet B is a useful and well-tolerated treatment for vitiligo. *J. Am. Acad. Dermatol.* **2001**, *44*, 999–1003. [CrossRef]
4. Dani, C.; Martelli, E.; Reali, M.F.; Bertini, G.; Panin, G.; Rubaltelli, F. Fiberoptic and conventional phototherapy effects on the skin of premature infants. *J. Pediatr.* **2001**, *138*, 438–440. [CrossRef] [PubMed]
5. Al Salman, M.; Ghiasi, M.; Farid, A.S.; Taraz, M.; Azizpour, A.; Mahmoudi, H. Oral simvastatin combined with narrowband UVB for the treatment of psoriasis: A randomized controlled trial. *Dermatol. Ther.* **2021**, *34*, e15075. [CrossRef] [PubMed]
6. Cheng, Z.; Liu, T.; Shen, M.; Peng, Y.; Yang, S.; Khan, W.U.; Zhang, Y. Luminescence and energy transfer of Ce^{3+}/Gd^{3+}/Tb^{3+}/Eu^{3+} doped hexagonal fluoride. *J. Lumin.* **2022**, *241*, 118477. [CrossRef]
7. Wang, X.; Chen, Y.; Kner, P.A.; Pan, Z. Gd^{3+}-activated narrowband ultraviolet-B persistent luminescence through persistent energy transfer. *Dalton Trans.* **2021**, *50*, 3499–3505. [CrossRef]
8. Reich, A.; Mędrek, K. Effects of narrow band UVB (311 nm) irradiation on epidermal cells. *Int. J. Mol. Sci.* **2013**, *14*, 8456–8466. [CrossRef]

9. Youssef, Y.E.; Eldegla, H.E.A.; Elmekkawy, R.S.M.; Gaballah, M.A. Evaluation of vitamin D receptor gene polymorphisms (ApaI and TaqI) as risk factors of vitiligo and predictors of response to narrowband UVB phototherapy. *Arch. Dermatol. Res.* **2022**, 1–8. [CrossRef]
10. Zhu, Q.; Wang, S.; Li, J.-G.; Li, X.; Sun, X. Compounds, Spherical engineering and space-group dependent luminescence behavior of YBO_3: Eu^{3+} red phosphors. *J. Alloys Compd.* **2018**, *731*, 1069–1079. [CrossRef]
11. Gao, Y.; Jiang, P.; Cong, R.; Yang, T. Photoluminescence of Bi^{3+} in $LiCaY_5(BO_3)_6$ and color-tunable emission through energy transfer to Eu^{3+}/Tb^{3+}. *J. Lumin.* **2022**, *251*, 119161. [CrossRef]
12. Wang, L.; Wang, Y. Enhanced photoluminescence of YBO_3:Eu^{3+} with the incorporation of Sc^{3+}, Bi^{3+} and La^{3+} for plasma display panel application. *J. Lumin.* **2007**, *122*, 921–923. [CrossRef]
13. Chen, L.; Yang, G.; Liu, J.; Shu, X.; Jiang, Y.; Zhang, G. Photoluminescence properties of Eu^{3+} and Bi^{3+} in YBO_3 host under VUV/UV excitation. *J. Appl. Phys.* **2009**, *105*, 013513. [CrossRef]
14. Zeng, X.; Im, S.-J.; Jang, S.-H.; Kim, Y.-M.; Park, H.-B.; Son, S.-H.; Hatanaka, H.; Kim, G.-Y.; Kim, S.-G. Luminescent properties of $(Y, Gd)BO_3$:Bi^{3+}, RE^{3+} (RE = Eu, Tb) phosphor under VUV/UV excitation. *J. Lumin.* **2006**, *121*, 1–6. [CrossRef]
15. Song, H.; Yu, H.; Pan, G.; Bai, X.; Dong, B.; Zhang, X.; Hark, S. Electrospinning preparation, structure, and photoluminescence properties of YBO_3:Eu^{3+} nanotubes and nanowires. *Chem. Mater.* **2008**, *20*, 4762–4767. [CrossRef]
16. Singh, V.; Sivaramaiah, G.; Rao, J.; Kim, S. Investigation of new UV-emitting, Gd-activated $Y_4Zr_3O_{12}$ phosphors prepared via combustion method. *J. Lumin.* **2015**, *157*, 82–87. [CrossRef]
17. Gupta, P.; Sahni, M.; Chauhan, S. Enhanced photoluminescence properties of rare earth elements doped $Y_{0.50}Gd_{0.50}BO_3$ phosphor and its application in red and green LEDs. *Optik* **2021**, *240*, 166810. [CrossRef]
18. Gawande, A.; Sonekar, R.; Omanwar, S. Combustion synthesis and energy transfer mechanism of $Bi^{3+}\rightarrow Gd^{3+}$ and $Pr^{3+}\rightarrow Gd^{3+}$ in YBO_3. *Combust. Sci. Technol.* **2014**, *186*, 785–791. [CrossRef]
19. Zhu, Q.; Wang, S.; Li, X.; Sun, X.; Li, J.-G. Well-dispersed $(Y_{0.95-x}Gd_xEu_{0.05})(B(OH)_4)CO_3$ colloidal spheres as a novel precursor for orthoborate red phosphor and the effects of Gd^{3+} doping on structure and luminescence. *CrystEngComm* **2018**, *20*, 4546–4555. [CrossRef]
20. Zhang, W.; Liu, S.; Hu, Z.; Liang, Y.; Feng, Z.; Sheng, X. Preparation of YBO_3: Dy^{3+}, Bi^{3+} phosphors and enhanced photoluminescence. *Mater. Sci. Eng. B* **2014**, *187*, 108–112. [CrossRef]
21. Liang, F.; Zhou, Y.-L.; Wang, S.-Q.; Zhang, Q.-P. Self-propagating High-temperature Synthesis and Photoluminescence Properties of $Bi_3B_5O_{12}$ Powders. *Chem. Lett.* **2015**, *44*, 571–573. [CrossRef]
22. Sun, S.; Zhang, Q.; Dai, Y.; Pei, X. Enhanced microwave dielectric properties of $Bi_6B_{10}O_{24}$ ceramics as ultra-low temperature co-fired ceramics materials. *J. Mater. Sci.-Mater. Electron.* **2022**, *33*, 13604–13613. [CrossRef]
23. Pianassola, M.; Stand, L.; Loveday, M.; Chakoumako, B.C.; Koschan, M.; Melcher, C.L.; Zhuravleva, M. Czochralski growth and characterization of the multicomponent garnet $(Lu_{1/4}Yb_{1/4}Y_{1/4}Gd_{1/4})_3Al_5O_{12}$. *Phys. Rev. Mater.* **2021**, *5*, 083401. [CrossRef]
24. Xu, H.; Wang, L.; Tan, L.; Wang, D.; Wang, C.; Shi, J. Sites occupancy preference of Bi^{3+} and white light emission through co-doped Sm^{3+} in $LiGd_5P_2O_8$. *J. Am. Ceram. Soc.* **2018**, *101*, 3414–3423. [CrossRef]
25. Accardo, G.; Audasso, E.; Yoon, S.P. Unravelling the synergistic effect on ionic transport and sintering temperature of nanocrystalline CeO_2 tri-doped with Li Bi and Gd as dense electrolyte for solid oxide fuel cells. *J. Alloys Compd.* **2022**, *898*, 162880. [CrossRef]
26. Wang, X.; Wang, Y.; Bu, Y.; Yan, X.; Wang, J.; Cai, P.; Vu, T.; Seo, H. Influence of doping and excitation powers on optical thermometry in Yb^{3+}-Er^{3+} doped $CaWO_4$. *Sci. Rep.* **2017**, *7*, 43383. [CrossRef] [PubMed]
27. Wang, X.; Bu, Y.; Xiao, Y.; Kan, C.; Lu, D.; Yan, X. Size and shape modifications, phase transition, and enhanced luminescence of fluoride nanocrystals induced by doping. *J. Mater. Chem. C* **2013**, *1*, 3158–3166. [CrossRef]
28. Yu, H.; Liu, J.; Zhang, W.; Zhang, S. Ultra-low sintering temperature ceramics for LTCC applications: A review. *J. Mater. Sci.-Mater. Electron.* **2015**, *26*, 9414–9423. [CrossRef]
29. Chadeyron, G.; El-Ghozzi, M.; Mahiou, R.; Arbus, A.; Cousseins, J. Revised structure of the orthoborate YBO_3. *J. Solid State Chem.* **1997**, *128*, 261–266. [CrossRef]
30. Zou, D.; Ma, Y.; Qian, S.; Huang, B.; Zheng, G.; Dai, Z. Improved luminescent properties of novel nanostructured Eu^{3+} doped yttrium borate synthesized with carbon nanotube templates. *J. Alloys Compd.* **2014**, *584*, 471–476. [CrossRef]
31. Armetta, F.; Saladino, M.L.; Martino, D.F.C.; Livreri, P.; Berrettoni, M.; Caponetti, E. Synthesis of yttrium aluminum garnet nanoparticles in confined environment II: Role of the thermal treatment on the composition and microstructural evolution. *J. Alloys Compd.* **2017**, *719*, 264–270. [CrossRef]
32. Solgi, S.; Ghamsari, M.S.; Tafreshi, M.J.; Karevane, R. Synthesis condition effects on the emission enhancement of YBO_3 powder. *Optik* **2020**, *218*, 165031. [CrossRef]
33. Velchuri, R.; Kumar, B.V.; Devi, V.R.; Prasad, G.; Prakash, D.J.; Vithal, M. Preparation and characterization of rare earth orthoborates, $LnBO_3$ (Ln = Tb, La, Pr, Nd, Sm, Eu, Gd, Dy, Y) and $LaBO_3$:Tb, Eu by metathesis reaction: ESR of $LaBO_3$:Gd and luminescence of $LaBO_3$:Tb, Eu. *Mater. Res. Bull.* **2011**, *46*, 1219–1226. [CrossRef]
34. Srivastava, S.; Behera, S.K.; Nayak, B. Effect of the Y:B ratio on phase purity and development of thermally stable nano-sized Eu^{3+}-doped YBO_3 red phosphor using sodium borohydride. *Dalton Trans.* **2015**, *44*, 7765–7769. [CrossRef] [PubMed]
35. Mokoena, P.; Gohain, M.; Bezuidenhoudt, B.; Swart, H.; Ntwaeaborwa, O. Luminescent properties and particle morphology of $Ca_3(PO_4)_2$: Gd^{3+}, Pr^{3+} phosphor powder prepared by microwave assisted synthesis. *J. Lumin.* **2014**, *155*, 288–292. [CrossRef]

36. Chen, L.; Zheng, H.; Cheng, J.; Song, P.; Yang, G.; Zhang, G.; Wu, C. Site-selective luminescence of Bi^{3+} in the YBO_3 host under vacuum ultraviolet excitation at low temperature. *J. Lumin.* **2008**, *128*, 2027–2030. [CrossRef]
37. Abo-Naf, S.M.; Abdel-Hameed, S.A.M.; Marzouk, M.A.; Elwan, R.L. Sol-gel synthesis, paramagnetism, photoluminescence and optical properties of Gd-doped and Bi-Gd-codoped hybrid organo-silica glasses. *J. Mater. Sci.-Mater. Electron.* **2015**, *26*, 2363–2373. [CrossRef]
38. De Hair, J.T.W.; Konijnend, W.L. The intermediate role of Gd^{3+} in the energy transfer from a sensitizer to an activator (especially Tb^{3+}). *J. Electrochem. Soc.* **1980**, *127*, 161. [CrossRef]
39. Carnall, W.; Fields, P.; Rajnak, K. Electronic energy levels of the trivalent lanthanide aquo ions. II. Gd^{3+}. *J. Chem. Phys.* **1968**, *49*, 4443–4446. [CrossRef]
40. Xia, Z.; Dong, B.; Liu, G.; Liu, D.; Chen, L.; Fang, C.; Liu, L.; Liu, S.; Tang, C.; Yuan, S. Effect of electronegativity on transport properties of $(La_{0.83}M_{0.17})_{0.67}Ca_{0.33}MnO_3$ (M = Y, Bi). *Phys. Status Solidi* **2005**, *202*, 113–119. [CrossRef]
41. Chen, L.; Luo, A.; Deng, X.; Xue, S.; Zhang, Y.; Liu, F.; Zhu, J.; Yao, Z.; Jiang, Y.; Chen, S. Luminescence and energy transfer in the Sb^{3+} and Gd^{3+} activated YBO_3 phosphor. *J. Lumin.* **2013**, *143*, 670–673. [CrossRef]
42. Yu, Z.; Luo, Z.; Liu, X.; Pun, E.Y.B.; Lin, H. Deagglomeration in Eu^{3+}-activated $Li_2Gd_4(MoO_4)_7$ polycrystalline incorporated polymethyl methacrylate. *Opt. Mater.* **2019**, *93*, 76–84. [CrossRef]
43. Blasse, G. Energy transfer between inequivalent Eu^{2+} ions. *J. Solid State Chem.* **1986**, *62*, 207–211. [CrossRef]
44. Kwon, I.-E.; Yu, B.-Y.; Bae, H.; Hwang, Y.-J.; Kwon, T.-W.; Kim, C.-H.; Pyun, C.-H.; Kim, S.-J. Luminescence properties of borate phosphors in the UV/VUV region. *J. Lumin.* **2000**, *87*, 1039–1041. [CrossRef]
45. Ajmal, M.; Atabaev, T.S. Facile fabrication and luminescent properties enhancement of bimodal Y_2O_3: Eu^{3+} particles by simultaneous Gd^{3+} codoping. *Opt. Mater.* **2013**, *35*, 1288–1292. [CrossRef]
46. Igashira, T.; Kawano, N.; Okada, G.; Kawaguchi, N.; Yanagida, T. Photoluminescence and scintillation properties of Ce-doped $Sr_2(Gd_{1-x}Lu_x)_8(SiO_4)_6O_2$ (x = 0.1, 0.2, 0.4, 0.5, 0.6) crystals. *Opt. Mater.* **2018**, *79*, 232–236. [CrossRef]

Disclaimer/Publisher's Note: The statements, opinions and data contained in all publications are solely those of the individual author(s) and contributor(s) and not of MDPI and/or the editor(s). MDPI and/or the editor(s) disclaim responsibility for any injury to people or property resulting from any ideas, methods, instructions or products referred to in the content.

Article

Effect of a-SiC$_x$N$_y$:H Encapsulation on the Stability and Photoluminescence Property of CsPbBr$_3$ Quantum Dots

Zewen Lin [1,2], Zhenxu Lin [1], Yanqing Guo [1], Haixia Wu [1], Jie Song [1], Yi Zhang [1], Wenxing Zhang [1], Hongliang Li [1], Dejian Hou [1] and Rui Huang [1,*]

[1] School of Materials Science and Engineering, Hanshan Normal University, Chaozhou 521041, China; 2636@hstc.edu.cn (Z.L.)
[2] National Laboratory of Solid State Microstructures/School of Electronics Science and Engineering/Collaborative Innovation Center of Advanced Microstructures, Nanjing University, Nanjing 210093, China
* Correspondence: rhuang@hstc.edu.cn

Abstract: The effect of a-SiC$_x$N$_y$:H encapsulation layers, which are prepared using the very-high-frequency plasma-enhanced chemical vapor deposition (VHF-PECVD) technique with SiH$_4$, CH$_4$, and NH$_3$ as the precursors, on the stability and photoluminescence of CsPbBr$_3$ quantum dots (QDs) were investigated in this study. The results show that a-SiCxNy:H encapsulation layers containing a high N content of approximately 50% cause severe PL degradation of CsPbBr$_3$ QDs. However, by reducing the N content in the a-SiCxNy:H layer, the PL degradation of CsPbBr$_3$ QDs can be significantly minimized. As the N content decreases from around 50% to 26%, the dominant phase in the a-SiCxNy:H layer changes from SiNx to SiCxNy. This transition preserves the inherent PL characteristics of CsPbBr$_3$ QDs, while also providing them with long-term stability when exposed to air, high temperatures (205 °C), and UV illumination for over 600 days. This method provided an effective and practical approach to enhance the stability and PL characteristics of CsPbBr$_3$ QD thin films, thus holding potential for future developments in optoelectronic devices.

Keywords: a-SiC$_x$N$_y$:H encapsulation; CsPbBr$_3$ QDs; stability; photoluminescence

1. Introduction

Recent studies have demonstrated that inorganic cesium lead halide perovskite (CsPbX$_3$, Cl, Br, and I) quantum dots (QDs) have the potential to be used in optoelectronic applications, such as light-emitting diodes (LEDs) and high-definition displays, due to their high quantum yields (QYs), ultralow-voltage operation, and ultra-narrow room-temperature emission [1–9]. However, for CsPbX$_3$ quantum dots, their crystal structure is inherently unstable, making them vulnerable to ion migration and decomposition in high-temperature, light, or humid conditions [10–15]. It has been found that oxygen and light play significant roles in the degradation of CsPbBr$_3$ quantum dots. They facilitate the maturation and growth of these quantum dots, which leads to a decrease in fluorescence quantum efficiency [14,15]. Under oxygen and light conditions, water vapor also acts as an ion transport channel, thus accelerating the degradation of CsPbBr$_3$ quantum dots [15]. Obviously, their poor stability when exposed to moist air, UV radiation, and high temperatures has been a barrier to their practical applications [10–14]. To overcome this, various strategies, such as doping engineering, surface engineering, and encapsulating engineering have been employed in an attempt to improve their stability [10,16–23]. For example, Zou et al. [16] utilized the substitution of Mn^{2+} to effectively stabilize perovskite lattices of CsPbX$_3$ QDs even under ambient air conditions with temperatures as high as 200 °C. Pan et al. [10] developed a postsynthesis passivation process for CsPbI$_3$ NCs by using a bidentate ligand, namely 2,2′-iminodibenzoic acid. This approach greatly enhanced the stability of red CsPbI$_3$ quantum dots, resulting in improved LED device performance. Kim et al. [17] reported highly efficient and stable CsPbBr$_3$ QDs, which retained more than 90% of the initial PLQY after 120 days of environmental storage by in situ surface reconstruction of $CsPbBr_3$-Cs_4PbBr_6 nanocrystals. In comparison to doping engineering and surface engineering, $CsPbX_3$ quantum dot composites produced through encapsulating engineering display higher stability [19–23]. Encapsulating engineering involves the use of an inert material layer to cover perovskite quantum dots, serving as a barrier against gas and ion diffusion and limiting the structure of the quantum dots. This method reduces the impact of air, light, water, and heat on the quantum dots, leading to improved stability. Moreover, the composite structure not only enhances the stability of perovskite quantum dots but also effectively passivates their surfaces, reducing surface defect states and improving photoluminescence (PL) efficiency. Therefore, utilizing encapsulating engineering for composite materials is a promising approach for enhancing the performance of $CsPbX_3$ quantum dots [19–23]. For example, $CsPbBr_3/SiO_2$ Janus nanocrystal films displayed higher photostability with only a slight drop (2%) in the PL intensity after nine hours of UV illumination [19]. Loiudice et al. [20] successfully prepared AlO_x on $CsPbX_3$ QD thin films using a low-temperature atomic layer deposition process, which improved their stability at high temperatures and under light exposure for hours. Despite the advances in encapsulation, the development of $CsPbX_3$ QDs with both high efficiency and excellent stability for practical applications remains challenging due to their sensitivity to the environment and the chemicals used in the encapsulation process, which may attack $CsPbX_3$ QDs and produce more defects, causing a marked deterioration of their PL intensity [14,22,23]. In our previous work [22], we developed a glow discharge plasma process combined with in situ real-time monitoring diagnosis to enhance the stability of $CsPbBr_3$ QDs, and demonstrated that an a-SiN_x:H encapsulating layer could significantly enhance the stability of $CsPbBr_3$ QDs under air exposure, UV illumination, and thermal treatment. However, the PL intensity was drastically reduced by 60% after being encapsulated by the a-SiN_x:H. Recently, to preserve the intrinsic photoluminescence (PL) characteristics of $CsPbBr_3$ QDs, we developed a damage-free plasma based encapsulation technique with real-time in situ diagnosis for $CsPbBr_3$ QD films. Our research revealed that the CH_4/SiH_4 plasma had negligible destructive effects on $CsPbBr_3$ QDs. Using low-temperature plasma-enhanced chemical vapor deposition, we fabricated a-SiC_x:H films that safeguarded the $CsPbBr_3$ QDs from surface damage during encapsulation, sustaining the PL efficiency. However, despite our efforts, the $CsPbBr_3$ QDs encapsulated by a-SiC_x:H still degraded after two months [23].

In this work, the effect of a-SiC_xN_y:H encapsulation layers on the stability and photoluminescence (PL) of $CsPbBr_3$ QDs was investigated. These layers were prepared using a very-high-frequency plasma-enhanced chemical vapor deposition (VHF-PECVD) technique with SiH_4, CH_4, and NH_3 as precursors. It is found that a-SiC_xN_y:H encapsulation layers with a high N content of ~50% cause a serious PL degradation of $CsPbBr_3$ QDs. However, by reducing the N content in the a-SiC_xN_y:H layer, the PL degradation of $CsPbBr_3$ QDs can be significantly minimized. As the N content decreases from around 50% to 26%, the dominant phase in the a-SiC_xN_y:H layer changes from SiN_x to SiC_xN_y, which not only makes $CsPbBr_3$ QDs retain their inherent PL characteristics but also endows $CsPbBr_3$ QDs with long-term stability when exposed to air, at a high temperature (205 °C), and under UV illumination for more than 600 days.

2. Materials and Methods

A very-high-frequency plasma-enhanced chemical vapor deposition (VHF-PECVD) technique was employed to prepare a-SiC_xN_y:H/$CsPbBr_3$ QDs/a-SiC_xN_y:H nanocomposite films with a sandwiched structure. The a-SiC_xN_y:H sublayer had a thickness of 15 nm and was firstly fabricated on silicon and quartz substrates using a mixture of SiH_4, CH_4, and NH_3. The flow rates of SiH_4 and CH_4 were set at 2.5 SCCM (short for 'standard cubic centimeters per minute') and 6 SCCM, respectively, while the NH_3 flow rate was different from 0 to 15 SCCM. The RF power, deposition pressure, and substrate temperature were maintained at 20 W, 20 Pa, and 150 °C, respectively. The 0.25 mg/mL $CsPbBr_3$ QDs solution was spin-coated on the substrates at a speed of 6000 rpm for 30 s, followed by the deposition of a 15 nm thick a-SiC$_x$N$_y$:H film. The CsPbBr$_3$ QDs were synthesized according to the procedures described by Protesescu et al. [1] To synthesize the CsPbBr$_3$ quantum dot, 5 mL of ODE and 0.188 mmol (0.069 g, ABCR, 98%) PbBr$_2$ were loaded into a 25 mL 3-neck flask and dried under vacuum at 120 °C for an hour. Then, 0.5 mL of dried oleylamine (OLA, Acros 80–90%) and 0.5 mL of dried OA were injected at 120 °C under a nitrogen atmosphere. Once the PbBr$_2$ salt was completely solubilized, the temperature was raised to 140–200 °C, and a Cs-oleate solution (0.4 mL, 0.125 M in ODE) was quickly injected. Five seconds later, the reaction mixture was cooled using an ice-water bath. The structure and composition of the CsPbBr$_3$ QDs/a-SiC$_x$N$_y$:H nanocomposite films were characterized using a Philips XL30 scanning electron microscope (SEM) and QUANTAX energy-dispersive X-ray spectroscope (EDS). The concentrations of Si, N, and C in the a-SiCxNy:H film were further determined by X-ray photoelectron spectroscopy (XPS). Optical absorption spectra were obtained via a Shimadzu UV-3600 spectrophotometer on quartz samples and used to estimate the optical bandgaps of a-SiC$_x$N$_y$:H films. For the acquisition of PL spectra, the Jobin Yvon FluoroLog-3 spectrophotometer was utilized, which is equipped with a 450 W continuous Xe lamp. Meanwhile, PL decay curves were recorded using an Edinburgh FLS980 spectrometer at room temperature. The temperature-dependent PL spectra were subsequently obtained by using a Raman Spectrometer that was equipped with a Linkam THMS 600 (Horiba LabRAM HR Evolution) from 25 to 205 °C.

3. Results and Discussion

Figure 1a shows the SEM image obtained from the CsPbBr$_3$ QDs/a-SiC$_x$N$_y$:H nanocomposite thin film, which indicates that the thin film was uniform. The EDS elemental maps displayed in Figure 1b,c reveal that the Cs, Pb, Br, Si, N, and C elements were well distributed. This confirms that CsPbBr$_3$ QDs are covered uniformly by an a-SiC$_x$N$_y$:H layer.

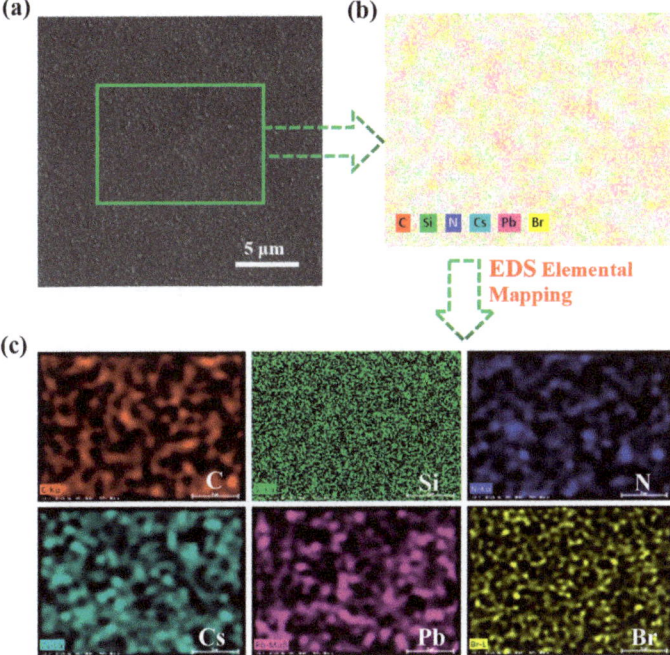

Figure 1. (**a**) SEM image of the CsPbBr$_3$ QDs/a-SiC$_x$N$_y$:H nanocomposite thin film. (**b**) Elemental mapping of the CsPbBr$_3$/a-SiC$_x$N$_y$:H nanocomposite thin film recorded by EDS. (**c**) EDS elemental mapping of Cs, Pb, Br, Si, N, and C in the thin film, respectively.

Figure 2 presents the PL from the $CsPbBr_3$ QDs before and after being encapsulated by a-SiC_xN_y:H layers, which were prepared with various NH_3 flow rates, respectively. As shown in Figure 2a, the PL intensity dropped remarkably by more than 40% after the $CsPbBr_3$ QDs were covered by the a-SiC_xN_y:H encapsulating layer prepared at a high NH_3 flow rate of 15 SCCM. With the decrease in NH_3 flow rate, the decline of PL from the $CsPbBr_3$ QDs was gradually reduced, as shown in Figure 2b–d. It was found that PL intensity from the $CsPbBr_3$ QDs remained nearly unchanged after being covered by the a-SiC_xN_y:H layer prepared at a low NH_3 flow rate of 5 SCCM. Obviously, PL intensity from $CsPbBr_3$ QDs is closely related to the NH_3 flow rate. As is demonstrated in our previous work [22], NH_3 can produce N-related reactive species in the glow discharge plasma through the collision of electrons and NH_3 molecules. Due to the high sensitivity of $CsPbX_3$ QDs to the environment, N-related reactive species interact with the atoms on the $CsPbX_3$ QDs surface and facilitatively create surface defects, leading to the degradation of the $CsPbBr_3$ QDs in the encapsulation process. The effect becomes more severe with increasing NH_3 flow rate. Therefore, the significant degradation of the $CsPbBr_3$ QDs after encapsulation by a-SiC_xN_y:H mainly stemmed from the high NH_3 flow rate used in the fabrication process. Figure 2c shows that a low NH_3 flow rate of 5 SCCM had little detrimental effect on the $CsPbBr_3$ QDs, suggesting that a low NH_3 flow rate is more suitable for forming a-SiC_xN_y:H encapsulating layers on the $CsPbBr_3$ QDs than a high NH_3 flow rate.

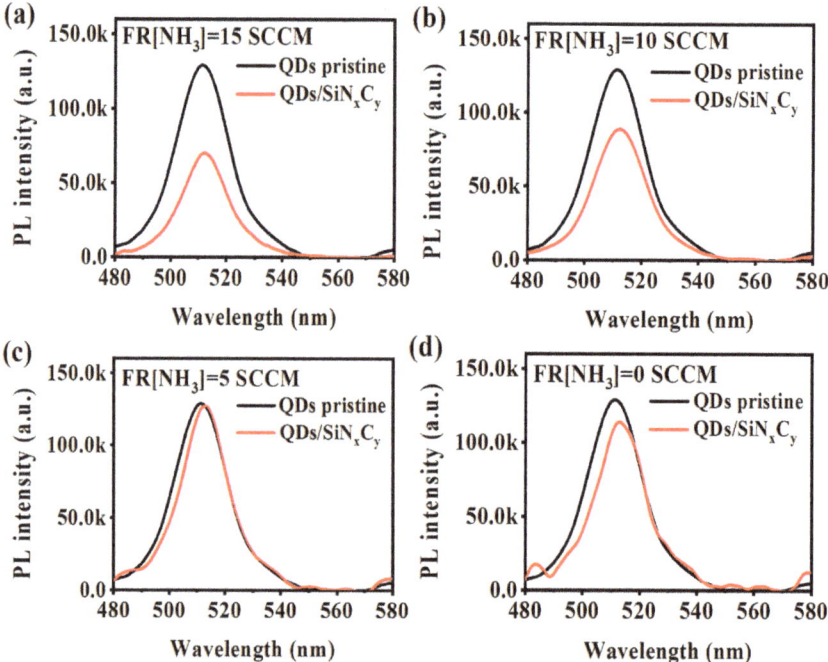

Figure 2. PL spectra acquired from the pristine $CsPbBr_3$ QD film and $CsPbBr_3$ QDs/a-SiC_xN_y:H nanocomposite thin films prepared with different NH_3 flow rates: (**a**) 15 SCCM; (**b**) 10 SCCM; (**c**) 5 SCCM; (**d**) 0 SCCM, respectively.

From Figure 2d, it can be observed that without the addition of ammonia, the luminescence intensity of the $CsPbBr_3$ QDs encapsulated with a-SiC_x:H was weaker than that of the uncoated $CsPbBr_3$ QDs. Additionally, the PL peak position of the SiC_x-encapsulated $CsPbBr_3$ QDs exhibited a redshift, and the full width at half maxima (FWHM) of the PL were narrower as compared to the uncoated $CsPbBr_3$ QDs. This phenomenon could possibly be attributed to the self-absorption effect of the a-SiC$_x$:H coating on the CsPbBr$_3$ QDs. To further test this hypothesis, the transmission spectra of a-SiC$_x$N$_y$:H films prepared using different NH$_3$ flows were measured, as depicted in Figure 3. As the NH$_3$ flow rate decreased from 15 SCCM to 0 SCCM, the absorption edge of the transmission spectrum progressively shifted towards the longer wavelength region. This suggests that the optical band gap of the film gradually decreased with the decrease of NH$_3$ flow rate. According to the formula [24]: $\alpha d = -\ln T$, where T is the transmittance and d is the film thickness, the absorption coefficient α of the film can be obtained. Thus, the Eopt can be calculated according to the Tauc equation $(\alpha h\nu)^{1/2} = B (h\nu - Eopt)$ [25], where α is the absorption coefficient and B is a constant. From the inset of Figure 3, it is evident that the E_{opt} of the a-SiC$_x$N$_y$:H decreased from 5.20 to 2.55 eV when the NH$_3$ flow rate was decreased from 15 to 0 SCCM. This variation indicates an evolution in the band structure of the a-SiC$_x$N$_y$:H attributed to the reduced N concentration as shown in the following Figure 5a. It is worth noting that the band gap value of the a-SiC$_x$:H film (2.55 eV) coincides with the short wave region of the luminescent peak of the CsPbBr$_3$ QDs at an NH$_3$ flow rate of 0, thereby explaining the PL decay due to the self-absorption effect of the a-SiC$_x$:H coating. It is therefore clear that the incorporation of N in the a-SiC$_x$:H coating plays a crucial role in increasing the optical band gap and avoiding PL decay caused by the self-absorption effect of the coating.

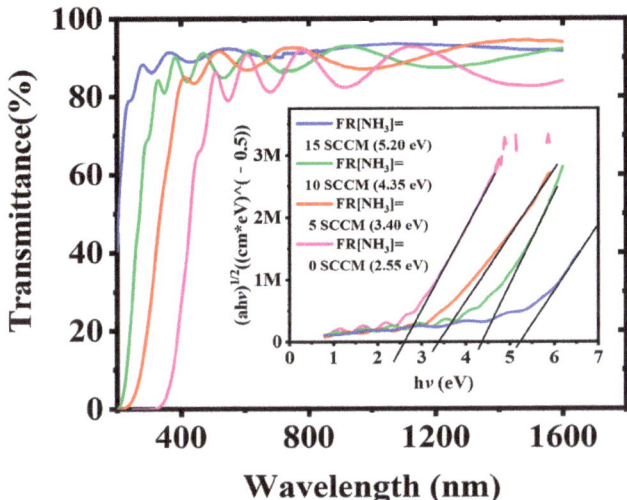

Figure 3. Transmission spectra of the a-SiC$_x$N$_y$:H films fabricated with different NH$_3$ flow rates. Inset shows the Tauc plots of the corresponding a-SiC$_x$N$_y$:H films.

To understand the PL characteristics, PL decay curves of the CsPbBr$_3$ QDs/a-SiC$_x$N$_y$:H nanocomposite thin films prepared with different NH$_3$ flow rates were measured under an excitation wavelength of 375 nm (375 nm, 70 ps excitation pulses LASER), as illustrated in Figure 4a. The PL decay curves were well fitted with a biexponential decay function [26–28]:

$$I(t) = I_0 + A_1 exp(\frac{-t}{\tau_1}) + A_2 exp(\frac{-t}{\tau_2}) \quad (1)$$

where I_0 is the background level; τ_1, τ_2 are the lifetime of each exponential decay component, and A_1, A_2 are the corresponding amplitudes, respectively. Thus, the average lifetime τ can be estimated as follows [27]: $\tau = (A_1 * \tau_1^2 + A_2 * \tau_2^2)/(A_1 * \tau_1 + A_2 * \tau_2)$.

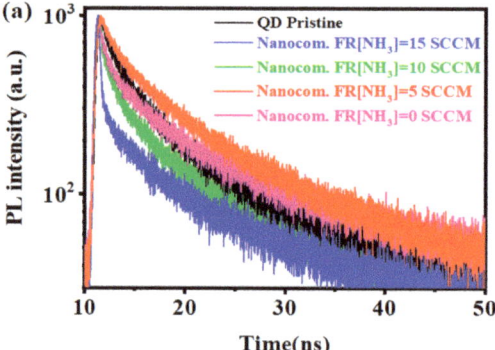

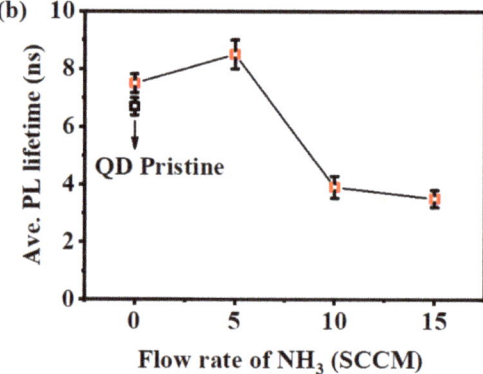

Figure 4. (a) PL decay traces and (b) lifetime of the pristine CsPbBr$_3$ QDs thin film and CsPbBr$_3$ QDs/a-SiC$_x$N$_y$:H nanocomposite thin films prepared with different NH$_3$ flow rates.

Figure 4b shows that the average lifetime (τ) increased gradually from 3.5 ns to 8.5 ns with the NH$_3$ flow rate decreasing from 15 SCCM to 5 SCCM. The increase in the average lifetime usually means that nonradiative decay is somewhat suppressed and more excitons tend to recombine along radiative paths [29], which is in good agreement with the improved PL shown in Figure 2. Therefore, the a-SiC$_x$N$_y$:H layers prepared with a low NH$_3$ flow rate are believed to effectively reduce the surface defect states of CsPbBr$_3$ QDs and thus achieve efficient photoluminescence. It seems that the PL intensity from CsPbBr$_3$ QDs is closely associated with the N content in the a-SiC$_x$N$_y$:H coatings.

The XPS spectra taken from the a-SiC$_x$N$_y$:H encapsulating layers were examined and the Si, N, and C contents in the films were estimated, as shown in Figure 5a. One can see that at the NH$_3$ flow rate of 15 SCCM, the contents of Si, N, and C in the a-SiC$_x$N$_y$:H encapsulating layer were around 48%, 43%, and 9%, respectively. By decreasing the NH$_3$ flow rate down to 5 SCCM, the N content sharply decreased to ~30%, while the Si and C contents rose to ~55% and ~15%, respectively. Without the addition of NH$_3$, the Si and C contents of the encapsulating layer were around 60% and 40%, respectively. To gain further insight into the a-SiC$_x$N$_y$:H encapsulating layers, we analyzed the Si 2p core level spectra, which is the symbol of the coexistence of different ionic states of Si atoms [30–32], as shown in Figure 5b. It is noted that the binding energy peaks are typically composed of various Si phases, namely SiN$_x$, SiC$_x$N$_y$, and SiC$_x$. At a high NH$_3$ flow rate of 15 SCCM, the a-SiC$_x$N$_y$:H encapsulating layer was dominated by the SiN$_x$ phase, which was attributed to the high content of nitrogen as illustrated in Figure 2a. With the decrease of nitrogen flow rate from 15 to 5 SCCM, the dominant phase in the encapsulating layer changed from the SiN_x phase into the SiC_xN_y phase. Without the addition of NH_3, encapsulating layers feature the SiC_x phase. Nitrogen plays a key role in the chemical bond reconstruction and the phase transformation in the a-SiC_xN_y:H encapsulating layer. Combined with the analysis of PL characteristics, it was found that the encapsulation layer with the SiC_xN_y phase was more beneficial to obtaining efficient PL than that with the SiN_x phase. As is illustrated in Figure 5b, the phase structure of nitride-rich a-SiC_xN_y:H film was dominated by the amorphous SiN_x phase, for which more N-active groups were generated in the preparation process. Excessive N-active precursors enhance the corrosion effect on the surface of $CsPbBr_3$ QDs, creating more surface defects, which in turn weaken the passivation effect of the encapsulation layer on $CsPbBr_3$ QDs. Accordingly, the PL intensity decreased significantly by more than 40% after the $CsPbBr_3$ QDs were covered by the encapsulating layer with the SiN_x phase. For the encapsulating layer with SiC_xN_y phase, fewer N active groups were produced in the preparation process, which weakened the corrosion effect of active groups on the surface of $CsPbBr_3$ QDs, helped to suppress the generation of non-radiative centers on the surface of $CsPbBr_3$ QDs, and thus enhanced the passivation effect of the encapsulation layer on $CsPbBr_3$ QDs. This is in line with the experimental phenomenon of the PL lifetime increasing with the decrease of nitrogen content. Therefore, the encapsulation layer with the SiC_xN_y phase is more favorable to obtaining efficient PL than that with the SiN_x phase.

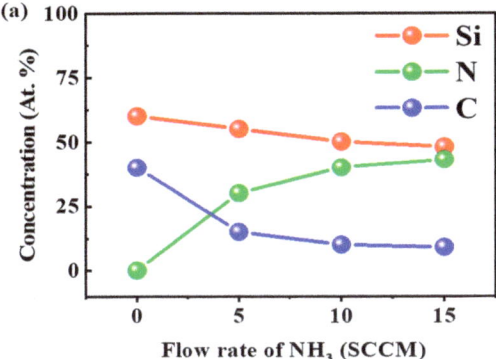

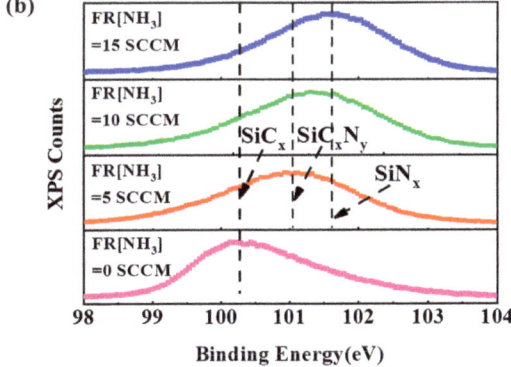

Figure 5. (a) The relative atomic concentration of Si, N, and C elements in a-SiC_xN_y:H encapsulating layers with different NH_3 flow rates. (b) Experimental Si 2p spectra for a-SiC_xN_y:H encapsulating layers fabricated at different NH_3 flow rates.

We assessed the stability of the CsPbBr$_3$ QDs encapsulated by the layers with SiC$_x$N$_y$ phase under various unfavorable environments. Figure 6a shows the change of PL intensity from CsPbBr$_3$ QDs/a-SiC$_x$N$_y$: H nanocomposites with the storage time in the air and under UV light illumination, respectively. After storing for more than 600 days in the air (humidity and temperature ranging from 40% to 70% and 15 to 30 °C, respectively), no significant PL intensity decline was observed from the CsPbBr$_3$ QDs/a-SiC$_x$N$_y$:H nanocomposites. They initially increased gradually and then stabilized after tripling. The enhanced PL was ascribed from the light irradiation during the PL measurement, which is referred to as photoactivation [22,33]. Similarly, the PL from CsPbBr$_3$ QDs/a-SiC$_x$N$_y$:H nanocomposites rapidly increased by more than 3 times after UV irradiation for 1 day. The enhanced PL remained stable during subsequent continuous UV illumination for at least 600 days. It is interesting that CsPbBr$_3$ QDs/a-SiC$_x$N$_y$:H nanocomposites, which were illuminated by continuous UV for at least 600 days, still showed strong green emissions even at the high temperature of 205 °C, as shown in Figure 6b. On the other hand, it is worth noting that the CsPbBr$_3$ QDs encapsulated by the layers with SiC$_x$ phase showed an obvious decline in PL after undergoing continuous UV light illumination for 2 months [23]. Evidently, the encapsulation layer with the SiC$_x$N$_y$ phase is more beneficial for obtaining stable and efficient PL than that with the SiC$_x$ phase.

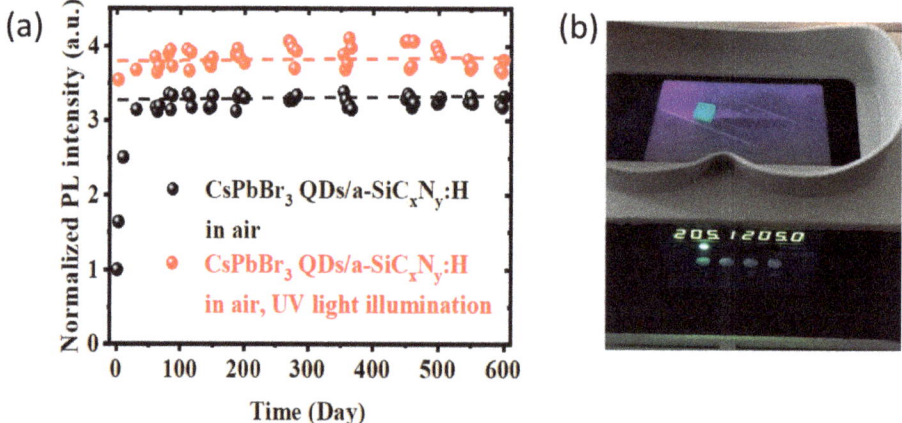

Figure 6. (a) PL intensity of the CsPbBr$_3$ QDs encapsulated by the layers with SiC$_x$N$_y$ phase versus time under ambient conditions and continuous UV (365 nm, 8 W) illumination time. (b) Photographs taken during UV illumination (λ = 365 nm, 8 W) at a high temperature of 205 °C.

To assess the thermal stability of CsPbBr$_3$ QDs/a-SiC$_x$N$_y$:H nanocomposite thin films, we monitored the integrated PL intensity as a function of temperature during thermal cycling, as shown in Figure 7. With increasing temperature from 25 to 205 °C, the PL was rapidly quenched. However, PL thermal quenching could be retrieved after the cooling process. Furthermore, the PL peak showed reversible modulation during the heating and cooling cycles. Compared to the irreversible PL deterioration of the original CsPbBr$_3$ QDs during thermal cycling [22], the thermal stability of CsPbBr$_3$ QDs/a-SiC$_x$N$_y$:H nanocomposite thin films was significantly improved, which can be ascribed to the constraint exerted by the CsPbBr$_3$ QDs/a-SiC$_x$N$_y$:H interface as demonstrated by our previous work [22]. These results indicate that the encapsulation layer with SiC$_x$N$_y$ phase not only provides nondestructive encapsulation for CsPbBr$_3$ quantum dots to achieve efficient PL, but also serves as a protective layer to realize the long-term stability of CsPbBr$_3$ QDs under harsh environment.

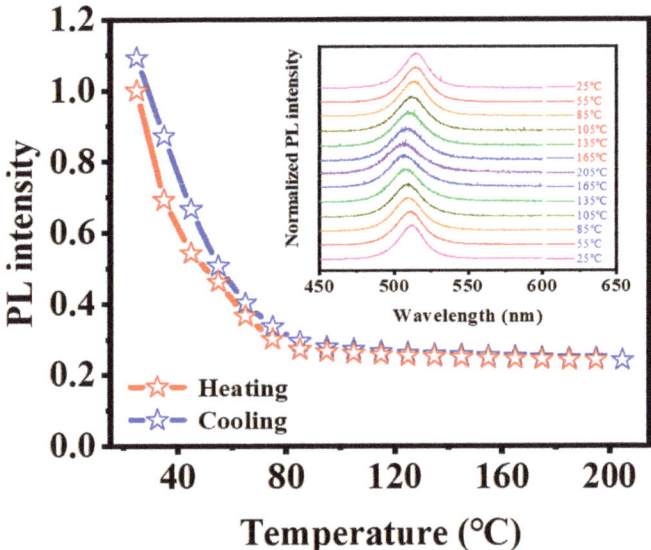

Figure 7. Heating and cooling cycling measurements of the CsPbBr$_3$ QDs encapsulated by the layers with SiC$_x$N$_y$ phase at various temperatures. The inset presents the PL spectra vs. temperature.

4. Conclusions

The effect of a-SiC$_x$N$_y$:H encapsulation on the stability and PL of CsPbBr$_3$ quantum dots were demonstrated. We found that a-SiC$_x$N$_y$:H encapsulation layer with a high N content of ~50% resulted in a serious PL degradation of CsPbBr$_3$ QDs. The PL degradation of CsPbBr$_3$ QDs can be significantly reduced by decreasing the N content in the a-SiC$_x$N$_y$:H encapsulation layer. With the N content decreasing from ~50% to ~26%, the dominant phase in the a-SiC$_x$N$_y$:H encapsulating layers transforms from the SiN$_x$ to the SiC$_x$N$_y$ states. The encapsulation layer with the SiC$_x$N$_y$ phase not only makes CsPbBr$_3$ QDs retain their inherent PL properties, but also makes CsPbBr$_3$ QDs have long-term stability when exposed to air, at a high temperature (205 °C), and under UV illumination for more than 600 days. Our results demonstrate an effective and practical approach to enhance the stability and PL characteristics of CsPbBr$_3$ QD thin films, with implications for the future development of optoelectronic devices.

Author Contributions: Z.L. (Zewen Lin): writing—review and editing, investigation, formal analysis. R.H.: writing—review and editing, formal analysis, funding acquisition. Z.L. (Zhenxu Lin): investigation, formal analysis. Y.G.: investigation, formal analysis. H.W.: investigation. J.S.: investigation. Y.Z.: investigation. W.Z.: formal analysis. H.L.: investigation. D.H.: investigation. All authors have read and agreed to the published version of the manuscript.

Funding: This research was funded by the Guangdong Basic and Applied Basic Research Foundation (2020A1515010432), Project of Guangdong Province Key Discipline Scientific Research Level Improvement (2021ZDJS039, 2022ZDJS067), Special Innovation Projects of Guangdong Provincial Department of Education (2019KTSCX096, 2020KTSCX076), Program of the Hanshan Normal University (QN202020), and Science and Technology Planning Projects of Chaozhou (2018GY18).

Data Availability Statement: Data underlying the results presented in this paper are not publicly available at this time but may be obtained from the authors upon reasonable request.

Conflicts of Interest: The authors declare no conflict of interest.

References

1. Protesescu, L.; Yakunin, S.; Bodnarchuk, M.I.; Krieg, F.; Caputo, R.; Hendon, C.H.; Yang, R.X.; Walsh, A.; Kovalenko, M.V. Nanocrystals of cesium lead halide perovskites (CsPbX$_3$, X = Cl, Br, and I): Novel optoelectronic materials showing bright emission with wide color gamut. *Nano Lett.* **2015**, *15*, 3692–3696. [CrossRef] [PubMed]
2. Kovalenko, M.V.; Protesescu, L.; Bodnarchuk, M.I. Properties and potential optoelectronic applications of lead halide perovskite nanocrystals. *Science* **2017**, *358*, 745–750. [CrossRef] [PubMed]
3. Lin, K.; Xing, J.; Quan, L.N.; de Arquer, F.P.; Gong, X.; Lu, J.; Xie, L.; Zhao, W.; Zhang, D.; Yan, C.; et al. Perovskite light-emitting diodes with external quantum efficiency exceeding 20 percent. *Nature* **2018**, *562*, 245–248. [CrossRef] [PubMed]
4. Liu, J.; Sheng, X.; Wu, Y.; Li, D.; Bao, J.; Ji, Y.; Lin, Z.; Xu, X.; Yu, L.; Xu, J.; et al. All-inorganic perovskite quantum dots/p-Si heterojunction light-emitting diodes under DC and AC driving modes. *Adv. Opt. Mater.* **2018**, *6*, 1700897. [CrossRef]
5. Han, T.H.; Tan, S.; Xue, J.; Meng, L.; Lee, J.K.; Yang, Y. Interface and defect engineering for metal halide perovskite optoelectronic devices. *Adv. Mater.* **2019**, *31*, 1803515. [CrossRef]
6. Zhang, Q.; Wang, B.; Zheng, W.; Kong, L.; Wan, Q.; Zhang, C.Y.; Li, Z.C.; Cao, X.Y.; Liu, M.M.; Li, L. Ceramic-like stable CsPbBr$_3$ nanocrystals encapsulated in silica derived from molecular sieve templates. *Nat. Commun.* **2020**, *11*, 31. [CrossRef]
7. Jiang, Y.; Sun, C.; Xu, J.; Li, S.; Cui, M.; Fu, X.; Liu, Y.; Liu, Y.; Wan, H.; Wei, K.; et al. Synthesis-on-substrate of quantum dot solids. *Nature* **2022**, *612*, 679–684. [CrossRef]
8. Zhu, Z.; Zhu, C.; Yang, L.; Chen, Q.; Zhang, L.; Dai, J.; Cao, J.; Zeng, S.; Wang, Z.; Wang, Z.; et al. Room-temperature epitaxial welding of 3D and 2D perovskites. *Nat. Mater.* **2022**, *21*, 1042–1049. [CrossRef]
9. Rainò, G.; Yazdani, N.; Boehme, S.C.; Kober-Czerny, M.; Zhu, C.; Krieg, F.; Rossell, M.D.; Erni, R.; Wood, V.; Infante, I.; et al. Ultra-narrow room-temperature emission from single CsPbBr$_3$ perovskite quantum dots. *Nat. Commun.* **2022**, *13*, 2587. [CrossRef]
10. Pan, J.; Shang, Y.; Yin, J.; Bastiani, M.D.; Peng, W.; Dursun, I.; Sinatra, L.; Zohry, A.M.E.; Hedhili, M.N.; Emwas, A.-H.; et al. Bidentate ligand-passivated CsPbI$_3$ perovskite nanocrystals for stable near-unity photoluminescence quantum yield and efficient red light-emitting diodes. *J. Am. Chem. Soc.* **2018**, *140*, 562–565. [CrossRef]
11. Wei, Y.; Cheng, Z.; Lin, J. An overview on enhancing the stability of lead halide perovskite quantum dots and their applications in phosphor-converted LEDs. *Chem. Soc. Rev.* **2019**, *48*, 310–350. [CrossRef] [PubMed]
12. Lv, W.; Li, L.; Xu, M.; Hong, J.X.; Tang, X.X.; Xu, L.G.; Wu, Y.H.; Zhu, R.; Chen, R.F.; Huang, W. Improving the stability of metal halide perovskite quantum dots by encapsulation. *Adv. Mater.* **2019**, *31*, 1900682. [CrossRef]
13. Liu, Y.; Chen, T.; Jin, Z.; Li, M.; Zhang, D.; Duan, L.; Zhao, Z.; Wang, C. Tough, stable and self-healing luminescent perovskite-polymer matrix applicable to all harsh aquatic environments. *Nat. Commun.* **2022**, *13*, 1338. [CrossRef] [PubMed]
14. Huang, S.; Li, Z.; Wang, B.; Zhu, N.; Zhang, C.; Kong, L.; Zhang, Q.; Shan, A.; Li, L. Morphology evolution and degradation of CsPbBr$_3$ nanocrystals under blue light-emitting diode illumination. *ACS Appl. Mater. Interfaces* **2017**, *9*, 7249–7258. [CrossRef] [PubMed]
15. Li, X.; Wang, Y.; Sun, H.; Zeng, H. Amino-mediated anchoring perovskite quantum dots for stable and low-threshold random lasing. *Adv. Mater.* **2017**, *29*, 1701185. [CrossRef]
16. Zou, S.; Liu, Y.; Li, J.; Liu, C.; Feng, R.; Jiang, F.; Li, Y.; Song, J.; Zeng, H.; Hong, M.; et al. Stabilizing cesium lead halide perovskite lattice through Mn(II) substitution for air-stable light-emitting diodes. *J. Am. Chem. Soc.* **2017**, *139*, 11443–11450. [CrossRef]
17. Kim, H.; Park, J.H.; Kim, K.; Lee, D.; Song, M.H.; Park, J. Highly emissive blue quantum dots with superior thermal stability via In situ surface reconstruction of mixed CsPbBr$_3$–Cs$_4$PbBr$_6$ nanocrystals. *Adv. Sci.* **2022**, *9*, 2104660.
18. Huang, H.; Chen, B.; Wang, Z.; Hung, T.F.; Susha, A.S.; Zhong, H.; Rogach, A.L. Water resistant CsPbX$_3$ nanocrystals coated with polyhedral oligomeric silsesquioxane and their use as solid state luminophores in all-perovskite white light-emitting devices. *Chem. Sci.* **2016**, *7*, 5699–5703. [CrossRef]
19. Hu, H.; Wu, L.; Tan, Y.; Zhong, Q.; Chen, M.; Qiu, Y.; Yang, D.; Sun, B.; Zhang, Q.; Yin, Y. Interfacial Synthesis of Highly Stable CsPbX$_3$/Oxide Janus Nanoparticles. *J. Am. Chem. Soc.* **2018**, *140*, 406–412. [CrossRef]
20. Loiudice, A.; Saris, S.; Oveisi, E.; Alexander, D.T.L.; Buonsanti, R. CsPbBr$_3$ QD/AlO$_x$ inorganic nanocomposites with exceptional stability in water, light, and heat. *Angew. Chem. Int. Ed.* **2017**, *56*, 10696–10701. [CrossRef]
21. Xu, L.; Chen, J.; Song, J.; Li, J.; Xue, J.; Dong, Y.; Cai, B.; Shan, Q.; Han, B.; Zeng, H. Double-protected all-inorganic perovskite nanocrystals by crystalline matrix and silica for triple-modal anti-counterfeiting codes. *ACS Appl. Mater. Interfaces* **2017**, *9*, 26556–26564. [CrossRef]
22. Lin, Z.; Huang, R.; Zhang, W.; Zhang, Y.; Song, J.; Li, H.; Hou, Z.; Guo, Y.; Song, C.; Wan, N.; et al. Highly luminescent and stable Si-based CsPbBr$_3$ quantum dot thin films prepared by glow discharge plasma with real-time and in situ diagnosis. *Adv. Funct. Mater.* **2018**, *28*, 1805214. [CrossRef]
23. Lin, Z.; Huang, R.; Song, J.; Guo, Y.; Lin, Z.; Zhang, Y.; Xia, L.; Zhang, W.; Li, H.; Song, C.; et al. Engineering CsPbBr$_3$ quantum dots with efficient luminescence and stability by damage-free encapsulation with a-SiC$_x$:H. *J. Lumin.* **2021**, *236*, 118086. [CrossRef]
24. Jha, H.S.; Yadav, A.; Singh, M.; Kumar, S.; Agarwal, P. Growth of wide-bandgap nanocrystalline silicon carbide films by HWCVD: Influence of filament temperature on structural and optoelectronic properties. *J. Electron. Mater.* **2015**, *44*, 922–928. [CrossRef]
25. Huang, R.; Xu, S.; Guo, Y.; Guo, W.; Wang, X.; Song, C.; Song, J.; Wang, L.; Ho, K.M.; Wang, N. Luminescence enhancement of ZnO-core/a-SiN$_x$:H-shell nanorod arrays. *Opt. Express* **2013**, *21*, 5891–5896. [CrossRef] [PubMed]
26. Lin, K.H.; Liou, S.C.; Chen, W.L.; Wu, C.L.; Lin, G.R.; Chang, Y.M. Tunable and stable UV-NIR photoluminescence from annealed SiO$_x$ with Si nanoparticles. *Opt. Express* **2013**, *21*, 23416–23424. [CrossRef] [PubMed]

27. Huang, R.; Lin, Z.; Guo, Y.; Song, C.; Wang, X.; Lin, H.; Xu, L.; Song, J.; Li, H. Bright red, orange-yellow and white switching photoluminescence from silicon oxynitride films with fast decay dynamics. *Opt. Mater. Express* **2014**, *4*, 205–212. [CrossRef]
28. Ma, Z.; Ji, X.; Wang, M.; Zhang, F.; Liu, Z.; Yang, D.; Jia, M.; Chen, X.; Wu, D.; Zhang, Y.; et al. Carbazole-containing polymer-assisted trap passivation and hole-injection promotion for efficient and stable $CsCu_2I_3$-based yellow LEDs. *Adv. Sci.* **2022**, *9*, 2202408. [CrossRef]
29. Li, J.; Xu, L.; Wang, T.; Song, J.; Chen, J.; Xue, J.; Dong, Y.; Cai, B.; Shan, Q.; Han, B.; et al. 50-fold EQE improvement up to 6.27% of solution-processed all-Inorganic perovskite $CsPbBr_3$ QLEDs via surface ligand density control. *Adv. Mater.* **2017**, *29*, 1603885. [CrossRef] [PubMed]
30. Singh, S.P.; Srivastava, P.; Ghosh, S.; Khan, S.A.; Oton, C.J.; Prakash, G.V. Phase evolution and photoluminescence in as-deposited amorphous silicon nitride films. *Scr. Mater.* **2010**, *63*, 605–608. [CrossRef]
31. Xu, Z.; Tao, K.; Jiang, S.; Jia, R.; Li, W.; Zhou, Y.; Jin, Z.; Liu, X. Application of polycrystalline silicon carbide thin films as the passivating contacts for silicon solar cells. *Sol. Energy Mater. Sol. Cells* **2020**, *206*, 110329. [CrossRef]
32. Lin, Z.; Huang, R.; Guo, Y.; Song, C.; Lin, Z.; Zhang, Y.; Wang, X.; Song, J.; Li, H.; Huang, X. Near-infrared light emission from Si-rich oxynitride nanostructures. *Opt. Mater. Express* **2014**, *4*, 816–822. [CrossRef]
33. Cordero, S.R.; Carson, P.J.; Estabrook, R.A.; Strouse, G.F.; Buratto, S.K. Photoactivated luminescence of CdSe quantum dot monolayers. *J. Phys. Chem. B* **2000**, *104*, 12137–12142. [CrossRef]

Disclaimer/Publisher's Note: The statements, opinions and data contained in all publications are solely those of the individual author(s) and contributor(s) and not of MDPI and/or the editor(s). MDPI and/or the editor(s) disclaim responsibility for any injury to people or property resulting from any ideas, methods, instructions or products referred to in the content.

Article

TiO₂/SnO₂ Bilayer Electron Transport Layer for High Efficiency Perovskite Solar Cells

Xiaolin Sun [1], Lu Li [2], Shanshan Shen [1] and Fang Wang [3,*]

[1] School of Aeronautical Engineering, Nanjing Vocational University of Industry Technology, Nanjing 210046, China
[2] School of Electrical Engineering, Nanjing Vocational University of Industry Technology, Nanjing 210046, China
[3] College of Electronic Engineering, Nanjing XiaoZhuang University, Nanjing 211100, China
* Correspondence: wangfang0182217@njxzc.edu.cn; Tel.: +86-153-4518-8763

Abstract: The electron transport layer (ETL) has been extensively investigated as one of the important components to construct high-performance perovskite solar cells (PSCs). Among them, inorganic semiconducting metal oxides such as titanium dioxide (TiO_2), and tin oxide (SnO_2) present great advantages in both fabrication and efficiency. However, the surface defects and uniformity are still concerns for high performance devices. Here, we demonstrated a bilayer ETL architecture PSC in which the ETL is composed of a chemical-bath-deposition-based TiO_2 thin layer and a spin-coating-based SnO_2 thin layer. Such a bilayer-structure ETL can not only produce a larger grain size of PSCs, but also provide a higher current density and a reduced hysteresis. Compared to the mono-ETL PCSs with a low efficiency of 16.16%, the bilayer ETL device features a higher efficiency of 17.64%, accomplished with an open-circuit voltage of 1.041 V, short-circuit current density of 22.58 mA/cm², and a filling factor of 75.0%, respectively. These results highlight the unique potential of TiO_2/SnO_2 combined bilayer ETL architecture, paving a new way to fabricate high-performance and low-hysteresis PSCs.

Keywords: perovskite solar cells; electron transport layer; low hysteresis; SnO_2; TiO_2

Citation: Sun, X.; Li, L.; Shen, S.; Wang, F. TiO₂/SnO₂ Bilayer Electron Transport Layer for High Efficiency Perovskite Solar Cells. *Nanomaterials* 2023, 13, 249. https://doi.org/10.3390/nano13020249

Academic Editor: Iván Mora-Seró

Received: 5 December 2022
Revised: 27 December 2022
Accepted: 30 December 2022
Published: 6 January 2023

Copyright: © 2023 by the authors. Licensee MDPI, Basel, Switzerland. This article is an open access article distributed under the terms and conditions of the Creative Commons Attribution (CC BY) license (https://creativecommons.org/licenses/by/4.0/).

1. Introduction

The high-efficiency, low-cost and facile fabrication process of halide perovskite solar cells (PSCs) have attracted tremendous attention in the field of photovoltaics in the past decade [1–5] and been regarded as the most promising substitute for traditional silicon (Si) and copper indium gallium selenide (CIGS) solar cells [6–8]. The sandwich structure of hybrid organic-inorganic based PSCs includes the electron transport layer (ETL), perovskite absorber layer, hole transport layer (HTL) and electrodes. Among them, ETL and HTL are used for the electron and hole extraction, respectively. However, the Spiro-OMeTAD are widely used as HTL in PSCs because of the simple synthesis, high carrier mobility and suitable valance band. The HTL are always fabricated by spin-coating on the top of a perovskite absorber layer with a dense and uniform film. In contrast, the ETL in PSCs is usually fabricated in a planar and/or mesoporous structure under the perovskite absorber layer [9–11]. The surface quality of ETL can substantially influence the deposition of perovskite film. Therefore, the electron transport layer and the corresponding interface of ETL/perovskite are significantly important parts to fabricate high-quality PSCs. Titanium dioxide (TiO_2) and/or tin oxide (SnO_2) thin films have been extensively investigated as an effective ETL in the PSCs, which can be fabricated by several different methods such as spin-coating, sputtering and chemical bath deposition (CBD) [12–16], to pursue a higher performance device.

Due to the facile planar configuration of PSCs, fabricating uniform, and compact ETL thin layer, it is imperative to pursue high performance. The conventional spin-coating method shows a facile and efficient way to fabricate the TiO_2-ETL. However, the uneven distribution of TiO_2 nanoparticles result in the carrier accumulation between perovskite (PVSK) and the ETL interface and an insufficient carrier extraction, leading to a low efficiency of resultant device [17,18]. Moreover, the large hysteresis of TiO_2-ETL also impedes the further application of TiO_2 in the PSCs [19]. Alternatively, SnO_2 presents a reduced hysteresis, high carrier mobility and good energy level towards perovskite, which can greatly improve the performance of PSCs [20–22]. For example, You et al. proposed SnO_2 as a planar ETL in the PSCs, which not only reduces the energy barrier between ETL/PVSK, but also reduces the hysteresis of devices, resulting in a high performance PSC with a champion PCE of 20.5% [21]. However, uniformity of SnO_2 nanoparticles is still a concern for the device fabrication because of its uneven distribution by spin-coating technique. Therefore, high-quality ETL plays a crucial role in the fabrication of devices, which paves a promising way for high-efficiency PSCs. To address this issue, Xu et al. introduced a bilayer ETL of TiO_2/ZnO thin layers into PSCs, which produces a compact interfacial layer to avoid direct contact between the FTO substrate and PVSK, leading to a reduced carrier accumulation at ETL/PVSK interface [23].

In this work, we propose a bilayer of ETLs that is composed of a CBD TiO_2 layer and a spin-coated SnO_2 layer. The presence of the SnO_2 thin layer on the top surface of CBD TiO_2 film can provide a higher current density and reduce the hysteresis of PSCs simultaneously. In addition, the diffusion of the K ion from SnO_2 can significantly improve the crystallinity of grains in the perovskite films. On the basis of this bilayer strategy, a higher power conversion efficiency (PCE) of 17.64% was achieved in comparison with the mono-TiO_2 ETL based PSCs with a PCE of 16.16%.

2. Materials and Methods

Materials: All reagents were used as received without further purification. Methylammonium iodide (MAI), methylammonium bromide (MABr), methylammonium chloride (MACl), formamidinium iodide (FAI), lead(II) iodide (PbI_2) and 2,2′,7,7′-tetrakis(N,N-di-p-methoxyphenylamine)9,9′-spirobifluorene (Spiro-OMeTAD) (99.5%) were purchased from Xi'an Polymer Light Technology (Xi'an, China). Dimethylformamide (DMF), dimethyl sulfoxide (DMSO), isopropanol (IPA), chlorobenzene (CB), and titanium tetrachloride ($TiCl_4$) were purchased from Sigma-Aldrich (Milwaukee, Germany).

Device Fabrication: The cleaned fluorine-doped tin oxide (FTO) substrates are treated using UV-ozone for 60 min. Then, the TiO_2 thin layer was prepared by using the CBD method and the SnO_2 thin layer was fabricated with spin-coating technologies, as shown in Figure 1. First, 2 M aqueous $TiCl_4$ mother solution was prepared by dropping $TiCl_4$ into distilled water. During the preparation, the mother solution was continuously stirred at a low temperature of around 0 °C. The as-prepared $TiCl_4$ mother solution was stored in the refrigerator (<10 °C). Second, the as-prepared $TiCl_4$ mother solution was diluted to a 0.2 M $TiCl_4$ solution. The cleaned FTO substrates were placed vertically in the glassware. Then, 300 mL of 0.2 M $TiCl_4$ solution was poured into the glassware. The glassware was put into an oven with a temperature of 75 °C. After 1 h heating, the glassware was taken out followed by rinsing the FTO substrates several times using distilled water. Finally, the FTO substrates were annealed at a high temperature of 450 °C for 30 min. The FTO substrates were washed by the acetone, distilled water, and ethyl alcohol for 20 min, respectively. Before the deposition of TiO_2 thin films, the FTO substrates are treated by using UV-ozone for 60 min. SnO_2 films were prepared by spin-coating Alfa Aesar SnO_2 (diluted by H_2O to 3%) at a speed of 3500 rpm for 30 s. The perovskite films were deposited by a two-step spin-coating method. Specifically, 1.35 M PbI_2 and 0.0675 M CsI were dissolved in organic solvent (DMF/DMSO = 19:1). The PbI_2 precursor solution was stirred at a temperature of 70 °C for 60 min. The mixed MAFA based organic cation precursor solution was prepared by dissolving 200 mg FAI, 100 mg MAI, 25 mg MABr and 25 mg MACl dissolved in 5 mL isopropanol. The PbI_2 precursor solution was first spin-coated at a speed of 3000 rpm for 30 s. The MA/FA cation solution was spin-coated at 3000 rpm for 30s. After annealing at

150 °C for 10 min, the perovskite film of $Cs_{0.05}FA_{0.54}MA_{0.41}Pb(I_{0.98}Br_{0.02})_3$ was obtained. The hole transport layer of the spiro-OMeTAD film was deposited by spin-coating the spiro-OMeTAD solution at a speed of 3500 rpm for 25 s. Finally, 80 nm Au film was deposited as a counter electrode by thermal evaporation.

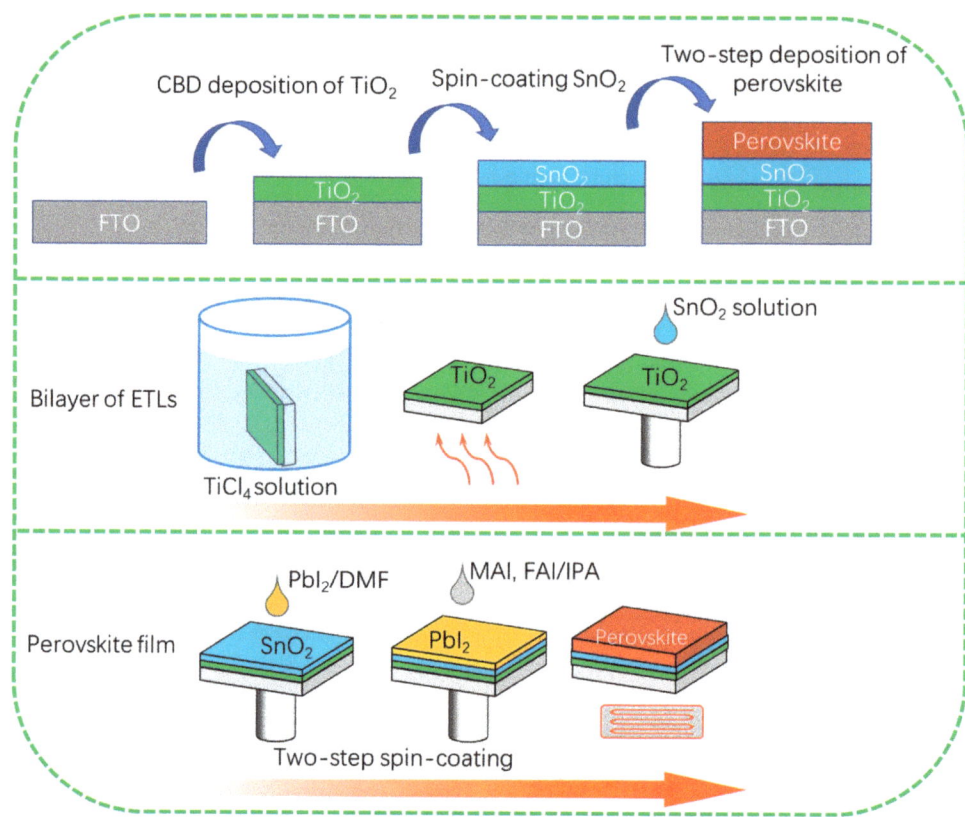

Figure 1. Schematic illustration of the bilayer of ETLs (TiO_2 and SnO_2 films) and perovskite films fabricated by chemical bath deposition and spin-coating.

Device Characterization: The diffraction data of perovskites are collected by using a Bruker D8 Discover diffractometer (Bruker AXS) from 10° to 60°. Surface and cross-section morphology images are recorded by a scanning electron microscope (SEM) (Helios NanoLab G3). The TRPL results were collected by using the Hamamatsu equipment which can provide an excitation wavelength of 450 nm. The photoluminescence (PL) spectra were acquired by a JASCO FP-8500 spectrometer with an excitation wavelength of 450 nm. The current-voltage (J-V) measurements were performed under one sun illumination (AM1.5G, 100 mW/cm^2) by using a Keithley 2420. The devices were test by using a metal shadow mask with a dimension of 0.3 × 0.3 cm^2. The EQE spectra of the devices were characterized by using Oriel IQE 200 equipment.

3. Results and Discussion

Figure 2a–d shows the top-view SEM images of the perovskite films fabricated on the TiO_2 and TiO_2/SnO_2 substrates, which clearly shows a larger grain size of perovskite thin film based on the TiO_2/SnO_2 substrates, compared with that on the TiO_2 substrates, with an average value changing from ~380 nm to ~540 nm, which can be verified by the statistics of perovskite grain size based on the TiO_2 and TiO_2/SnO_2 substrates, as presented in Figure 2e,f. As is well-known, the commercial SnO_2 colloid precursor is stabilized by incorporating potassium hydroxide (KOH) [24]. The presence of K ion in the SnO_2 will diffuse into the perovskite thin film during the annealing process, which greatly enhances the crystallinity of perovskite grains, and reduces the hysteresis of resultant devices [25–27].

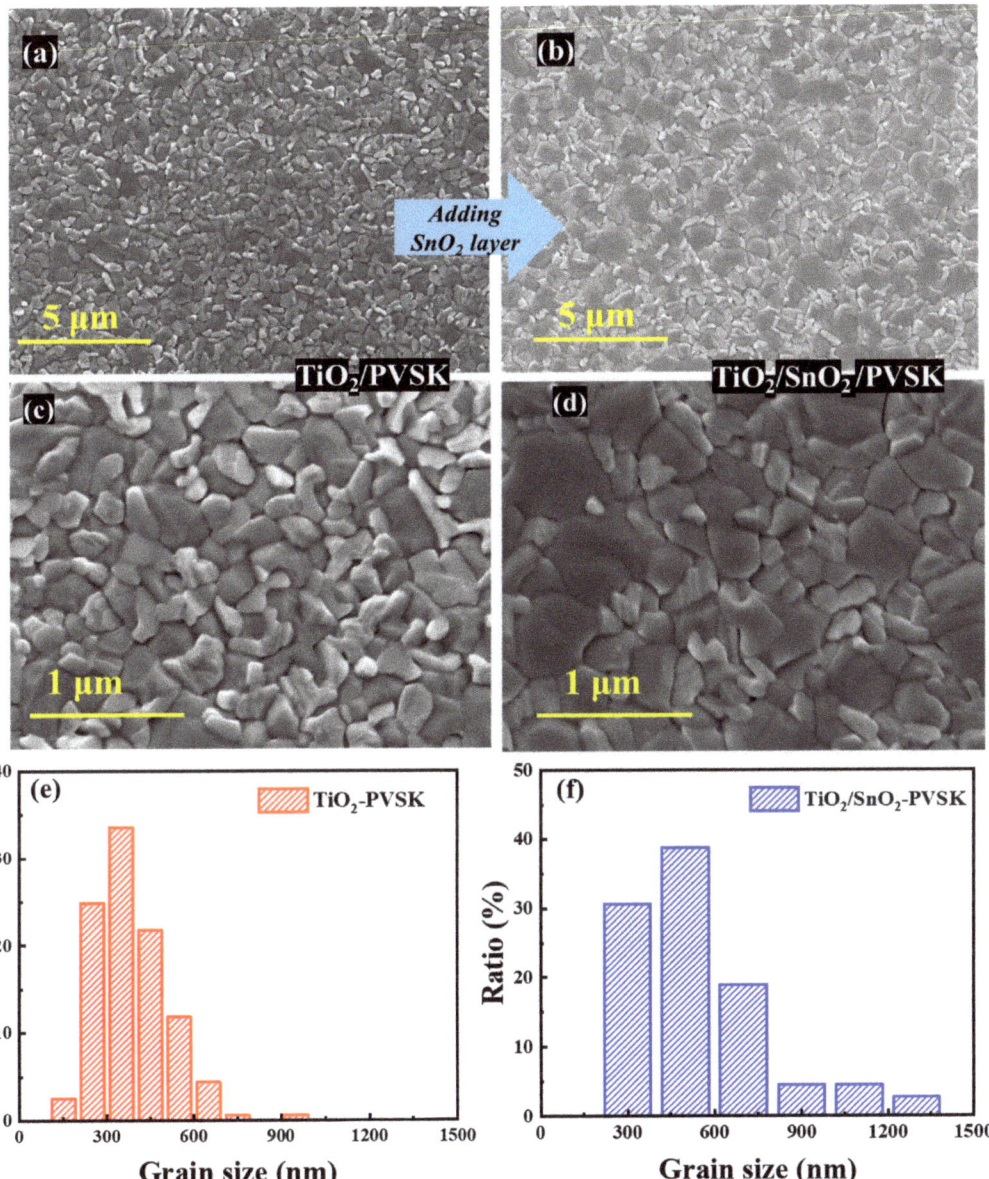

Figure 2. (**a**) Top-view SEM images of monolayer TiO_2-ETL PSCs and bilayer TiO_2/SnO_2-ETL PSCs (**a–d**), and their corresponding statistic of grain size (**e,f**).

Furthermore, the phase structure of perovskite thin film deposited on the TiO$_2$ and TiO$_2$/SnO$_2$ substrates was investigated by X-ray diffraction (XRD), as presented in Figure 3a. The increase of XRD intensity (on the TiO$_2$/SnO$_2$ substrate) verifies that the improved crystallinity of perovskites is accomplished with high absorption in a short-wavelength region (as shown in Figure 3b).

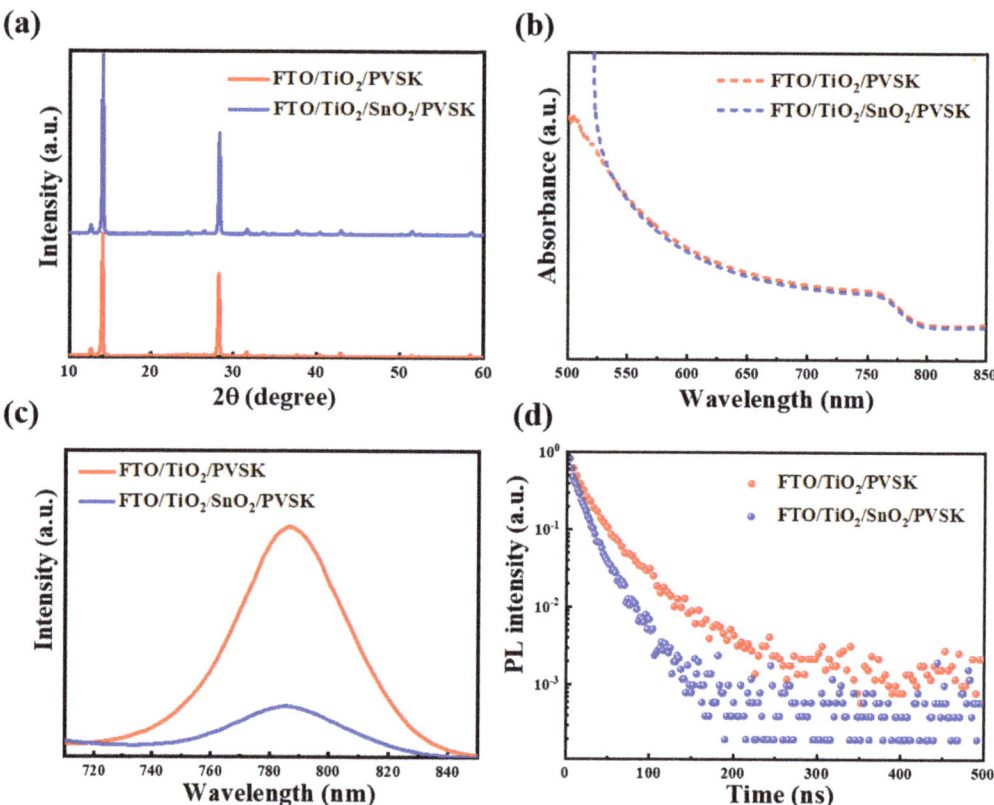

Figure 3. (a) XRD patterns, (b) absorption spectra, (c) PL spectra and (d) TRPL curves of the perovskite films deposited on FTO/TiO$_2$ and FTO/TiO$_2$/SnO$_2$ substrates.

In addition, the steady-state photoluminescence (PL) and time-resolved photoluminescence (TRPL) experiments were carried out to investigate carrier transport behavior. As seen in Figure 3c,d, the faster PL quenching of the perovskite thin film on the TiO$_2$/SnO$_2$ substrate indicates an enhanced electron extraction capability [28]. Moreover, the lifetimes of the corresponding perovskite thin films were fitted by a biexponential decay function [29,30]. The lifetime of the TiO$_2$/SnO$_2$-based sample is 15.4 ns, which is shorter than that of the TiO$_2$-based sample (22.2 ns), indicating a faster carrier extraction from the perovskite thin film to TiO$_2$/SnO$_2$ electron transport layer [31].

Figure 4a,b shows the cross-section SEM images of devices fabricated on TiO$_2$ and TiO$_2$/SnO$_2$ substrates. The uniform and dense perovskite absorber layers not only ensure the light harvest, but also effectively impede the carrier recombination in the devices. The current density-voltage (J-V) curves of the devices were measured under standard AM 1.5 G illumination and are shown in Figure 5a and Table 1, while the key performance parameters of open-circuit voltage (V_{OC}), short-circuit current (J_{SC}), fill factor (FF), power conversion efficiency (PCE) and their statistical analyses are displayed in Figure 6a–d and Table 2, respectively. The PCE of 16.16% (V_{OC} = 1.012 V, J_{SC} = 22.06 mA/cm^2 and FF = 72.4%) and 10.37% (V_{OC} = 0.905 V, J_{SC} = 22.06 mA/cm^2 and FF = 51.9%) under reverse scan (RS) and forward scan (FS) indicate large hysteresis in the TiO$_2$-based devices. In contrast, the high PCE of 17.64% (V_{OC} = 1.041 V, J_{SC} = 22.58 mA/cm^2 and FF = 75.0%) and 15.29% (V_{OC} = 1.001 V, J_{SC} = 22.73 mA/cm^2 and FF = 67.2%) under RS and FS were obtained for TiO$_2$/SnO$_2$-based solar cells. The improved efficiency of TiO$_2$/SnO$_2$-based solar cells can be attributed to a higher crystallinity of perovskite grains, which enhances light capture and reduces the defects at grain boundaries [14,25]. The EQE spectra of the corresponding devices were presented in Figure 5b. The improved EQE in the short wavelength in terms of TiO$_2$/SnO$_2$-based device indicates faster carrier extraction and reduced recombination at the TiO$_2$/SnO$_2$/PVSK interface [32]. Similarly, the enhanced EQE at the long wavelength region also suggests that reduced defects and carrier recombination in the perovskite bulk film, which can be explained by the enlarged grain size and improved crystallinity of the perovskite grains [32]. As a result, the integrated J_{SC} from EQE of the TiO$_2$/SnO$_2$-based device is 21.59 mA/cm^2, which is higher than that of the TiO$_2$ based device (21.17 mA/cm^2). Furthermore, the TiO$_2$/SnO$_2$-based device exhibited a stable output (under initial maximum power point (MPP) voltage) with a PCE of 17.65%. In contrast, the TiO$_2$ based solar cell shows a poor output under MPP, yielding a low PCE of 15.74% (Figure 5c). More importantly, the hysteresis (hysteresis index (HI) = PCE$_{RS}$/PCE$_{FS}$) of the TiO$_2$/SnO$_2$-based devices is also reduced, compared to TiO$_2$-based devices [32–34]. The HI of the TiO$_2$-based PSC is 1.56, which is decreased to 1.15 by incorporating SnO$_2$ into devices to construct the TiO$_2$/SnO$_2$ bilayer ETL. Compared to TiO$_2$ based devices with a large hysteresis of 1.51, the improved efficiency and reduced HI of 1.18 for TiO$_2$/SnO$_2$-based PSCs indicates the bilayer ETL can improve the reproducible fabrication and the device performance.

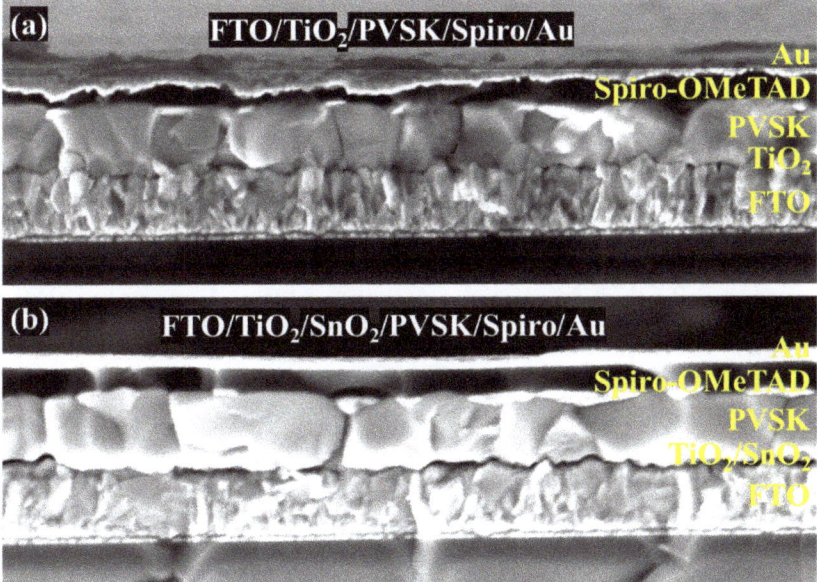

Figure 4. Cross-section images of PSCs in (**a**) TiO$_2$-PSCs and (**b**) TiO$_2$/SnO$_2$ PSCs.

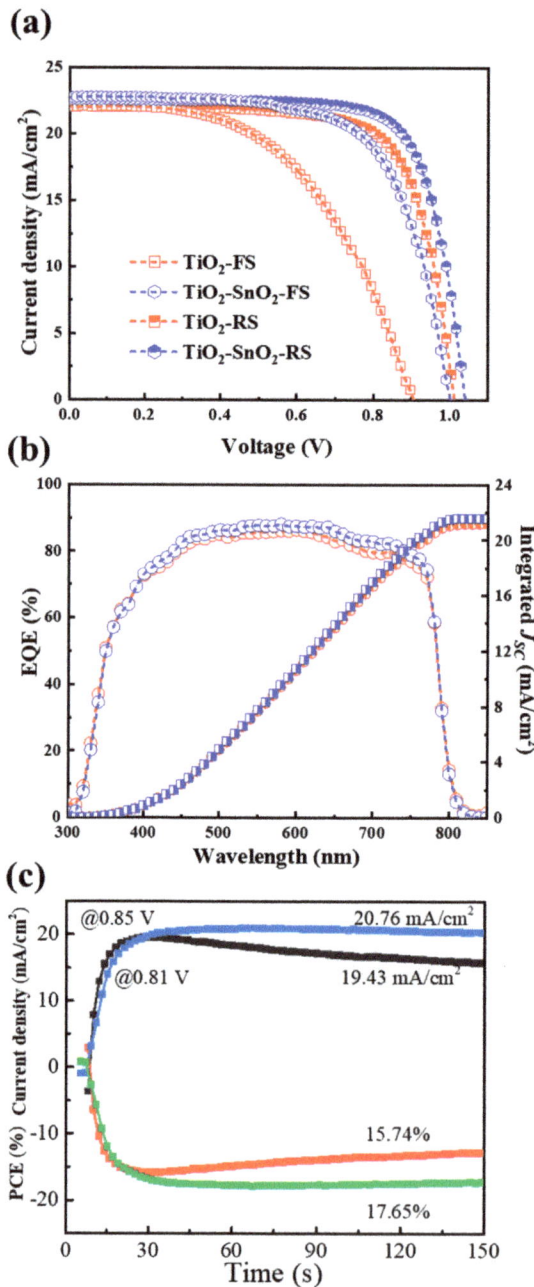

Figure 5. (**a**) Measured current density-voltage curves of champion devices. (**b**) Corresponding EQE spectra and their integrated current density. (**c**) Stable output of perovskite solar cells based on based on TiO$_2$ and TiO$_2$/SnO$_2$ ETLs.

Table 1. Photovoltaics parameters of PSCs based on TiO_2 and TiO_2/SnO_2 ETLs.

Sample	Scan Direction	V_{oc} (V)	J_{sc} (mA/cm^2)	FF	PCE (%)	HI
TiO_2/SnO_2 PSC	FS.	1.001	22.73	0.672	15.29	1.18
	RS.	1.041	22.58	0.750	17.64	
TiO_2-PSCs	FS.	0.905	22.06	0.519	10.37	1.51
	RS.	1.012	22.06	0.724	16.16	

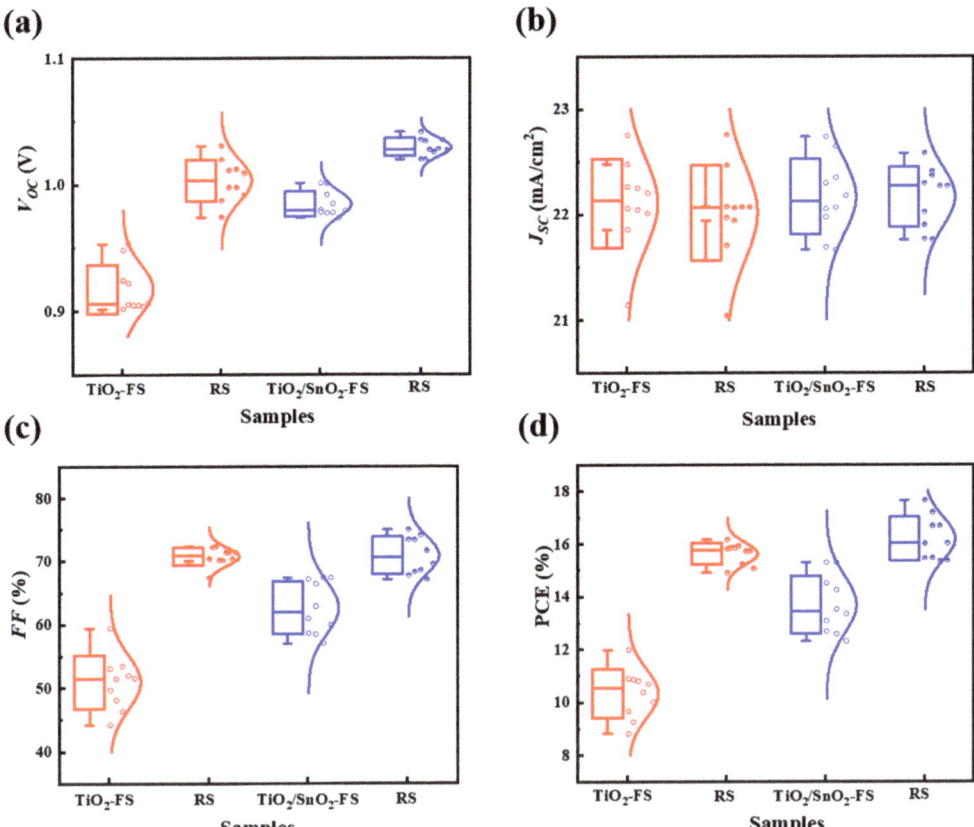

Figure 6. Statistical distribution of TiO_2 and TiO_2/SnO_2 based PSCs (10 devices). (a) V_{OC}, (b) J_{SC}, (c) FF and (d) PCE.

Table 2. Average photovoltaics parameters of PSCs based on TiO_2 and TiO_2/SnO_2 ETLs.

Sample	Scan Direction	V_{oc} (V)	J_{sc} (mA/cm^2)	FF	PCE (%)
TiO_2/SnO_2 PSC	FS.	0.985 ± 0.010	22.17 ± 0.34	0.626 ± 0.039	13.68 ± 1.04
	RS.	1.029 ± 0.007	22.16 ± 0.27	0.709 ± 0.028	16.18 ± 0.78
TiO_2-PSCs	FS.	0.917 ± 0.018	22.11 ± 0.40	0.5090 ± 0.040	10.33 ± 0.88
	RS.	1.003 ± 0.016	22.01 ± 0.43	0.707 ± 0.014	15.61 ± 0.40

4. Conclusions

In summary, we developed a bilayer electron transport layer by combining CBD-TiO_2 and spin-coated SnO_2 in the perovskite solar cells. The TiO_2/SnO_2 bilayer ETLs provide not only a compact electron transport layer, but also accelerate the carrier transport in the solar cells. Furthermore, the presence of K ion from SnO_2 can greatly improve the crystallinity of perovskite thin film and significantly reduce the hysteresis of resultant devices. Compared with the TiO_2-based solar cells, the TiO_2/SnO_2-based solar cells demonstrate a higher PCE of 17.64% and a lower hysteresis index. These results highlight the potential fabrication of the TiO_2/SnO_2 bilayer electron transport layers and will be a beneficial strategy to fabricate a high-quality perovskite thin film solar cell.

Author Contributions: Conceptualization, X.S. and L.L.; methodology, X.S. and F.W.; software, X.S.; validation, X.S. and S.S.; formal analysis, X.S. and F.W.; investigation, X.S. and L.L.; resources, X.S.; data curation, S.S.; writing—original draft preparation, X.S.; writing—review and editing, L.L. and S.S.; project administration, F.W.; funding acquisition, X.S. and F.W. All authors have read and agreed to the published version of the manuscript.

Funding: This research was funded by the High-level Scientific Research Foundation for the introduction of talent of Nanjing Vocational University of Industry Technology under Nos. YK20-03-03 and YK20-03-02. The 2021 High-end training of professional leaders of teachers in higher vocational colleges in Jiangsu Province under No. 2021GRGDYX009. The National Natural Science Foundation of China under No. 32101535. The Jiangsu Postdoctoral Research Foundation under No. 2021K112B. The National Natural Science Foundation of China under No. 62205144. The Qing Lan project of Jiangsu Universities under No. QL078.

Data Availability Statement: The data is available on reasonable request from the corresponding author.

Conflicts of Interest: The authors declare no conflict of interest. The funders had no role in the design of the study; in the collection, analyses, or interpretation of data; in the writing of the manuscript; or in the decision to publish the results.

References

1. Huang, H.-H.; Liu, Q.-H.; Tsai, H.; Shrestha, S.; Su, L.-Y.; Chen, P.-T.; Chen, Y.-T.; Yang, T.-A.; Lu, H.; Chuang, C.-H.; et al. A Simple One-Step Method with Wide Processing Window for High-Quality Perovskite Mini-module Fabrication. *Joule* **2021**, *5*, 958–974. [CrossRef]
2. Tong, G.; Li, H.; Li, D.; Zhu, Z.; Xu, E.; Li, G.; Yu, L.; Xu, J.; Jiang, Y. Dual-phase CsPbBr3-CsPb2Br5 Perovskite Thin Films via Vapour Deposition for High-performance Rigid and Flexible Photodetectors. *Small* **2018**, *14*, 1702523–1702530. [CrossRef]
3. Chao, L.; Niu, T.; Gao, W.; Ran, C.; Song, L.; Chen, Y.; Huang, W. Solvent Engineering of the Precursor Solution toward Large-Area Production of Perovskite Solar Cells. *Adv. Mater.* **2021**, *33*, 2005410. [CrossRef] [PubMed]
4. Yoo, J.; Seo, G.; Chua, M.; Park, T.; Lu, Y.; Rotermund, F.; Kim, Y.-K.; Moon, C.; Jeon, N.; Correa-Baena, J.-P.; et al. Efficient Perovskite Solar Cells via Improved Carrier Management. *Nature* **2021**, *590*, 587–593. [CrossRef]
5. Yang, H.; Xu, E.; Wu, C.; Li, J.; Liu, B.; Hong, F.; Zhang, L.; Chang, Y.; Zhang, Y.; Tong, G.; et al. Bifunctional Interface Engineering by Oxidating Layered TiSe2 for High-Performance CsPbBr3 Solar Cells. *ACS Appl. Energy Mater.* **2022**, *5*, 8254–8261. [CrossRef]
6. Tong, G.; Son, D.; Ono, L.; Liu, Y.; Hu, Y.; Zhang, H.; Jamshaid, A.; Qiu, L.; Liu, Z.; Qi, Y. Scalable Fabrication of >90 cm^2 Perovskite Solar Modules with 1000 h Operational Stability Based on the Intermediate Phase Strategy. *Adv. Energy Mater.* **2021**, *11*, 2003712. [CrossRef]
7. Werner, J.; Boyd, C.C.; Moot, T.; Wolf, E.J.; France, R.M.; Johnson, S.A.; van Hest, M.F.A.M.; Luther, J.M.; Zhu, K.; Berry, J.J.; et al. Learning from Existing Photovoltaic Technologies to Identify Alternative Perovskite Module Designs. *Energy Environ. Sci.* **2020**, *13*, 3393. [CrossRef]
8. Liu, Z.; Qiu, L.; Ono, L.; He, S.; Hu, Z.; Jiang, M.; Tong, G.; Wu, Z.; Jiang, Y.; Son, D.-Y.; et al. A Holistic Approach to Interface Stabilization for Efficient Perovskite Solar Modules with over 2,000-hour Operational Stability. *Nat. Energy* **2020**, *5*, 596–604. [CrossRef]
9. Wu, T.; Liu, X.; Luo, X.; Segawa, H.; Tong, G.; Zhang, Y.; Ono, L.K.; Qi, Y.B.; Han, L. Heterogeneous FASnI3 Absorber with Enhanced Electric Field for High-Performance Lead-Free Perovskite Solar Cells. *Nano-Micro Lett.* **2022**, *14*, 99. [CrossRef]
10. Jeon, N.; Na, H.; Jung, E.; Yang, T.-Y.; Lee, Y.; Kim, G.; Shin, H.-W.; Seok, S.I.; Lee, J.; Seo, J. A Fluorene-Terminated Hole-transporting Material for Highly Efficient and Stable Perovskite Solar Cells. *Nat. Energy* **2018**, *3*, 682–689. [CrossRef]
11. Lin, L.; Jones, T.; Wang, J.; Cook, A.; Pham, N.; Duffy, N.; Mihaylov, B.; Grigore, M.; Anderson, K.; Duck, B.; et al. Strategically Constructed Bilayer Tin (IV) Oxide as Electron Transport Layer Boosts Performance and Reduces Hysteresis in Perovskite Solar Cells. *Small* **2020**, *16*, 1901466. [CrossRef] [PubMed]
12. Chen, T.; Tong, G.; Xu, E.; Li, H.; Li, P.; Zhu, Z.; Tang, J.; Qi, Y.B.; Jiang, Y. Accelerating Hole Extraction by Inserting 2D Ti3C2-MXene Interlayer to All Inorganic Perovskite Solar Cells with Long-Term Stability. *J. Mater. Chem. A* **2019**, *7*, 20597–20603. [CrossRef]
13. Wu, W.-Q.; Chen, D.; Caruso, R.; Cheng, Y.-B. Recent Progress in Hybrid Perovskite Solar Cells Based on N-type Materials. *J. Mater. Chem. A* **2017**, *5*, 10092–10109. [CrossRef]

14. Tong, G.; Ono, L.; Liu, Y.; Zhang, H.; Bu, T.; Qi, Y. Up-Scalable Fabrication of SnO_2 with Multifunctional Interface for High Performance Perovskite Solar Modules. *Nano-Micro Lett.* **2021**, *13*, 155. [CrossRef]
15. Paik, M.; Lee, Y.; Yun, H.; Lee, S.; Hong, S.; Seok, S. TiO_2 Colloid-Spray Coated Electron-Transporting Layers for Efficient Perovskite Solar Cells. *Adv. Energy Mater.* **2020**, *10*, 2001799. [CrossRef]
16. Li, H.; Tong, G.; Chen, T.; Zhu, H.; Li, G.; Chang, Y.; Wang, L.; Jiang, Y. Interface Engineering Using Perovskite Derivative-Phase for Efficient and Stable $CsPbBr_3$-Solar Cells. *J. Mater. Chem. A* **2018**, *6*, 14225. [CrossRef]
17. Wang, P.; Li, R.; Chen, B.; Hou, F.; Zhang, J.; Zhao, Y.; Zhang, X. Gradient Energy Alignment Engineering for Planar Perovskite Solar Cells with Efficiency Over 23%. *Adv. Mater.* **2020**, *32*, 1905766. [CrossRef] [PubMed]
18. Wojciechowski, K.; Stranks, S.; Abate, A.; Sadoughi, G.; Sadhanala, A.; Kopidakis, N.; Rumbles, G.; Li, C.-Z.; Friend, R.; Jen, A.-Y.; et al. Heterojunction Modification for Highly Efficient Organic–Inorganic Perovskite Solar Cells. *ACS Nano* **2014**, *8*, 12701–12709. [CrossRef]
19. Shin, S.; Yeom, E.; Yang, W.; Hur, S.; Kim, M.; Im, J.; Seo, J.; Noh, J.; Seok, S. Colloidally Prepared La-doped $BaSnO_3$ Electrodes for Efficient, Photostable Perovskite Solar Cells. *Science* **2017**, *356*, 167–171. [CrossRef] [PubMed]
20. Jiang, Q.; Zhang, X.; You, J. SnO_2: A Wonderful Electron Transport Layer for Perovskite Solar Cells. *Small* **2018**, *14*, 1801154. [CrossRef]
21. Jiang, Q.; Zhang, L.; Wang, H.; Yang, X.; Meng, J.; Liu, H.; Yin, Z.; Wu, J.; Zhang, X.; You, J. Enhanced Electron Extraction Using SnO_2 for High-Efficiency Planar-Structure $HC(NH_2)_2PbI_3$-Based Perovskite Solar Cells. *Nat. Energy* **2016**, *2*, 16177. [CrossRef]
22. Deng, K.; Chen, Q.; Li, L. Modification Engineering in SnO_2 Electron Transport Layer toward Perovskite Solar Cells: Efficiency and Stability. *Adv. Funct. Mater.* **2020**, *30*, 2004209. [CrossRef]
23. Xu, X.; Zhang, H.; Shi, J.; Dong, J.; Luo, Y.; Li, D.; Meng, Q. Highly Efficient Planar Perovskite Solar Cells with a TiO_2/ZnO Electron Transport Bilayer. *J. Mater. Chem. A* **2015**, *3*, 19288–19293. [CrossRef]
24. Bu, T.; Li, J.; Zheng, F.; Chen, W.; Wen, X.; Ku, Z.; Peng, Y.; Zhong, J.; Cheng, Y.; Huang, F. Universal Passivation Strategy to Slot-Die Printed SnO_2 for Hysteresis-Free Efficient Flexible Perovskite Solar Module. *Nat. Commun.* **2018**, *9*, 4609. [CrossRef]
25. Zhu, P.; Gu, S.; Luo, X.; Gao, Y.; Li, S.; Zhu, J.; Tan, H. Simultaneous Contact and Grain-Boundary Passivation in Planar Perovskite Solar Cells Using SnO_2-KCl Composite Electron Transport Layer. *Adv. Energy Mater.* **2019**, *10*, 1903083. [CrossRef]
26. Bu, T.; Li, J.; Li, H.; Tian, C.; Su, J.; Tong, G.; Ono, L.K.; Wang, C.; Lin, Z.; Chai, N.; et al. Lead Halide-Templated Crystallization of Methylamine-Free Perovskite for Efficient Photovoltaic Modules. *Science* **2021**, *378*, 1327–1332. [CrossRef] [PubMed]
27. Bi, H.; Liu, B.; He, D.; Bai, L.; Wang, W.; Zang, Z.; Chen, J. Interfacial Defect Passivation and Stress Release by Multifunctional KPF_6 Modification for Planar Perovskite Solar Cells with Enhanced Efficiency and Stability. *Chem. Eng. J.* **2021**, *418*, 129375. [CrossRef]
28. Liu, Z.; Deng, K.; Hu, J.; Li, L. Coagulated SnO_2 Colloids for High-Performance Planar Perovskite Solar Cells with Negligible Hysteresis and Improved Stability. *Angew. Chem.* **2019**, *58*, 11497–11504. [CrossRef]
29. Tong, G.; Jiang, M.; Son, D.; Ono, L.; Qi, Y. 2D Derivative Phase Induced Growth of 3D All Inorganic Perovskite Micro–Nanowire Array Based Photodetectors. *Adv. Funct. Mater.* **2020**, *30*, 2002526. [CrossRef]
30. Tong, G.; Chen, T.; Li, H.; Qiu, L.; Liu, Z.; Dang, Y.; Song, W.; Ono, L.K.; Jiang, Y.; Qi, Y.B. Phase Transition Induced Recrystallization and Low Surface Potential Barrier Leading to 10.91%-Efficient $CsPbBr_3$ Perovskite Solar Cells. *Nano Energy* **2018**, *65*, 536–542. [CrossRef]
31. Wang, Z.; Wu, T.; Xiao, L.; Qin, P.; Yu, X.; Ma, L.; Xiong, L.; Li, H.; Chen, X.; Wang, Z.; et al. Multifunctional Potassium Hexafluorophosphate Passivate Interface Defects for High Efficiency Perovskite Solar Cells. *Power Sources* **2021**, *488*, 229451. [CrossRef]
32. Tong, G.; Son, D.-Y.; Ono, L.; Kang, H.-B.; He, S.; Qiu, L.; Zhang, H.; Liu, Y.; Hieulle, J.; Qi, Y. Removal of Residual Compositions by Powder Engineering for High Efficiency Formamidinium-Based Perovskite Solar Cells with Operation Lifetime over 2000 h. *Nano Energy* **2021**, *87*, 106152. [CrossRef]
33. Domanski, K.; Alharbi, E.; Hagfeldt, A.; Grätzel, M.; Tress, W. Systematic Investigation of the Impact of Operation Conditions on the Degradation Behaviour of Perovskite Solar Cells. *Nat. Energy* **2018**, *3*, 61–67. [CrossRef]
34. Habisreutinger, S.; Noel, N.; Snaith, H. Hysteresis Index: A Figure without Merit for Quantifying Hysteresis in Perovskite Solar Cells. *ACS Energy Lett.* **2018**, *3*, 2472–2476. [CrossRef]

Disclaimer/Publisher's Note: The statements, opinions and data contained in all publications are solely those of the individual author(s) and contributor(s) and not of MDPI and/or the editor(s). MDPI and/or the editor(s) disclaim responsibility for any injury to people or property resulting from any ideas, methods, instructions or products referred to in the content.

Article

Nanocrystalline ZnSnN₂ Prepared by Reactive Sputtering, Its Schottky Diodes and Heterojunction Solar Cells

Fan Ye *, Rui-Tuo Hong †, Yi-Bin Qiu †, Yi-Zhu Xie, Dong-Ping Zhang, Ping Fan and Xing-Min Cai *

Key Laboratory of Optoelectronic Devices and Systems of Ministry of Education and Guangdong Province, and Shenzhen Key Laboratory of Advanced Thin Films and Applications, College of Physics and Optoelectronic Engineering, Shenzhen University, Shenzhen 518060, China
* Correspondence: yefan@szu.edu.cn (F.Y.); caixm@szu.edu.cn (X.-M.C.)
† These authors contributed equally to this work.

Abstract: AbstractZnSnN$_2$ has potential applications in photocatalysis and photovoltaics. However, the difficulty in preparing nondegenerate ZnSnN$_2$ hinders its device application. Here, the preparation of low-electron-density nanocrystalline ZnSnN$_2$ and its device application are demonstrated. Nanocrystalline ZnSnN$_2$ was prepared with reactive sputtering. Nanocrystalline ZnSnN$_2$ with an electron density of approximately 10^{17} cm^{-3} can be obtained after annealing at 300 °C. Nanocrystalline ZnSnN$_2$ is found to form Schottky contact with Ag. Both the current I vs. voltage V curves and the capacitance C vs. voltage V curves of these samples follow the related theories of crystalline semiconductors due to the limited long-range order provided by the crystallites with sizes of 2–10 nm. The $I-V$ curves together with the nonlinear $C^{-2}-V$ curves imply that there are interface states at the Ag-nanocrystalline ZnSnN$_2$ interface. The application of nanocrystalline ZnSnN$_2$ to heterojunction solar cells is also demonstrated.

Keywords: ZnSnN$_2$; nanocrystalline; Schottky diode; heterojunction; solar cell

Citation: Ye, F.; Hong, R.-T.; Qiu, Y.-B.; Xie, Y.-Z.; Zhang, D.-P.; Fan, P.; Cai, X.-M. Nanocrystalline ZnSnN$_2$ Prepared by Reactive Sputtering, Its Schottky Diodes and Heterojunction Solar Cells. *Nanomaterials* **2023**, *13*, 178. https://doi.org/10.3390/nano13010178

Academic Editor: Shunri Oda

Received: 14 November 2022
Revised: 15 December 2022
Accepted: 26 December 2022
Published: 30 December 2022

Copyright: © 2022 by the authors. Licensee MDPI, Basel, Switzerland. This article is an open access article distributed under the terms and conditions of the Creative Commons Attribution (CC BY) license (https:// creativecommons.org/licenses/by/ 4.0/).

1. Introduction

The family of Zn-IV-N$_2$ (IV = Si, Ge and Sn) is an analogue of the III nitrides [1]. As a member of this family, ZnSnN$_2$ has many advantages, which include a direct band gap, earth-abundance of constituent elements, no toxicity, a high absorption coefficient, a low fabrication cost, etc. [1–6]. These make ZnSnN$_2$ a highly competent candidate in photocatalytic, photovoltaic and light-emitting applications. Recent theoretical work shows that the photon-to-electron conversion efficiency of ZnSnN$_2$/CuCrO$_2$ heterojunction solar cells is approximately 22% [2].

ZnSnN$_2$ has been fabricated by various methods such as sputtering [3–14], molecular beam epitaxy (MBE) [15], high pressure metathesis reaction [16] and vapor-liquid-solid growth [17]. During the preparation of ZnSnN$_2$, Sn atoms can easily occupy the positions of Zn atoms, and this results in the formation of the intrinsic antisite defect Sn$_{Zn}$ which is a divalent donor and which almost has the lowest formation energy in ZnSnN$_2$ [18,19]. Therefore, the antisite defect Sn$_{Zn}$ is considered to be the major donor in most cases, and the prepared ZnSnN$_2$ usually has a very high density of approximately 10^{19} cm^{-3}. Preparing ZnSnN$_2$ with an electron density of approximately or below 10^{17} cm^{-3} is still challenging due to the facile formation of the donor defect Sn$_{Zn}$.

Compared with its polycrystalline, crystalline or microcrystalline counterparts [3–17], nanocrystalline ZnSnN$_2$ whose crystalline grains or crystallites are smaller than 10 nm has never been studied. In this paper, we approach the high electron density problem of ZnSnN$_2$ with nanocrystallization. We show that nanocrystalline ZnSnN$_2$ with crystallites of approximately 2 nm in the amorphous matrix can be deposited by sputtering an alloy target of Zn and Sn (the Zn/Sn atomic ratio to be 3:1) in an atmosphere of N$_2$ and Ar. Annealing at 300 °C makes the reactively deposited ZnSnN$_2$ turn *n*-type conductive with an electron density of approximately 10^{17} cm^{-3} and the annealed ZnSnN$_2$ remains nanocrystalline with a grain size of 2–10 nm. The device applications, such as Ag-ZnSnN$_2$ Schottky diodes and Cu$_2$O-ZnSnN$_2$ heterojunction solar cells, are studied.

2. Materials and Methods

2.1. Preparation and Characterization of ZnSnN$_2$

Reactive radio frequency (RF) magnetron sputtering was used to deposit the samples. The substrates were K9 glasses. The substrates were ultrasonically cleaned in acetone, ethanol and deionized water, and the washing time in each liquid was 15 min. The target was the alloy of Zn (99.99%) and Sn (99.99%), with the Zn/Sn atomic ratio to be 3:1, and the gases were Ar (99.99%) and N$_2$ (99.999%). The flow rate of Ar was 20 standard cubic centimeters per min (sccm), and that of N$_2$ was 6 sccm. The base pressure of the sputtering chamber was 5.4×10^{-4} Pa. During sputtering, the chamber pressure was 5 Pa and the RF sputtering power was 35 W. The substrates rotated with the substrate holder at 0.6π rad/s. Before film deposition, the target was sputtered for 5 min to clean the surface. The time for film deposition was 60 min. Some deposited samples were annealed for 1 h in a vacuum chamber in the flow of Ar (20 sccm) and at a pressure of 5 Pa. The annealing temperatures were 300 °C, 350 °C, 400 °C and 450 °C.

The films without annealing or annealed at different temperatures were characterized with X-ray diffraction (XRD, D/max 2500 PC, 18 kW, Cu Kα radiation) and transmission electron microscopy (TEM, FEI Titan Cubed Themis G2 300). The samples for TEM observation were prepared by using the mechanical stripping method. The film thickness was measured with the surface profiler (Veeco Dektak 3ST), and the surface of the samples was characterized with scanning electron microscopy (SEM, Supra55 Sapphire). The samples were also characterized with atomic force microscopy (AFM, Oxford Instrument, MFP-3D Infinity, noncontact mode), room temperature Hall effect (HL 5500 PC, Van der Pauw electrode was used, and melted alloy of In and Sn was dropped at the four corners of a square sample to act as the electrodes) as well as X-ray photoelectron spectroscopy (XPS, Thermo Fisher, Thermo escalab 250Xi, Al Kα = 1486.6 eV; The C 1 s was calibrated to be 284.8 eV; the calculation of the atomic ratio is elaborated elsewhere [14]).The Seebeck coefficient S_e of the samples was measured with $\Delta V / \Delta T$ where ΔV is the voltage difference (ΔV is obtained with a voltmeter (Keithley 2400)) and ΔT is the temperature difference between the hot side and the cold side.

2.2. Preparation and Characterization of the Schottky Diodes and Heterojunctions

The steps to prepare the Schottky diodes with the structure of Ag\ITO\ZnSnN$_2$\Ag (ITO refers to indium tin oxide, and its resistivity is approximately 6.22×10^{-4} $\Omega \cdot$cm) are as follows (Supplementary Materials Figure S1): ZnSnN$_2$ was first deposited on ITO (whose substrate is glass) without annealing, and the thickness was 95 nm. During film deposition, some surface area of the ITO was covered with a mask in order to produce electrodes later. Secondly, a silver paste of 0.1 cm × 0.1 cm was dropped on ZnSnN$_2$ and the surface of ITO to act as electrodes. To compare the effect of annealing, ZnSnN$_2$ deposited on ITO under the same conditions was annealed for 1 h at 300 °C in the flow of Ar before making the electrodes. During annealing, the vacuum pressure was also 5 Pa, and then the second step of dropping silver paste was repeated to produce electrodes for another Schottky diode.

To prepare Cu$_2$O-ZnSnN$_2$ heterojunctions (Supplementary Materials Figure S1), Cu$_2$O was first deposited on ITO with reactive DC sputtering (part of the ITO was also masked for making an electrical connection later). The sputtering voltage and current were 289 V and 35 mA. The target was copper (99.999%), and the gases were Ar (30 sccm) and O$_2$ (1.8 sccm). The work pressure was 0.6 Pa, and the substrate temperature was 400 °C. The sputtering time was 10 h, and the thickness of Cu$_2$O was 1672 nm (to obtain the electrical properties of Cu$_2$O, Cu$_2$O was also deposited on K9 glass, and Hall measurements showed that Cu$_2$O is p-type conductive with a hole density of 1.02×10^{15} cm^{-3}). After Cu$_2$O cooled to room temperature naturally, ZnSnN$_2$ was deposited on Cu$_2$O and the parameters were the same as those mentioned previously except that the sputtering time was 3 h here (the thickness of ZnSnN$_2$ was 236 nm here). Silver paste was dropped on the surface of ITO and ZnSnN$_2$ to produce one solar cell with the structure of Ag\ITO\Cu$_2$O\ZnSnN$_2$\Ag. Additionally, with silver paste dropped on the surface of ITO to produce an electrode, Au was evaporated on ZnSnN$_2$ to produce another solar cell with the structure of Ag\ITO\Cu$_2$O\ZnSnN$_2$\Au. The base pressure for evaporating Au was 5×10^{-4} Pa. Au wire (99.999%) was heated first with a current of 75 A for 10 min and then at a current of 90 A for 5 min. The solar cells had an area of 0.04 cm^2.

The current I-voltage V curves of the devices were measured at room temperature with a source measurement unit instrument (Keithley Standard Series 2400). The capacitance C-voltage V curves of the devices were measured at room temperature with a semiconductor characterization system (Keithley 4200-SCS), and the frequency used was 10 kHz. For the Schottky diodes, the forward-bias is defined as the state at which the Ag electrode, in direct contact with the ZnSnN$_2$ layer, is connected with the anode of the voltage. The solar cells are forward-biased when the electrostatic potential of Cu$_2$O is higher than that of ZnSnN$_2$. The current density-voltage V curves of the two Cu$_2$O-ZnSnN$_2$ heterojunction solar cells were measured at room temperature with a multi-meter (Keithley, 2400 Series) under AM1.5 light illumination (which is from an AAA solar simulator with the intensity calibrated to 100 mW/cm^2 through a Si reference cell).

3. Results and Discussion

The XRD patterns of the samples deposited at 35 W and room temperature (the substrates were not intentionally heated) without annealing and those deposited at the same conditions but annealed at 300–450 °C are presented in Figure 1a. No diffraction peaks were observed in the XRD patterns of these samples. The samples deposited at 35 W and room temperature without annealing were amorphous, and they were still amorphous after annealing at 300–450 °C. In our equipment, when the sputtering power and other parameters were kept unchanged, the samples deposited at the substrate temperature of 100–300 °C became polycrystalline, and the preferred orientation changed when the substrate temperature was over 200 °C (Figure 1b, also Supplementary Materials Figure S2). However, when the sputtering power is 50 W, even the samples deposited at room temperature are polycrystalline (Figure 1c; also Supplementary Materials Figure S3). This implies that relatively higher sputtering power favors crystallization, and this agrees with F. Alnjiman et al., who showed that polycrystalline ZnSnN$_2$ can be deposited without heating the substrate intentionally [7]. Though the samples deposited at the substrate temperature of 100–300 °C were polycrystalline, the samples deposited at room temperature and then annealed at 300 °C or above after deposition remained amorphous, and post-deposition annealing failed to make the amorphous samples turn to polycrystalline. This is different from amorphous ZnSnN$_2$ prepared by direct current sputtering, which turns into polycrystalline after annealing [12]. Therefore, amorphous ZnSnN$_2$ is relatively stable since it remains amorphous even if the post-deposition annealing temperature is higher than 300 °C. It is reported that reducing the Zn/Sn ratio to a certain degree leads to microcrystalline and amorphous ZnSnN$_2$ [13], while in our work, reducing the substrate temperature to fabricate amorphous ZnSnN$_2$ was used. Later, we will show that both the ZnSnN$_2$ samples without annealing and the ZnSnN$_2$ samples annealed at 300 °C are actually nanocrystalline with crystallites of 2–10 nm. The Raman scattering spectra of these samples show that the obtained ZnSnN$_2$ is phonon-glass-like (Supplementary Materials Figure S4).

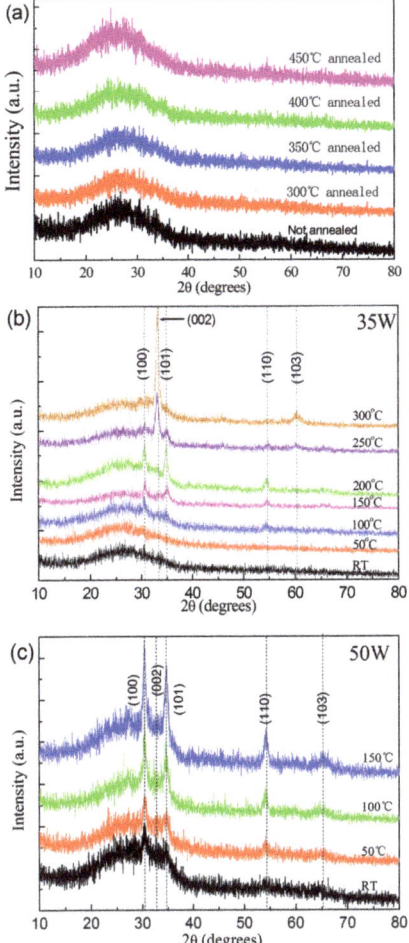

Figure 1. (a) XRD patterns of the samples without annealing or annealed at 300–450 °C; (b) XRD patterns of the samples deposited at 35 W and under different substrate temperatures; (c) XRD patterns of the samples deposited at 50 W and under different substrate temperatures.

The transmittance (T_r) and reflectance (R) spectra of the samples were measured with an UV/VIS spectrophotometer. The absorption coefficient α is equal to $t^{-1}\ln[(1-R)T_r^{-1}]$, where t is the thickness of the film (cm). The thickness of the films before and after annealing is approximately 95 nm. For semiconductors with a direct band gap, the relation between the absorption coefficient α and the optical band gap E_g^{opt} is as follows:

$$(\alpha h\nu)^2 = A(h\nu - E_g^{opt}), \tag{1}$$

where A is a constant, h is the Plank constant and ν is the photon frequency. The optical bandgap E_g^{opt} can be obtained by extrapolating the linear region to intercept the $h\nu$ axis in the plot of $(\alpha h\nu)^2$ vs. $h\nu$. With the assumption that nanocrystalline ZnSnN$_2$ has a direct band gap, the optical bandgap E_g^{opt} of all the samples can be obtained, as shown in Figure 2. The optical bandgap of nanocrystalline ZnSnN$_2$ is approximately 2.75 eV, which is larger than the experimental band gap of crystalline or polycrystalline ZnSnN$_2$

(0.94–2.38 eV [15,20]), and this is in agreement with the fact that the band gap of amorphous Si is larger than its crystalline counterparts. The optical band gaps of the samples annealed at 300 °C, 350 °C, 400 °C and 450 °C were 2.40 eV, 2.35 eV, 2.11 eV and 2.15 eV, respectively. In amorphous and microcrystalline ZnSnN$_2$ fabricated by reducing the Zn/Sn ratio, the optical bandgap is also approximately 2.53–2.59 eV [13], and our work agrees with this. It has been found that annealing reduces the band gap.

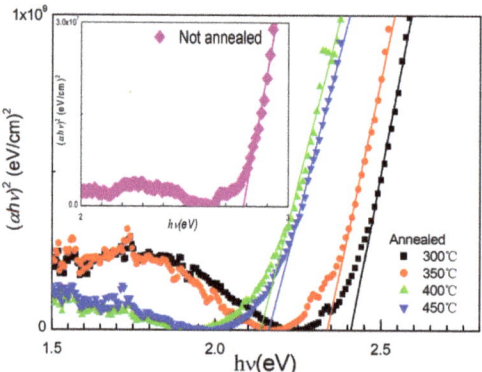

Figure 2. Tauc plots of ZnSnN$_2$ without annealing or annealed at 300–450 °C.

Hall effect measurements were conducted, and the results of the samples are listed in Table 1. The samples without annealing are almost insulating, and the resistivity is beyond the measurement scope of the Hall effect instrument. The samples without annealing are still n-type conductive since their Seebeck coefficients S_e were measured to be negative [21], and no reversal in the conduction type is observed here. After being annealed at 300–450 °C, the samples turn conductive. The electron density n of the samples annealed at 300 °C is 5.54×10^{17} cm^{-3} and this is much lower than N_c, the room temperature density of the states of the conduction band of ZnSnN$_2$ (1.04×10^{18} cm^{-3}) [4,14]. The gap between the conduction band minimum E_c and the Fermi level E_F was calculated to be 0.0165 eV with

$$n = N_c \exp[-(E_c - E_F)/(kT)], \quad (2)$$

where k is the Boltzmann constant and T is the absolute temperature (300 K here). The electron density increases to ~10^{18} cm^{-3} when the annealing temperature is 350, 400 or 450 °C. The mobility is similar to that of the amorphous and microcrystalline samples deposited by reducing the Zn/Sn ratio [13].

Table 1. The mobility μ (cm^{2}V^{-1}s^{-1}) and carrier concentration n (cm^{-3}) of the samples annealed at 300–450 (°C) (The resistivity of the samples without annealing is beyond the scope of the instrument).

°C	μ (cm^{2}V^{-1}s^{-1})	n (cm^{-3})
300	0.988	5.34×10^{17}
350	5.54	8.37×10^{18}
400	2.96	6.48×10^{18}
450	1.22	8.11×10^{18}

To reveal the microscopic change of the structure, the samples without annealing and the samples annealed at 300 °C were studied with TEM (Figure 3). Crystallites with a size slightly larger than 2 nm can be observed, and no diffraction patterns are observed in the samples without annealing (Figure 3a). The fact that no diffraction patterns are observed possibly results from the fact that the density of the crystallites or the volume ratio of the crystalline component to the amorphous component is small (Figure 3a). After being annealed at 300 °C, crystallites with bigger sizes, larger densities and diffraction spots are observed in the high-resolution image and diffraction patterns, and the sizes of the crystallites are 5–10 nm (Figure 3b). Annealing is found to increase the size of crystallites. In amorphous Si with crystallites smaller than 20 Å, a reversal in the conduction type can be observed in the Hall effect measurement [22,23]. The fact that no reversal in the conduction type is observed in the Hall effect measurement of our samples very possibly results from the nanocrystalline grains, which provide limited long-range order [22].

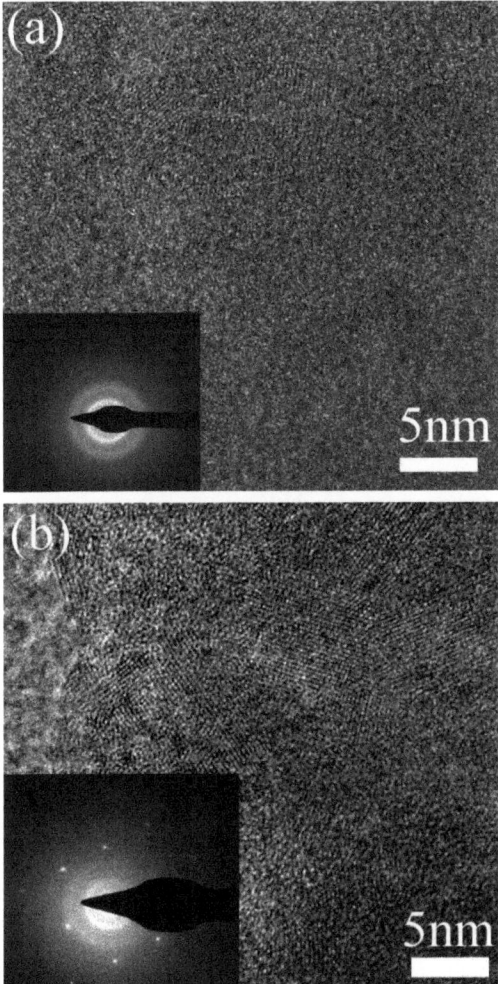

Figure 3. TEM images and the diffraction patterns of ZnSnN$_2$. (a) Without annealing; (b) annealed at 300 °C.

The AFM images of the samples fabricated at 50 °C, 100 °C, 250 °C and 300 °C and at 35 W are presented in Figure 4. The root mean square (RMS) roughness of the samples deposited at 50 °C, 100 °C, 250 °C and 300 °C is 0.77 nm, 1.27 nm, 0.92 nm and 2.35 nm, respectively. The amorphous samples deposited at 50 °C are found to have the lowest RMS roughness. The morphology of polycrystalline samples is similar to those obtained by others [24].

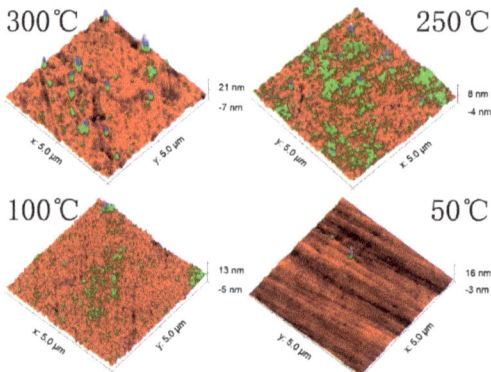

Figure 4. AFM images of the ZnSnN$_2$ samples deposited at 35 W under different substrate temperatures.

The samples were measured with XPS. A survey in 0–1400 eV shows that the samples contain Zn, Sn and N, together with adventitious O. The high-resolution binding energy spectra of Zn, Sn, N and O of the samples without annealing and those annealed at 300 °C are presented in Figure 5a,b.

For the ZnSnN$_2$ samples without annealing, Zn 2p3/2 and Zn2p1/2 are at 1021.38 and 1044.45 eV, while for the ZnSnN$_2$ samples annealed at 300 °C, these peaks shift to 1021.57 and 1044.64 eV. Therefore, Zn is at +2 [24]. The peak at approximately 498 eV present in the Sn spectrum is due to the Zn L3M45M45 Auger peak [4,25]. Sn is at +4 since Sn 3d5/2 and 3d3/2 are at 486.11 and 494.52 before annealing, and these peaks shift to 486.18 and 494.58 after annealing [13,24]. In the high resolution XPS spectrum of N 1s, a peak at approximately 403 eV is observed and it is from absorbed γ-type nitrogen molecules labeled as γ-N$_2$ (N ≡ N). The possible origin of γ-N$_2$ is nitrogen molecules, which are not decomposed but absorbed directly during sputtering [26,27]. The peak below 400 eV can be deconvoluted into two peaks. In the samples without annealing, the two N 1 s peaks are at 396.31 and 397.80, and they shift to 396.90 and 398.35 eV after annealing. The lower energy peak with a much stronger intensity at 396.31 and 396.90 eV implies nitrogen is at −3, while the weaker higher energy peak at 397.80 and 398.35 eV is from organic nitrogen [7,13].

The O 1s peak can also be decomposed into two peaks. The peaks of the samples without annealing are at 529.90 and 531.18 eV, and after annealing, the peaks shift to 529.99 and 530.58 eV. The exact origin of the oxygen present in ZnSnN$_2$ is still under debate. Some researchers think that the oxygen present in ZnSnN$_2$ is completely from post-deposition adsorption in air [7]. In our work, we think that the oxygen has two possible origins: the low energy peak is possibly from residual oxygen in the sputtering chamber, which was incorporated during sputtering and which substitutes nitrogen in the lattice [13]; the high energy peak is due to post-deposition adsorption in air.

The atomic ratio of Zn to Zn + Sn is 0.80 before annealing and 0.78 after annealing. In the literature, ZnSnN$_2$ samples with Zn/(Zn + Sn) = 0.72 and an electron density of 2.7×10^{17} cm^{-3} were reported [5]. The decrease is mainly due to the loss of Zn, since Zn has much higher vapour pressure than Sn at the annealing temperature of 300 °C. Most possibly, Sn atoms, which substitute the positions of Zn atoms, are the major donors in these nanocrystalline samples, and this is similar to its polycrystalline or crystalline counterpart [18,19].

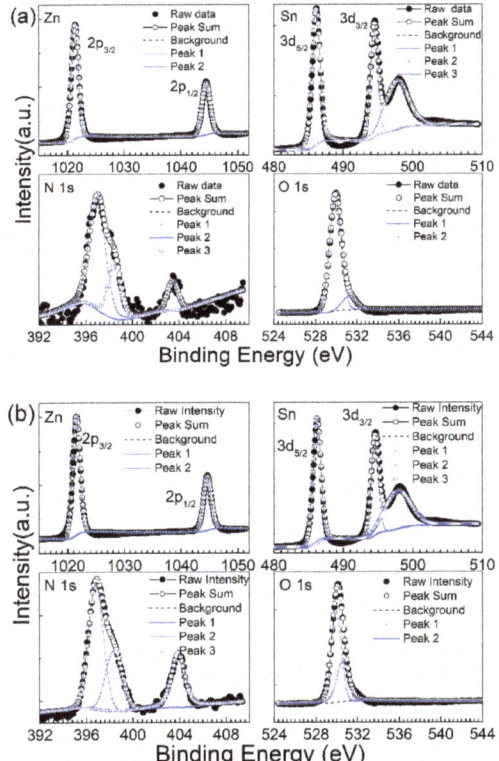

Figure 5. High resolution binding energy peak of Zn, Sn, N and O in ZnSnN$_2$. (**a**) Without annealing; (**b**) annealed at 300 °C.

Figure 6 shows the current density J vs. voltage V curves of the Schottky diodes with the structure of Ag\ITO\ZnSnN$_2$\Ag. The diode is forward-biased when the silver metal directly on the ZnSnN$_2$ layer is connected with the anode of the voltage. Rectification is observed, and obviously the rectification is from the Schottky contact formed between ZnSnN$_2$ and Ag since the structure of Ag\ITO\ZnSnN$_2$\ITO\Ag shows linear JV curves (Supplementary Materials Figure S5). The transport properties of our nanocrystalline ZnSnN$_2$ very possibly follow those of crystalline semiconductors since the transport behaviour of amorphous and microcrystalline Si with crystallites of or over 20 Å follows that of crystalline Si [22,23], and the crystallite of our nanocrystalline ZnSnN$_2$ meets this size requirement. In the following, theories or models based on crystalline semiconductors will be used. Under the thermal emission-diffusion model, the current through a Schottky barrier diode at a voltage of V [28,29] is as follows:

$$J = J_s \exp[-\xi q(V - IR_s)/(kT)]\{\exp[q(V - IR_s)/(kT)] - 1\}, \tag{3}$$

where $J_s = I_s/S = A^{**}T^2 \exp[-q\phi_{b0}/(kT)]$, J_s is the current density, I_s is the saturation current at zero bias, S is the diode area (0.10 × 0.10 cm^2 here), A^{**} is the effective Richardson constant ($A^{**} = 14.40$ A·cm^{-2}·K^{-2} for ZnSnN$_2$ since the Richardson constant A^* for a free electron is 120 A·cm^{-2}·K^{-2}, and the effective electron mass of ZnSnN$_2$ is 0.12 times that of a free electron [14]), q is the elementary charge, ϕ_{b0} is the barrier height at zero bias (the energy needed by an electron at the Fermi level in the metal to enter the conduction band of the semiconductor at zero bias), R_s is the series resistance due to the neutral region of the semiconductors, ξ is a factor without unit, ξ is equal to $1 - \eta^{-1}$, where η is the ideality factor, and other symbols have the same meaning as in previous formulas. From the forward-biased experimental data, the three unknown parameters ϕ_{b0}, ξ, and R_S can be fitted with the least square method, and the results, together with the calculated η, are presented in Figure 6a.

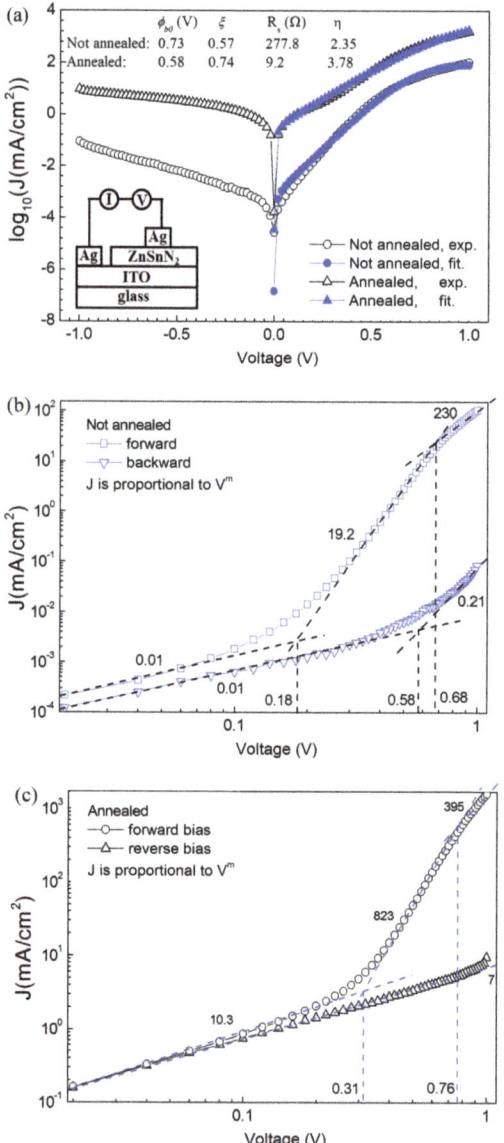

Figure 6. (a) The $J-V$ curves of Ag\ITO\ZnSnN$_2$\Ag Schottky contact (The Schottky contact is forward-biased when the Ag electrode in direct contact with ZnSnN$_2$ is connected with the anode of the voltage. The inset shows the structure of the Schottky diodes). (b) log J–log V curves of the samples without annealing; (c) log J–log V curves of the samples annealed at 300 °C.

At lower forward biased voltages, the fitted current is slightly larger than the experimental data, while good fitting is observed when the forward bias is over ~0.3 V in both samples. Since the mobility of both samples is low, diffusion is most likely the major transport mechanism at lower forward biased voltages, and thermal emission becomes obvious with an increase in the forward biased voltage. The fitted series resistance R_s of ZnSnN$_2$ without annealing is 277.8 Ω, which is larger than that of ZnSnN$_2$ annealed at 300 °C (9.2 Ω) and this agrees with the fact that the former has poorer conductivity than the latter.

The theoretical Schottky barrier height ϕ_{b0} at zero bias for ideal or intimate Ag-ZnSnN$_2$ is calculated to be 0.36 eV (the ideal band diagram is shown in Supplementary Materials Figure S6a) since theoretical ϕ_{b0} is equal to $W_m - \chi$, where W_m is the work function of Ag (W_m = 4.26 eV for Ag) and χ is the electron affinity of ZnSnN$_2$ (χ = 3.9 eV for ZnSnN$_2$ [30,31]). The fitted ϕ_{b0} is larger than the theoretical value, and the Schottky barrier height is enhanced. Schottky barrier height enhancement can result from surface Fermi-level pinning [32], a surface inversion layer (which possibly results from the diffusion of Ag into ZnSnN$_2$, and Ag atoms substituting Zn or Sn are acceptors) [33,34] or interface oxide states [32]. Surface Fermi-level pinning is less likely here since the forbidden band gap E_g calculated from the fitted ϕ_{b0} will be approximately 1.0 eV (under surface Fermi-level pinning, ϕ_{b0} is equal to $E_g - \phi_0$, where ϕ_0 is the surface neutral energy level measured from the edge of the valence band and is usually approximately $E_g/3$ [32]), which is less than half of those obtained from the Tauc plot. The fact that the calculated ideality factor η is over two also suggests the existence of interface oxide states since the ideality factor of a Schottky diode whose barrier height is enhanced by a surface inversion layer is usually below two [33].

A. M. Cowley and S. M. Sze [32] proposed that when there is interfacial oxide (the band diagram of the Schottky contact with interface states is shown in Supplementary Materials Figure S6b), the Schottky barrier height is expressed as:

$$\phi_{b0} = \gamma(W_m - \chi) + (1 - \gamma)(E_g - \phi_0)/q - \Delta\varphi_n \quad (4)$$

and the interface states $D_s = (1 - \gamma)\varepsilon_i/(\gamma q^2 \delta)$, where γ is a weighting factor (without unit), $\Delta\varphi_n$ is the image force barrier lowering, ε_i is the dielectric constant of the interfacial layer, δ is its thickness and other symbols have the same meaning as previously defined. If we suppose $\Delta\varphi_n = 0$, $\phi_0 = E_g/3$, where E_g is chosen to be 2.5 eV and ϕ_{b0} = the fitted value, γ will be 0.71 for the samples without annealing and 0.83 for the samples annealed at 300 °C. With the assumption that $\varepsilon_i = \varepsilon_0$ (free space dielectric constant) and δ = 5 Å [32], the interface states D_s will be 4.42 $\times$ 10^{12} states·eV^{-1}·cm^{-2} for the samples without annealing and 2.19 $\times$ 10^{12} states·eV^{-1}·cm^{-2} for the samples annealed. The samples without annealing have larger interface states than the samples annealed. The interface states of our samples are in the same magnitude with that of the Al/n-Si system (where Si is single-crystalline) [35], but roughly one magnitude smaller than that (3 $\times$ 10^{13} states·eV^{-1}·cm^{-2}) of metals and amorphous Si systems which have the same oxide thickness of 5 Å and whose Fermi-level were pinned [36].

Figure 6b,c shows the logarithmic J-V curves of the samples without annealing or annealed at 300 °C. For both samples, both the forward- and reverse-biased data obey the power law ($J \propto V^m$), but m varies in different voltage ranges. For the samples without annealing, below 0.18 V, m is less than one, and this suggests that the current is controlled by the diode. Between 0.18 V and 0.68 V, m is approximately 19.2. After 0.68 V, m is approximately 230. This implies that after 0.18 V, the current is bulk-controlled, and there is an exponential distribution of trap-levels within the forbidden gap of ZnSnN$_2$. The reverse-biased data of the samples without annealing is diode-controlled since m is below one in the whole measurement range. For the samples annealed at 300 °C, m is well over two in both the forward and backward bias, and this suggests that the current is bulk-controlled, and annealing reduces the resistivity of ZnSnN$_2$.

Figure 7a,b show the curves of the inverse of the square of the capacitance C (per unit area) vs. voltage V for the samples without annealing or annealed at 300 °C. For both samples, nonlinear C^{-2}-V curves are observed. The peak in Figure 7b is possibly due to the series resistance [37,38]. Considering the nonlinearity of the C^{-2}-V curves and the fact that the ideality factor η obtained from the IV curves is over two, we think the possibility that the barrier height enhancement results from a surface inversion layer can be excluded [33]. The nonlinearity of the C^{-2}-V curves suggests the existence of excess capacitance. When there is an interfacial layer in the Schottky contact, the capacitance C of a Schottky diode [39] is equal to $[C_i^{-1} + (C_D + C_{ex})^{-1}]^{-1}$, where C_i is the capacitance due to the interfacial layer, C_D is the depletion capacitance of the junction and C_{ex} is the excess capacitance due to surface states or deep traps. For thin interfacial layers, C_i is very large and can be ignored. Therefore, C is equal to $C_D + C_{ex}$. Though C^{-2} vs. V curves are not linear, $(C - C_{ex})^{-2}$ vs. V or C_D^{-2} vs. V will be linear.

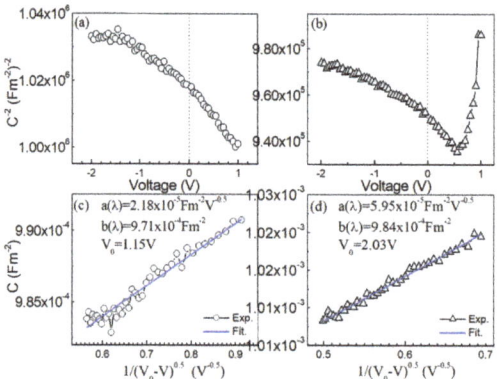

Figure 7. The C^{-2}–V curves of the samples without annealing (**a**) or annealed at 300 °C (**b**); $C - (V_0 - V)^{-0.5}$ curves of the samples without annealing (**c**) or annealed at 300 °C (**d**).

For the Schottky barrier where there is an oxide between the metal and the semiconductor, J. Szatkowski et al. [40] further proposed that

$$C = a(\lambda)(V_0 - V)^{-0.5} + b(\lambda), \quad (5)$$

where $a(\lambda)$ is $(1 - \lambda + \lambda\gamma)(q\varepsilon_0\varepsilon_r N_D\gamma/2)^{-0.5}$, N_D is the doping concentration, $b(\lambda)$ is $\lambda\gamma q^2 D_s$, V_0 and λ are constants, ε_r is the relative dielectric constant of ZnSnN$_2$ (ε_r = 11 [20]) and other symbols have the same meaning as in previous formulas. From the data about the capacitance C and reverse-biased voltage V, $a(\lambda)$, $b(\lambda)$ and V_0 can be fitted with the least square method. Both the experimental and fitted capacitance C vs. $(V_0 - V)^{-0.5}$ curves (where only the reverse-biased data are used) of the samples without annealing or with annealing are shown in Figure 7c,d. The fitted $a(\lambda)$, $b(\lambda)$ and V_0 are also presented in Figure 7c,d. As can been seen from Figure 7c,d, the relation between the capacitance C and $(V_0 - V)^{-0.5}$ is indeed linear.

If N_D is known, λ can be calculated from $a(\lambda)$ and then D_s from $b(\lambda)$, since γ has been obtained from the JV curves, and $a(\lambda)$ and $b(\lambda)$ have been obtained from the CV curves. For the samples annealed at 300 °C, λ and D_s are calculated to be 5.86 and 1.25×10^{11} states·eV^{-1}·cm^{-2} if we suppose that $N_D = n = 5.34 \times 10^{17}$ cm^{-3}. The samples without annealing are almost insulating, and we can suppose that $N_D = 10^{15}$ cm^{-3} [36]. λ and D_s are then calculated to be 2.49 and 3.40×10^{11} states·eV^{-1}·cm^{-2}. The interface states extracted from the capacitance-voltage of the samples without annealing are larger than those of the samples annealed at 300 °C, and this is in agreement with that obtained from I-V curves.

In addition, the interface states extracted from the C-V curves are roughly one-magnitude smaller than those from the J-V curves. The J-V and C-V curves of the Schottky diodes formed between nanocrystalline ZnSnN$_2$ and Ag are found to follow the theories based on crystalline semiconductors, in agreement with amorphous or microcrystalline Si with crystallites between 20–130 Å [22,23].

Figure 8a shows the current density J vs. the voltage V curves of the Ag\ITO\Cu$_2$O\ZnSnN$_2$\Ag and Ag\ITO\Cu$_2$O\ZnSnN$_2$\Au heterojunction solar cells under the illumination of AM1.5. The only difference between the two solar cells is that the electrode in connection with ZnSnN$_2$ is Ag for one solar cell and Au for the other. The open-circuit voltage V_{oc} (V), short-circuit current density J_{sc} (mA/cm^2), fill factor FF and power conversion efficiency PCE (%) are listed in the inset. The statistical distribution of these solar cells is presented in Figure S7 (the statistics are based on nine samples for the Ag\ITO\Cu$_2$O\ZnSnN$_2$\Au solar cell, while the statistics are based on eight samples for the Ag\ITO\Cu$_2$O\ZnSnN$_2$\Ag solar cell). The average V_{oc}, J_{sc}, FF and PCE of the solar cell with Au as the electrode are 0.22 V, 1.75 mA/cm^2, 42% and 0.15%, while those of the solar cell with Ag as the electrode are 0.19 V, 0.34 mA/cm^2, 22% and 0.02%, respectively. When the electrode in connection with ZnSnN$_2$ changes from Ag to Au, the four parameters increase greatly, implying better device performance. The measured series resistance of the solar cell with Au as the electrode is 0.83 kΩ, and that with Ag as the electrode is 4.19 kΩ. The improvement in performance mainly results from the change in the electrode. Previous work shows that Au can react with Sn to form AuSn even at room temperature, and the contact between AuSn and n-type GaAs is ohmic [41]. Though the contact between Au and ZnSnN$_2$ is theoretically non-ohmic (the work function of Au is 5.1 eV and the Fermi level of Au is much lower than that of ZnSnN$_2$), the Au electrode was thermally evaporated, and it is possible that there are interfacial alloys formed such as AuSn that reduce the Au-ZnSnN$_2$ contact resistance greatly.

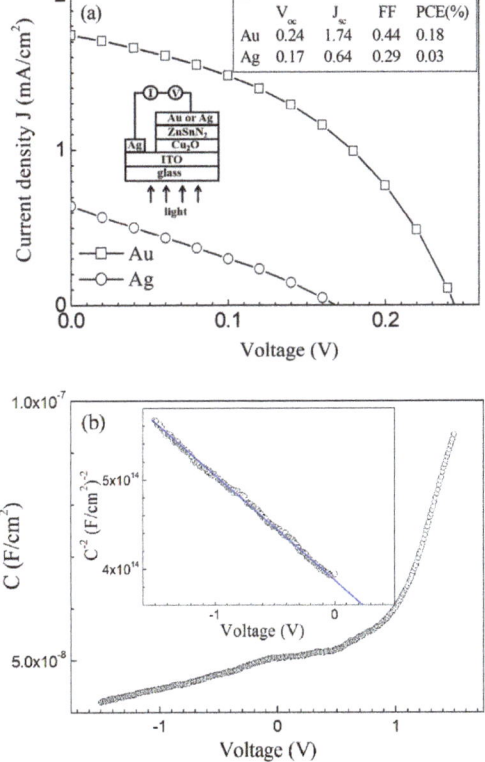

Figure 8. *Cont.*

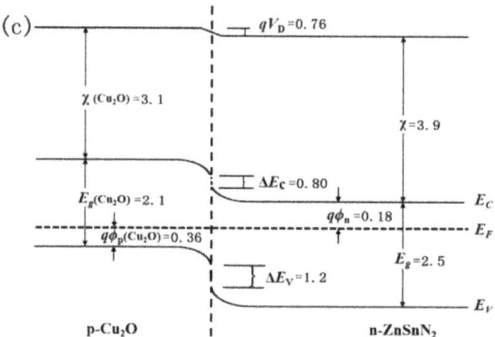

Figure 8. (a) The current density J vs. the voltage V curves of the Ag\ITO\Cu$_2$O\ZnSnN$_2$\Ag and Ag\ITO\Cu$_2$O\ZnSnN$_2$\Au solar cells (the inset shows the structure of these heterojunction solar cells); (b) the CV curve of the Ag\ITO\Cu$_2$O\ZnSnN$_2$\Au solar cell; (c) the energy band diagram of Cu$_2$O−ZnSnN$_2$ heterojunction at zero bias.

Figure 8b shows the capacitance C-Voltage V curve of the Ag\ITO\Cu$_2$O\ZnSnN$_2$\Au heterojunction solar cell (measured under darkness), with the inset showing the C^{-2}-V curve under reverse bias. It can be found that C^{-2}-V in the reverse bias is linear, and the intercept between the C^{-2}-V line and the voltage axis is approximately 0.24 V. This suggests that Cu$_2$O-ZnSnN$_2$ heterojunction is abrupt, with a built-in potential of 0.24 V. The built-in potential of 0.24 V is very near to the average open-circuit voltage (0.22 V) of the solar cell with Au as the electrode. Figure 8c shows the energy band diagram of the Cu$_2$O-ZnSnN$_2$ heterojunction at zero bias. The diagram shows a staggered, type II heterojunction. The built-in potential V_D, conduction band and valence band discontinuities, ΔE_c and ΔE_v, are estimated to be 0.76 V, 0.80 eV and 1.2 eV, respectively (Supplementary Materials Figure S8). The obtained V_{oc} needs great improvement as compared with V_D. The presence of the conduction band and valence band discontinuities, ΔE_c and ΔE_v, accounts partially for the low PCE.

In addition, after being stored in air without encapsulation for 11 months, the JV curves of the Ag\ITO\Cu$_2$O\ZnSnN$_2$\Au solar cells were measured under the AM1.5 illumination and the average V_{oc}, J_{sc}, FF and PCE (%) were 0.06 V, 0.56 mA/cm^2, 27% and 0.01%, respectively (the statistical data obtained after the storage for 11 months are also presented in Figure S7). The performance degradation might be due to the storage in air without encapsulation. Though all performance factors decrease as compared with those obtained 11 months ago, the photovoltaic effect still exists.

Figure 9a shows the dark JV curve on a log-log scale for the Ag\ITO Cu$_2$O\ZnSnN$_2$\Au heterojunction. The forward-biased JV curve obeys the power law ($J \propto V^m$) with different exponent values m at different voltage ranges (the junction is forward-biased when the electrostatic potential of Cu$_2$O is higher than that of ZnSnN$_2$). In 0−0.19 V, m is smaller than unity, implying that the current is controlled by the heterojunction diode. This also suggests that the resistance of the heterojunction is not large, since large resistance usually results in ohmic behaviour under relatively low voltage. In 0.19–0.98 V, m is approximately two while over 0.98 V m is over three. This implies that the regime in 0.19–0.98 V is the trap-filled limited (TFL) region where only partial traps are filled and the current is due to a space charge limited current (SCLC) [42]. In the region over 0.98 V, all the traps are filled, strong injection happens and the current increases sharply with an increase in the voltage. If we suppose that V_{tr} (0.19 V in Figure 9a) is the turn-on voltage at which the space charge limited conduction takes place in ZnSnN$_2$ and V_{TFL} (0.98 V in Figure 9a) is the voltage required to fill the traps of ZnSnN$_2$, the trap density N_t and the trap energy level E_t are calculated to be 2.14×10^{16} cm^{-3} and 0.1 eV, since V_{tr} equals $16qnt^2 N_t / \{9\varepsilon_r \varepsilon_0 N_c \exp[(E_t - E_c)/(kT)]\}$ and V_{TFL} equals $qN_t t^2/(2\varepsilon_r \varepsilon_0)$, where n is the free carrier density ($n = N_D = 10^{15}$ cm^{-3}), t is the film thickness (236 nm) and the other symbols have the same meaning as previously defined.

The heterojunction parameters, including the diode current J_0, the ideality factor η, the series resistance R_s and the shunt conductance G can be extracted based on a single diode model [43,44]:

$$J = J_0 \exp[\frac{q}{\eta kT}(V - R_s J)] + GV - J_L, \quad (6)$$

where J_L is the light-induced current ($J_L = J_{SC} = 1.74$ mA/cm^2 here), while the other parameters have the same meaning as previously defined. Figure 9b shows the dJ/dV vs. V curve, and the shunt conductance G is found to be 0.43 mS/cm^2 from the nearly flat area in the reverse bias. The dV/dJ vs. $[J + J_{SC}]^{-1}$ curve is presented in Figure 9c, and the series resistance R_s is calculated to be 1.26 Ω·cm^2. From the $J + J_{SC} - GV$ vs. $V - R_s J$ curve in Figure 9d, the diode current J_0 and the ideality factor η are calculated to be 7.76×10^{-2} mA/cm^2 and 2.90, respectively. The ideality factor η larger than two implies that the current transport is recombination-limited [45]. To improve the conversion efficiency, the series resistance needs to be reduced and the shunt resistance needs to be improved. The possible methods include better ohmic contact, optimum film thickness, etc. The band gap needs to be reduced to improve the short-circuit current density J_{sc}. The theoretical open-circuit voltage V_{oc} of the solar cell is estimated to be 0.24 V since V_{oc} approximately equal $(\eta kT/q) \ln[(J_{sc}/J_0) + 1]$ if the series resistance and shunt resistance are negligible and the photogenerated current equals the short-circuit current density J_{sc}. The diode current J_0 needs to be reduced to further improve the open-circuit voltage V_{oc}.

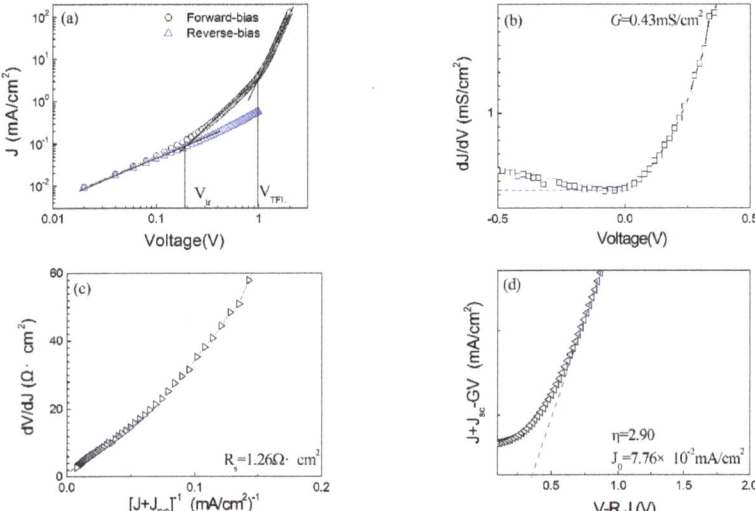

Figure 9. (a) The log–log dark JV curve; (b) dJ/dV vs. V curve; (c) dV/dJ vs. $[J + J_{SC}]^{-1}$ curve; (d) $J + J_{SC} - GV$ vs. $V - RJ$ curve.

The major structure and power conversion efficiency of several heterojunction solar cells, such as polycrystalline ZnSnN$_2$, Si, GaAs, etc [12,30,46–49], are listed in Table 2. The PCE (%) of Cu$_2$O-ZnSnN$_2$ needs great improvement compared with these heterojunction solar cells. The low PCE of Cu$_2$O-ZnSnN$_2$ possibly mainly results from the band gap of nanocrystalline ZnSnN$_2$ and the energy band diagram of the heterojunction. The band gap of nanocrystalline ZnSnN$_2$ is approximately 2.5 eV and this is much wider than the optimum band gap to yield the highest PCE (%). A recent report shows that polycrystalline ZnSnN$_2$ with a band gap of 1.43 eV can be prepared by sputtering under bias [50], and this provides a clue about reducing the band gap of nanocrystalline ZnSnN$_2$. One possible method to reduce the conduction and valence band discontinuities (Figure 8c) is to deposit a buffer between Cu_2O and $ZnSnN_2$. Replacing Cu_2O with other *p*-type semiconductors such as $CuCrO_2$ is another method since theoretical work shows that the PCE (%) of $CuCrO_2$-$ZnSnN_2$ is approximately 22% [2]. $ZnSnN_2$ belongs to the few solar cell absorption materials that can meet tetra Watt power needs at very low cost, and this necessitates further work in improving its material properties and device performance.

Table 2. The major structure and power conversion efficiency (PCE) of several heterojunction solar cells (pc refers to polycrystalline).

Layer/Stack	PCE (%)
pc-$ZnSnN_2$/SnO and	0.37 [12]
pc-$ZnSnN_2$/Al_2O_3/SnO	and 1.54 [30]
α-C/Si	1.5 [46]
ZnO/ZnGeO/Cu_2O:Na	8.1 [47]
MXene/GaAs	9.69 [48]
α-Si:H/c-Si	25.6 [49]

4. Conclusions

The reactive preparation and properties of nanocrystalline $ZnSnN_2$, together with its Schottky diodes and heterojunction solar cells, were investigated. $ZnSnN_2$ reactively deposited at 35 W and room temperature is nanocrystalline with crystalline grains of 2 nm. Annealing increases the grain size and the volume ratio of the crystalline component to the amorphous component. Annealing reduces the optical band gap and improves the electrical conductivity of $ZnSnN_2$. Nanocrystalline $ZnSnN_2$ annealed at 300 °C in Ar has an electron concentration of 5.54×10^{17} cm^{-3} and mobility of 0.998 $cm^2 \cdot V^{-1} \cdot s^{-1}$, which is much better than that of polycrystalline or crystalline $ZnSnN_2$, whose electron density in most cases is usually approximately 10^{19} cm^{-3}. The thermal emission-diffusion model was used to analyze the *I-V* curves of the Ag-nanocrystalline $ZnSnN_2$ Schottky diode. At lower forward biased voltages, diffusion is the major transport mechanism, and thermal emission becomes obvious at higher forward biased voltages. Both the *I-V* and *C-V* curves suggest the existence of interface states, which are calculated to be approximately 10^{12} states$\cdot eV^{-1} \cdot cm^{-2}$ from the *IV* curves and approximately 10^{11} states$\cdot eV^{-1} \cdot cm^{-2}$ from the *CV* curves. Nanocrystalline $ZnSnN_2$-based heterojunctions were successfully fabricated. The heterojunction interface is abrupt. The performance of the Ag\ITO\$Cu_2O$\$ZnSnN_2$\Au heterojunction solar cell is much better than that of the Ag\ITO\$Cu_2O$\$ZnSnN_2$\Ag solar cell. The current transport of the Ag\ITO\$Cu_2O$\$ZnSnN_2$\Au solar cell is recombination-limited. The fact that nanocrystallization can greatly reduce the electron density of $ZnSnN_2$, the *IV* and *CV* curves of the Schottky contact formed between nanocrystalline $ZnSnN_2$ and Ag obey the theories based on crystalline semiconductors and nanocrystalline $ZnSnN_2$-based heterojunction solar cells are successfully prepared suggest that nanocrystallization can pave a new way for the device application of $ZnSnN_2$.

Supplementary Materials: The following supporting information can be downloaded at: https://www.mdpi.com/article/10.3390/nano13010178/s1, Figure S1: the schematic diagram to show the fabrication steps for the Schottky diodes and heterojunction solar cells; Figure S2: XRD patterns of the samples deposited at 35 W and different substrate temperatures; Figure S3: XRD patterns of the samples deposited at 50 W and different substrate temperatures; Figure S4: Raman scattering spectra of the samples without annealing or with annealing; Figure S5: IV curve of the structure of Ag\ITO\$ZnSnN_2$\ITO\Ag; Figure S6: the ideal and nonideal band diagram of Ag-$ZnSnN_2$ at zero bias; Figure S7: the statistical distribution of the solar cell performance parameters; Figure S8: the energy band diagram for the Cu_2O-$ZnSnN_2$ heterojunction [4,17,24,29,51–53].

Author Contributions: Conceptualization, F.Y. and X.-M.C.; funding acquisition, F.Y. and X.-M.C.; investigation, R.-T.H. and Y.-B.Q.; project administration, F.Y. and X.-M.C.; resources, F.Y., Y.-Z.X., D.-P.Z., P.F. and X.-M.C.; supervision, F.Y.; validation, F.Y., R.-T.H., Y.-B.Q. and X.-M.C.; writing—

original draft, F.Y., R.-T.H., Y.-B.Q. and X.-M.C.; writing—review and editing, F.Y. and X.-M.C. All authors have read and agreed to the published version of the manuscript.

Funding: This work was financially supported by the Natural Science Foundation of China (No.:61674107), the Shenzhen Municipal Science and Technology Funding of for Innovation (Project No.: 20200812151132002).

Data Availability Statement: The data presented in this study are available upon request from the corresponding author. The data are not publicly available due to privacy.

Conflicts of Interest: The authors declare to have no competing interests.

References

1. Punya, A.; Paudel, T.R.; Lambrecht, W. Electronic and lattice dynamical properties of II-IV-N$_2$ semiconductors. *Phys. Status Solidi C* **2011**, *8*, 2492–2499. [CrossRef]
2. Laidouci, A.; Aissat, A.; Vilcot, J. Numerical study of solar cells based on ZnSnN$_2$ structure. *Sol. Energy* **2020**, *211*, 237–243. [CrossRef]
3. Chinnakutti, K.; Panneerselvam, V.; Salammal, S. Ba-acceptor doping in ZnSnN$_2$ by reactive RF magnetron sputtering: (002) faceted Ba-ZnSnN$_2$ films. *J. Alloys Compd.* **2021**, *855*, 157380. [CrossRef]
4. Ye, F.; Chen, Q.Q.; Cai, X.M.; Xie, Y.Z.; Ma, X.F.; Vaithinathan, K.; Zhang, D.P.; Fan, P.; Roy, V. Improving the chemical potential of nitrogen to tune the electron density and mobility of ZnSnN$_2$. *J. Mater. Chem. C* **2020**, *8*, 4314–4320. [CrossRef]
5. Wang, Y.; Ohsawa, T.; Meng, X.; Alnjiman, F.; Pierson, J.; Ohashi, N. Suppressing the carrier concentration of zinc tin nitride thin films by excess zinc content and low temperature growth. *Appl. Phys. Lett.* **2019**, *115*, 232104. [CrossRef]
6. Gogova, D.; Olsen, V.; Bazioti, C.; Lee, I.; Vines, L.; Kuznetsov, A. High electron mobility single-crystalline ZnSnN$_2$ on ZnO (0001) substrates. *CrystEngComm* **2020**, *22*, 6268–6274. [CrossRef]
7. Alnjiman, F.; Diliberto, S.; Ghanbaja, J.; Haye, E.; Kassavetis, S.; Patsalas, P.; Gendarme, C.; Bruyère, S.; Cleymand, F.; Miska, P.; et al. Chemical environment and functional properties of highly crystalline ZnSnN$_2$ thin films deposited by reactive sputtering at room temperature. *Sol. Energy Mater. Sol. Cells* **2018**, *182*, 30–36. [CrossRef]
8. Shing, A.; Tolstova, Y.; Lewis, N.; Atwater, H. Effects of surface condition on the work function and valence-band position of ZnSnN$_2$. *Appl. Phys. A-Mater.* **2017**, *123*, 735. [CrossRef]
9. Fioretti, A.; Stokes, A.; Young, M.; Gorman, B.; Toberer, E.; Tamboli, A.; Zakutayev, A. Effects of Hydrogen on Acceptor Activation in Ternary Nitride Semiconductors. *Adv. Electron. Mater.* **2017**, *3*, 1600544. [CrossRef]
10. Chinnakutti, K.; Panneerselvam, V.; Salammal, S. Investigation on structural and optoelectronic properties of in-situ post growth annealed ZnSnN$_2$ thin films. *Mater. Sci. Semicond. Process.* **2019**, *89*, 234–239. [CrossRef]
11. Qin, R.; Cao, H.; Liang, L.; Xie, Y.; Zhuge, F.; Zhang, H.; Gao, J.; Javaid, K.; Liu, C.; Sun, W. Semiconducting ZnSnN$_2$ thin films for Si/ZnSnN$_2$ p-n junctions. *Appl. Phys. Lett.* **2016**, *108*, 142104. [CrossRef]
12. Javaid, K.; Wu, W.; Wang, J.; Fang, J.; Zhang, H.; Gao, J.; Liang, L.; Cao, H. Band Offset Engineering in ZnSnN$_2$-Based Heterojunction for Low-Cost Solar Cells. *ACS Photonics* **2018**, *5*, 2094–2099. [CrossRef]
13. Wu, X.; Meng, F.; Chu, D.; Yao, M.; Guan, K.; Zhang, D.; Meng, J. Carrier Tuning in ZnSnN$_2$ by Forming Amorphous and Microcrystalline Phases. *Inorg. Chem.* **2019**, *58*, 8480–8485. [CrossRef]
14. Cai, X.; Wang, B.; Ye, F.; Vaithinathan, K.; Zeng, J.; Zhang, D.; Fan, P.; Roy, V. Tuning the photoluminescence, conduction mechanism and scattering mechanism of ZnSnN$_2$. *J. Alloys Compd.* **2019**, *779*, 237–243. [CrossRef]
15. Veal, T.; Feldberg, N.; Quackenbush, N.; Linhart, W.; Scanlon, D.; Piper, L.; Durbin, S. Band Gap Dependence on Cation Disorder in ZnSnN$_2$ Solar Absorber. *Adv. Energy Mater.* **2015**, *5*, 1501462. [CrossRef]
16. Kawamura, F.; Yamada, N.; Imai, M.; Taniguchi, T. Synthesis of ZnSnN$_2$ crystals via a high-pressure metathesis reaction. *Cryst. Res. Technol.* **2016**, *51*, 220–224. [CrossRef]
17. Quayle, P.; Junno, G.; He, K.; Blanton, E.; Shan, J.; Kash, K. Vapor-liquid-solid synthesis of ZnSnN$_2$. *Phys. Status Solidi B* **2017**, *254*, 1600718. [CrossRef]
18. Chen, S.; Narang, P.; Atwater, H.A.; Wang, L. Phase Stability and Defect Physics of a Ternary ZnSnN$_2$ Semiconductor: First Principles Insights. *Adv. Mater.* **2014**, *26*, 311–315. [CrossRef]
19. Pan, J.; Cordell, J.; Tucker, G.J.; Tamboli, A.; Zakutayev, A.; Lany, S. Interplay between composition, electronic structure, disorder, and doping due to dual sublattice mixing in non-equilibrium synthesis of ZnSnN$_2$:O. *Adv. Mater.* **2019**, *31*, 1807406. [CrossRef]
20. Cao, X.; Kawamura, F.; Ninomiya, Y.; Taniguchi, T.; Yamada, N. Conduction-band effective mass and bandgap of ZnSnN$_2$ earth-abundant solar absorber. *Sci. Rep.* **2017**, *7*, 14987. [CrossRef]
21. Jones, D.; Spear, W.; Lecomber, P. Transport properties of amorphous-germanium prepared by glow-discharge technique. *J. Non-Cryst. Solids* **1976**, *20*, 259–270. [CrossRef]
22. Spear, W.; Willeke, G.; Le Comber, P. Electronic properties of microcrystalline silicon prepared in the glow discharge plasma. *Physica* **1983**, *117–118*, 908–913.
23. Willeke, G.; Spear, W.; Jones, D.; Le Comber, P. Thermoelectric power, Hall effect and density-of-states measurements on glow-discharge microcrystalline silicon. *Philos. Mag. B* **1982**, *46*, 177–190. [CrossRef]

24. Chinnakutti, K.; Panneerselvam, V.; Salammal, S. Tailoring optoelectronic properties of earth abundant ZnSnN$_2$ by combinatorial RF magnetron sputtering. *J. Alloys Compd.* **2019**, *772*, 348–358. [CrossRef]
25. Zatsepin, D.; Boukhvalov, D.; Kurmaev, E.; Zhidkov, I.; Kim, S.; Cui, L.; Gavrilov, N.; Cholakh, S. XPS and DFT study of Sn incorporation into ZnO and TiO$_2$ host matrices by pulsed ion implantation. *Phys. Status Solidi B* **2015**, *252*, 1890–1896. [CrossRef]
26. Ye, F.; Zeng, J.; Qiu, Y.; Cai, X.; Wang, B.; Wang, H.; Zhang, D.; Fan, P.; Roy, V. Deposition-rate controlled nitrogen-doping into cuprous oxide and its thermal stability. *Thin Solid Films* **2019**, *674*, 44–51. [CrossRef]
27. Asahi, R.; Morikawa, T.; Ohwaki, T.; Aoki, K.; Taga, Y. Visible-Light Photocatalysis in Nitrogen-Doped Titanium Oxides. *Science* **2001**, *293*, 269–271. [CrossRef] [PubMed]
28. Chand, S.; Kumar, J. Current-voltage characteristics and barrier parameters of Pd$_2$Si/p-Si (111) Schottky diodes in a wide temperature range. *Semicond. Sci. Technol.* **1995**, *10*, 1680–1688. [CrossRef]
29. Sze, S.; Ng, K. *Physics of Semiconductor Devices*, 3rd ed.; Wiley Interscience: Hoboken, NJ, USA, 2006; p. 159.
30. Javaid, K.; Yu, J.; Wu, W.; Wang, J.; Zhang, H.; Gao, J.; Zhuge, F.; Liang, L.; Cao, H. Thin Film Solar Cell Based on ZnSnN$_2$/SnO Heterojunction. *Phys. Status Solidi RRL* **2018**, *12*, 1700332. [CrossRef]
31. Punya, A.; Lambrecht, W. Band offsets between ZnGeN$_2$, GaN, ZnO, and ZnSnN$_2$ and their potential impact for solar cells. *Phys. Rev. B* **2013**, *88*, 075302-1–075302-6. [CrossRef]
32. Cowley, A.M.; Sze, S.M. Surface States and Barrier Height of Metal-Semiconductor Systems. *J. Appl. Phys.* **1965**, *36*, 3212–3220. [CrossRef]
33. Eglash, S.J.; Newman, N.; Pan, S.; Mo, D.; Shenai, K.; Spicer, W.; Collins, D.M. Engineered Schottky barrier diodes for themodification and control of Schottky barrier heights. *J. Appl. Phys.* **1987**, *61*, 5159–5169. [CrossRef]
34. Shannon, J.M. Increasing the effective height of a Schottky barrier using low-energy ion implantation. *Appl. Phys. Lett.* **1974**, *25*, 75–77. [CrossRef]
35. Turut, A.; Saglam, M.; Efeoglu, H.; Yalcin, N.; Yildirim, M.; Abay, B. Interpreting the nonideal reverse bias C-V characteristicsand importance of the dependence of Schottky barrier heighton applied voltage. *Phys. B* **1995**, *205*, 41–50. [CrossRef]
36. Wronski, C.; Carlson, D. Surface states and barrier heights of metal-amorphous silicon Schottky barriers. *Solid State Commun.* **1977**, *23*, 421–424. [CrossRef]
37. Chattopadhyay, P.; Raychaudhuri, B. Origin of the anomalous peak in the forward capacitance-voltage plot of a Schottky barrier diode. *Solid State Electron.* **1992**, *35*, 875–878.
38. Chattopadhyay, P.; Raychaudhuri, B. Frequency dependence of forward capacitance-voltage characteristics of Schottky barrier diodes. *Solid State Electron.* **1993**, *36*, 605–610. [CrossRef]
39. Vasudev, P.; Mattes, B.; Pietras, E.; Bube, R. Excess capacitance and non-ideal Schottky barriers on GaAs. *Solid State Electron.* **1976**, *19*, 557–559. [CrossRef]
40. Szatkowski, J.; Sierański, K. Simple interface-layer model for the nonideal characteristics of the Schottky-barrier diode. *Solid State Electron.* **1992**, *35*, 1013–1015. [CrossRef]
41. Buene, L.; Jacobsen, S.-T. Room Temperature Interdiffusion in Evaporated Au-Sn Films. *Phys. Scr.* **1978**, *18*, 397. [CrossRef]
42. Chiu, F.-C. A review on conduction mechanism in dielectric films. *Adv. Mater. Sci. Eng.* **2014**, *2014*, 578168. [CrossRef]
43. Hegedus, S.; Shafarman, W. Thin-film solar cells: Device measurements and analysis. *Prog. Photovolt Res. Appl.* **2004**, *12*, 155–176. [CrossRef]
44. Chen, S.; Ishap, M.; Xiong, W.; Shah, U.; Farooq, U.; Luo, J.; Zheng, Z.; Su, Z.; Fan, P.; Zhang, X.; et al. Improved open-circuit voltage of Sb$_2$Se$_3$ thin film solar cells via interfacial sulfur diffusion-induced gradient bandgap engineering. *Solar RRL* **2021**, *5*, 2100419. [CrossRef]
45. Carlson, D.E.; Wronski, C.R. Amorphous silicon solar cell. *Appl. Phys. Lett.* **1976**, *28*, 671–673. [CrossRef]
46. Li, X.; Zhu, H.; Wang, K.; Cao, A.; Wei, J.; Li, C.; Jia, Y.; Li, Z.; Li, X.; Wu, D. Graphene-On-Silicon Schottky Junction Solar Cells. *Adv. Mater.* **2010**, *22*, 2743–2748. [CrossRef] [PubMed]
47. Minami, T.; Nishi, Y.; Miyata, T. Efficiency enhancement using a Zn$_{1-x}$Ge$_x$-O thin film as an n-type window layer in Cu$_2$O-based heterojunction solar cells. *Appl. Phys. Express* **2016**, *9*, 052301-1–052301-4. [CrossRef]
48. Zhang, Z.; Lin, J.; Sun, P.; Zeng, Q.; Deng, X.; Mo, Y.; Chen, J.; Zheng, Y.; Wang, W.; Li, G. Air-stable MXene/GaAs heterojunction solar cells with a high initial efficiency of 9.69%. *J. Mater. Chem. A* **2021**, *9*, 16160–16168. [CrossRef]
49. Masuko, K.; Shigematsu, M.; Hashiguchi, T.; Fujishima, D.; Kai, M.; Yoshimura, N.; Yamaguchi, T.; Ichihashi, Y.; Mishima, T.; Matsubara, N.; et al. Achievement of More Than 25% Conversion Efficiency With Crystalline Silicon Heterojunction Solar Cell. *IEEE J. Photovolt.* **2014**, *4*, 1433–1435. [CrossRef]
50. Virfeu, A.; Alnjiman, F.; Ghanbaja, J.; Borroto, A.; Gendarme, C.; Migot, S.; Kopprio, L.; Gall, S.L.; Longeaud, C.; Vilcot, J.-P. Approaching Theoretical Band Gap of ZnSnN$_2$ Films via Bias Magnetron Cosputtering at Room Temperature. *ACS Appl. Electron. Mater.* **2021**, *3*, 3855–3866. [CrossRef]
51. Quayle, P.C.; He, K.; Shan, J.; Kash, K. Synthesis, lattice structure, and band gap of ZnSnN$_2$. *MRS Commun.* **2013**, *3*, 135–138. [CrossRef]

52. Quayle, P.C.; Blanton, E.W.; Punya, A.; Junno, G.T.; He, K.; Han, L.; Zhao, H.; Shan, J.; Lambrecht, W.R.L.; Kash, K. Charge-neutral disorder and polytypes in heterovalent wurtzite-based ternary semiconductors: The importance of the octet rule. *Phys. Rev. B* **2015**, *91*, 205207. [CrossRef]
53. Meyer, B.K.; Polity, A.; Reppin, D.; Becker, M.; Hering, P.; Klar, P.J.; Sander, T.; Reindl, C.; Benz, J.; Eickhoff, M.; et al. Binary copper oxide semiconductors: From materials towards devices, *Phys. Status Solidi B* **2012**, *249*, 1487–1509. [CrossRef]

Disclaimer/Publisher's Note: The statements, opinions and data contained in all publications are solely those of the individual author(s) and contributor(s) and not of MDPI and/or the editor(s). MDPI and/or the editor(s) disclaim responsibility for any injury to people or property resulting from any ideas, methods, instructions or products referred to in the content.

Article

Structural, Electronic and Optical Properties of Some New Trilayer Van de Waals Heterostructures

Beitong Cheng [1], Yong Zhou [1,2], Ruomei Jiang [1], Xule Wang [1], Shuai Huang [1], Xingyong Huang [3], Wei Zhang [1], Qian Dai [1], Liujiang Zhou [4], Pengfei Lu [5,*] and Hai-Zhi Song [1,4,6,*]

[1] Quantum Research Center, Southwest Institute of Technical Physics, Chengdu 610041, China; beitong20@163.com (B.C.)
[2] School of Electronic Engineering, Chengdu Technological University, Chengdu 611730, China
[3] Faculty of Science, Yibin University, Yibin 644007, China
[4] Institute of Fundamental and Frontier Sciences, University of Electronic Science and Technology of China, Chengdu 610054, China
[5] State Key Laboratory of Information Photonics and Optical Communications, Beijing University of Posts and Telecommunications, Beijing 100876, China
[6] State Key Laboratory of High Power Semiconductor Lasers, Changchun University of Science and Technology, Changchun 130013, China
* Correspondence: lupengfei@bupt.edu.cn (P.L.); hzsong@uestc.edu.cn (H.-Z.S.); Tel.: +86-158-2823-9155 (H.-Z.S.)

Citation: Cheng, B.; Zhou, Y.; Jiang, R.; Wang, X.; Huang, S.; Huang, X.; Zhang, W.; Dai, Q.; Zhou, L.; Lu, P.; et al. Structural, Electronic and Optical Properties of Some New Trilayer Van de Waals Heterostructures. *Nanomaterials* 2023, 13, 1574. https://doi.org/10.3390/nano13091574

Academic Editor: Iván Mora-Seró

Received: 7 April 2023
Revised: 30 April 2023
Accepted: 5 May 2023
Published: 8 May 2023

Copyright: © 2023 by the authors. Licensee MDPI, Basel, Switzerland. This article is an open access article distributed under the terms and conditions of the Creative Commons Attribution (CC BY) license (https://creativecommons.org/licenses/by/4.0/).

Abstract: Constructing two-dimensional (2D) van der Waals (vdW) heterostructures is an effective strategy for tuning and improving the characters of 2D-material-based devices. Four trilayer vdW heterostructures, BP/BP/MoS$_2$, BlueP/BlueP/MoS$_2$, BP/graphene/MoS$_2$ and BlueP/graphene/MoS$_2$, were designed and simulated using the first-principles calculation. Structural stabilities were confirmed for all these heterostructures, indicating their feasibility in fabrication. BP/BP/MoS$_2$ and BlueP/BlueP/MoS$_2$ lowered the bandgaps further, making them suitable for a greater range of applications, with respect to the bilayers BP/MoS$_2$ and BlueP/MoS$_2$, respectively. Their absorption coefficients were remarkably improved in a wide spectrum, suggesting the better performance of photodetectors working in a wide spectrum from mid-wave (short-wave) infrared to violet. In contrast, the bandgaps in BP/graphene/MoS$_2$ and BlueP/graphene/MoS$_2$ were mostly enlarged, with a specific opening of the graphene bandgap in BP/graphene/MoS$_2$, 0.051 eV, which is much larger than usual and beneficial for optoelectronic applications. Accompanying these bandgap increases, BP/graphene/MoS$_2$ and BlueP/graphene/MoS$_2$ exhibit absorption enhancement in the whole infrared, visible to deep ultraviolet or solar blind ultraviolet ranges, implying that these asymmetrically graphene-sandwiched heterostructures are more suitable as graphene-based 2D optoelectronic devices. The proposed 2D trilayer vdW heterostructures are prospective new optoelectronic devices, possessing higher performance than currently available devices.

Keywords: two-dimensional material; van der Waals heterostructure; trilayer; the first-principles calculation

1. Introduction

Van der Waals (vdW) heterostructures [1–3] containing different two-dimensional (2D) material monolayers have recently attracted interest for fundamental physics in low-dimensional systems [4–8] and for applications in microelectronic and optoelectronic devices [9–12]. Several kinds of typical 2D materials such as graphene [13–15], transition metal dichalcogenide (TMD) [16,17] and phosphorene [18–20] have been incorporated into this construction. In graphene/MoS$_2$, good thermal stability with high atom cohesive energy was confirmed [21] and found to mainly be associated with vdW coupling [22], and an electric field applied at the hetero-interface can control the Schottky barriers and Ohmic contacts [23]. Bilayer graphene/black-phosphorous (BP) was proposed to protect

BP from its structural and chemical degradation with a slightly opened graphene band gap while still maintaining the electronic characteristics of graphene and BP monolayers [24,25]. Graphene/hexagonal-boron-nitride (hBN) appears to have good electric-field-driven switching, which is beneficial to the field-effect transistor [26]. MoS_2/MX_2 (M = Mo, Cr, W; X = S, Se) bilayer heterostructures exhibit semiconductor properties with an indirect band gap, except for MoS_2/WSe_2 [27], and all of them can undergo transiting from semiconductors to metals with the strain and band gap decreasing with the electric field [28], showing prospects for construction of ultrathin flexible devices and for application in viable optoelectronic fields. The TMD/BN heterostructures were explored for application in photocatalytic fields owing to the direct bandgap and powerful built-in electric field across the interface [29]. The type-II band alignments of MoS_2/ZnO and WS_2/ZnO help to separate the photo-generated charge effectively, and all TMD/graphene-like-ZnO structures show optical absorption in the visible and infrared regions [30]. Blue-phosphorous (BlueP)/BP possesses a tunable bandgap, band edges and electron-hole behavior by applying perpendicular electric field, and it thus shows potential for use in novel optoelectronic devices [31]. BlueP/GaN vdW heterostructure was found active for facilitating charge injection and thus promising for unipolar electronic device applications [32]. The photoelectric performances of BP/TMD heterostructures can be improved by applying compressive stress [33] or an electric field [34]. BP/MoS_2 demonstrated an improved response and absorption characteristics in photodetectors [35]. Many BP/XT_2 (X = Mo, W; T = S, Se, Te) heterostructures with type-II band alignments would be prospective as spin-filter devices [36]. $BlueP/MoS_2$ has been considered the next generation of photovoltaic devices and water-splitting materials due to its wide optical response range and good light absorption ability [37]. In addition to the theoretical research such as that mentioned above, there have been experimental studies on bilayer vdW heterostructures to explore their practical prospects. The BP/MoS_2 heterostructure photodetector was prepared and demonstrated with a wide range of current-rectifying behavior, microsecond response speed and high detectivity [38]. Graphene/BP heterostructure phototransistors exhibited good photoconductive gain and thus an ultrahigh photoresponse at near-infrared wavelength, implying the potential applications in remote sensing and environmental monitoring [39]. It is apparent that building vdW heterostructures is an effective way to achieve higher performance from devices based on graphene, phosphorenes and TMDs.

As an extension to the above idea, trilayer vdW heterostructures have recently been attracting more and more attention, since one more layer might open more space for tuning the properties further [40,41]. Simulation by Datta et al. demonstrated that the electronic properties of $MoS_2/MX_2/MoS_2$ exhibit a smaller electron effective mass and a semiconductor-to-metal transition under tensile strain [42]. Device level performance of $MoS_2/MX_2/MoS_2$ were further calculated, showing that these trilayer heterostructures would be prospective to construct highly sensitive field effect transistors for nano-biomolecules sensing as well as for pH sensing [43]. Bafekry et al. theoretically analyzed trilayer vdW heterostructures in which MoS_2 monolayer is encapsulated by two graphitic carbon nitride monolayers, revealing their magnetic metallic characteristics [44] probably applicable in nano-spintronic devices. Liu et al. proposed trilayer heterostructures $MoS_2/SiC/MoS_2$ and $SiC/MoS_2/SiC$, and predicted their strain-induced tunable band structure promising for future optoelectronic devices [45]. Calculation on $BlueP/MoX_2$-based trilayer heterostructures demonstrated the excellent ability to accelerate the separation of photo-generated electron-hole pairs and the controllable power conversion efficiency indicating great potential for photocatalysis and photovoltaics [46]. BN/graphene/BN was proposed and simulated to open the graphene bandgap and enhance the application potential for radio-frequency devices and switching transistors [47,48]. Xu et al. theoretically argued that MoS_2-graphene-based trilayer heterostructures are of unique optoelectronic properties tunable by external electric field and thus of great potential for solar energy harvesting and conversion [49]. External electric field modulation effects were also simulated on $PtSe_2$/graphene/graphene and graphene/$PtSe_2$/graphene, revealing their usefulness to guide the design of high-performance field effect transistors [50]. In an experimental work, Long et al. fabricated WSe$_2$/graphene/MoS$_2$ p-g-n heterostructure photodetector, which achieved a wide detection wavelength range of 400–2400 nm and a short rise/fall time [51]. Li et al. fabricated photodetectors based on dielectric shielded MoTe$_2$/graphene/SnS$_2$ p–g–n junctions, and achieved extraordinary responsivity as high as 2600 A/W, detectivity as good as 10^{13} Jones with fast photoresponse in the range of 405–1550 nm [52]. It is thus clear that constructing trilayer vdW heterostructures can improve further the performance of electronic and optoelectronic devices based on graphene, phosphorenes and TMDs. However, the reported trilayer heterostructures used mainly symmetrical structures and/or contained two types of materials. Further efforts should be made to construct new vdW heterostructures.

In this work, we attempt to construct and study trilayer vdW heterostructures with asymmetric structures and/or three different types of 2D materials. Four kinds of trilayer heterostructures containing MoS$_2$, phosphorene and/or graphene are simulated in terms of structural, optical and optoelectronic properties. It is suggested that sophisticated 2D trilayer vdW heterostructures can provide further optimized characters for a new generation of optoelectronic devices.

2. Calculation Methods

In this study, we designed four trilayer vdW heterostructures BP/BP/MoS$_2$, BlueP/BlueP/MoS$_2$, BP/graphene/MoS$_2$ and BlueP/graphene/MoS$_2$, which are structurally asymmetric and/or composed of more than two different types of 2D materials. The first-principles calculation was performed using the tool of Vienna ab initio simulation package (VASP) [53] within Density Functional Theory (DFT) and the projector-augmented wave (PAW) method [54]. The generalized gradient approximation (GGA) with the Perdew–Burke–Ernzerhof (PBE) method was applied for calculating electron exchange and correlation potentials [54,55]. Due to the weak vdW interaction between the monolayers, the opt-PBE vdW method [55,56] was adopted to modify the vdW dispersion. To avoid the interaction between adjacent slab models, the thickness of the vacuum layer was set to 20 Å. The energy cutoff, convergence criteria of total energy and force were set to 450 eV, 10^{-5} eV and 0.01 eV/A, respectively. When optimizing the trilayer atomic structures, $1 \times 7 \times 1$, $3 \times 3 \times 1$, $2 \times 2 \times 1$ and $3 \times 3 \times 1$ k-point sets were adopted for BP/BP/MoS$_2$, BlueP/BlueP/MoS$_2$, BP/graphene/MoS$_2$ and BlueP/graphene/MoS$_2$, respectively. When calculating the electronic structures, $5 \times 21 \times 1$, $16 \times 16 \times 1$, $5 \times 5 \times 1$ and $8 \times 8 \times 1$ k-point sets were adopted for the four structures, respectively. When calculating the optical properties, $8 \times 21 \times 1$, $20 \times 20 \times 1$, $10 \times 10 \times 1$ and $13 \times 13 \times 1$ k-point sets were adopted for the four structures, respectively. The above-mentioned k-point sets were actually determined by convergence tests: increasing the k-point set scale until a state in which the calculation results do not change beyond the convergence standard any longer. It reflects that calculations for structural, electronic and optical properties need more and more computational cost and accuracy.

To establish the stability of the proposed vdW heterostructures, we should in principle evaluate their vibrational frequencies to seek real and positive phonon parameters. Nevertheless, most of the first-principles simulations on vdW heterostructures took the binding energy as the figure of merit to characterize the structure stability [21,27,30,36], because the vibrations are so computationally expensive that a complex structure would be technically hard to deal with. In our case, the proposed atomic structures are even larger and complicated, so we judge the structure stability and thus the experimental feasibility by calculating the binding energy, which would be negative for a stable system [23]. The binding energy (E_b) is calculated using the following equation:

$$E_b = E_{hts} - E_A - E_B - E_C \qquad (1)$$

where E_{hts} represents the overall energy of the heterostructure and E_A, E_B, and E_C represent the energy of each monolayer in its free state. The work function, used as the key parameter to deduce the band alignment, is described by

$$W = E_{vac} - E_f \tag{2}$$

where E_{vac} is the energy of a stationary electron in the vacuum and E_f is the Fermi level.

The macroscopic optical response function is usually measured by the complex dielectric function $\varepsilon(\omega) = \varepsilon_1(\omega) + i\varepsilon_2(\omega)$, where ω represents the circular frequency of light. The real part $\varepsilon_1(\omega)$ can be processed via the Kramers–Kronig relation and the imaginary part $\varepsilon_2(\omega)$ is produced by Fermi's golden rule [57]. Based on these parameters, we can obtain the absorption coefficient by [58],

$$\alpha(\omega) = \frac{\sqrt{2}\omega}{c}\{[\varepsilon_1{}^2(\omega) + \varepsilon_2{}^2(\omega)]^{1/2} - \varepsilon_1(\omega)\}^{1/2} \tag{3}$$

3. Results and Discussion

3.1. Structural Parameters

Due to the surface relaxation and interfacial mismatch, the trilayer vdW heterostructure supercells must be carefully set, for which we here extend monolayer cells individually to satisfy the length matching between different layers. The BP/BP/MoS$_2$ heterostructure is constructed from a $4\sqrt{3} \times 1$ supercell MoS$_2$ monolayer and a 5×1 supercell bilayer-BP, while the BlueP/BlueP/MoS$_2$ heterostructure is formed by stacking a 3×3 supercell bilayer-BlueP on a 3×3 supercell MoS$_2$ monolayer. After structure optimization calculation, we obtain the rectangular lattice constants $a = 22.227$ and $b = 3.278$ Å for BP/BP/MoS$_2$, and the hexagonal lattice constants $a = b = 9.642$ Å for BlueP/BlueP/MoS$_2$. For BP/graphene/MoS$_2$, bilayer BP/graphene cells are first constructed by matching 3×1 BP with $4\sqrt{3} \times 1$ supercell graphene, and then the 1×2 supercell of the bilayer BP/graphene is set to match the $2\sqrt{3} \times 3$ supercell MoS$_2$. For BlueP/graphene/MoS$_2$, a 3×3 supercell monolayer BlueP, 4×4 supercell graphene, and 3×3 supercell monolayer MoS$_2$ are successively stacked. The optimized rectangular lattice constants of BP/graphene/MoS$_2$ are $a = 10.256$ Å and $b = 8.907$ Å, and the hexagonal ones of BlueP/graphene/MoS$_2$ are $a = b = 9.776$ Å. The simulated models of BP/BP/MoS$_2$, BlueP/BlueP/MoS$_2$, BP/graphene/MoS$_2$ and BlueP/graphene/MoS$_2$ are depicted in Figure 1.

The trilayer vdW heterostructures will have different energies with different layer spacing. Figure 1 gives the layer spacing when the system reaches the final equilibrium. The calculated binding energy of the four trilayer heterostructures is -21.12, -65.41, -21.92 and -65.78 eV, respectively. The negative binding energies indicate that all the four proposed vdW heterostructures are structurally highly stable like the reported bilayer heterostructures [21,36]. The structure stability may imply the fabrication feasibility to some degree. As is known, BP/MoS$_2$ [38], graphene/MoS$_2$ [59], graphene/BP [39] bilayer heterostructures, and TMD/graphene/MoS$_2$ [51] trilayer heterostructures have been successfully fabricated using physical and chemical methods including mechanical stripping, liquid-phase stripping, hydrothermal and chemical vapor deposition. BlueP 2D monolayer has been well fabricated by molecular beam epitaxy [60], and the possibility to form its heterostructure with other 2D materials has been stated by a few studies [37,46]. Combining together the above fabrication abilities and considering the confirmed structure stability, the preparation of our proposed vdW heterostructures will be experimentally feasible. Noting the binding energy differences among these heterostructures, BlueP/BlueP/MoS$_2$ and BlueP/graphene/MoS$_2$ might be easier to be fabricated than BP/BP/MoS$_2$ and BP/graphene/MoS$_2$, although BlueP-related 2D materials seem currently farther away from practical manufacture than BP-related ones.

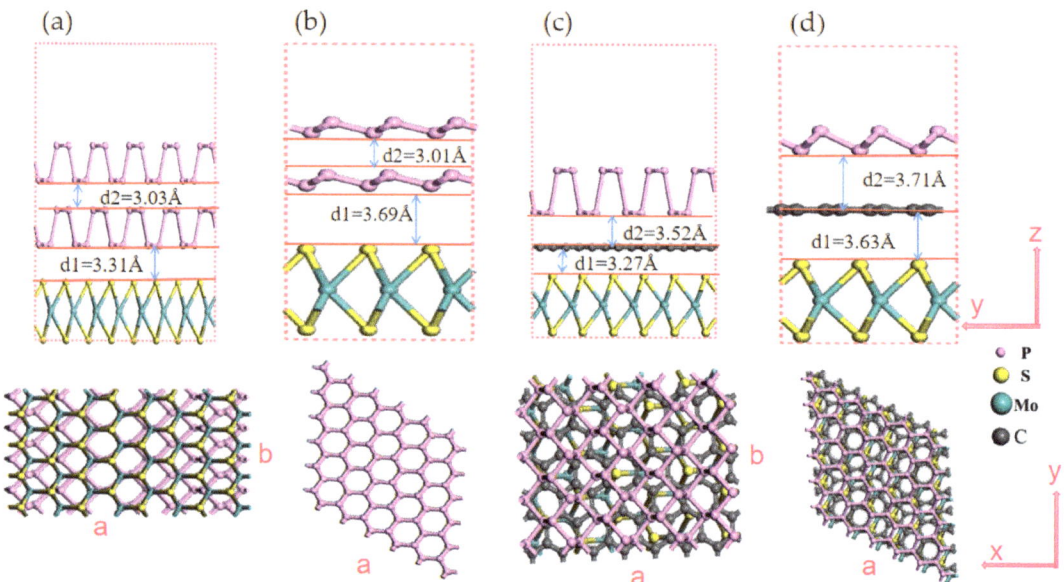

Figure 1. The models of trilayer 2D vdW heterostructures. (**a**) BP/BP/MoS$_2$; (**b**) BlueP/BlueP/MoS$_2$; (**c**) BP/graphene/MoS$_2$; (**d**) BlueP/graphene/MoS$_2$.

3.2. Electronic Properties

3.2.1. Band Structure

Figure 2a,b demonstrate the projected band structures of BP/BP/MoS$_2$ and BlueP/BlueP/MoS$_2$ calculated using the PBE method. BP/BP/MoS$_2$ exhibits the conduction band minimum (CBM) located between the Y and G points and mostly contributed by MoS$_2$ (red lines) as well as the valance band maximum (VBM) located at the high-symmetry G point and mostly contributed by BP (blue lines). These are qualitatively consistent with our calculation of the bilayer counterpart BP/MoS$_2$ (in agreement with [34]), suggesting that the high hole mobility of BP and high electron mobility of MoS$_2$ are still preserved. The indirect bandgap of the $4\sqrt{3} \times 1$ supercell MoS$_2$, 1.338 eV, and the direct bandgap of the 5×1 supercell bilayer-BP, 0.706 eV, are reduced compared with those in bilayer BP/MoS$_2$ (MoS$_2$ 1.398 eV, BP 0.876 eV), which are already lower than the free monolayer ones (MoS$_2$ 1.69 eV, BP 0.9 eV [36]). Significantly, the whole BP/BP/MoS$_2$ heterostructure possesses an indirect band gap of 0.167 eV, nearly half that of bilayer BP/MoS$_2$, 0.326 eV. It is seen that this trilayer heterostructure effectively tunes the bandgap further. For BlueP/BlueP/MoS$_2$, the CBM is located between the G and M points and is also mostly contributed by MoS$_2$ (red lines), whereas the VBM is located at the high-symmetry K point and is mostly contributed by BlueP (blue lines), consistent with the calculated results of the bilayer counterpart BlueP/MoS$_2$ (in agreement with [37]). Similar to the above BP/BP/MoS$_2$, the direct bandgap of the 3×3 supercell MoS$_2$ monolayer, 1.298 eV, and the indirect bandgap of the 3×3 supercell BlueP monolayer, 1.624 eV, are smaller than those in the bilayer BlueP/MoS$_2$ (MoS$_2$ 1.328 eV, BlueP 1.698 eV), which are already decreased compared with the corresponding free monolayer ones (MoS$_2$ 1.69 eV, BlueP 1.908 eV [2,37]). The whole trilayer heterostructure possesses an indirect band gap of 1.2632 eV, smaller than that of the bilayer BlueP/MoS$_2$, 1.3746 eV. It falls in between the BlueP/MoS$_2$/BlueP of 1.013 eV and MoS$_2$/BlueP/MoS$_2$ of 1.561 eV reported by Han et al. [46], again implying the further effective tuning by such an asymmetric trilayer heterostructure. The BP(BlueP) and MoS$_2$ bandgap decrease as they compose bilayer or trilayer heterostructures, which can be attributed to ~2.5% and ~2% lattice mismatch and a weak vdW force [35]. A whole-bandgap decrease with the addition of a BP(BlueP) monolayer was discovered by Dong et al. and Qiao et al. in their studies on BP [57,61]; bandgap engineering by means of the simple addition of monolayers appears thus to be effective. Moreover, it is interesting here that the VBM of BlueP and MoS$_2$ are becoming much closer, which supports application with the function of carrier transferring or tunneling.

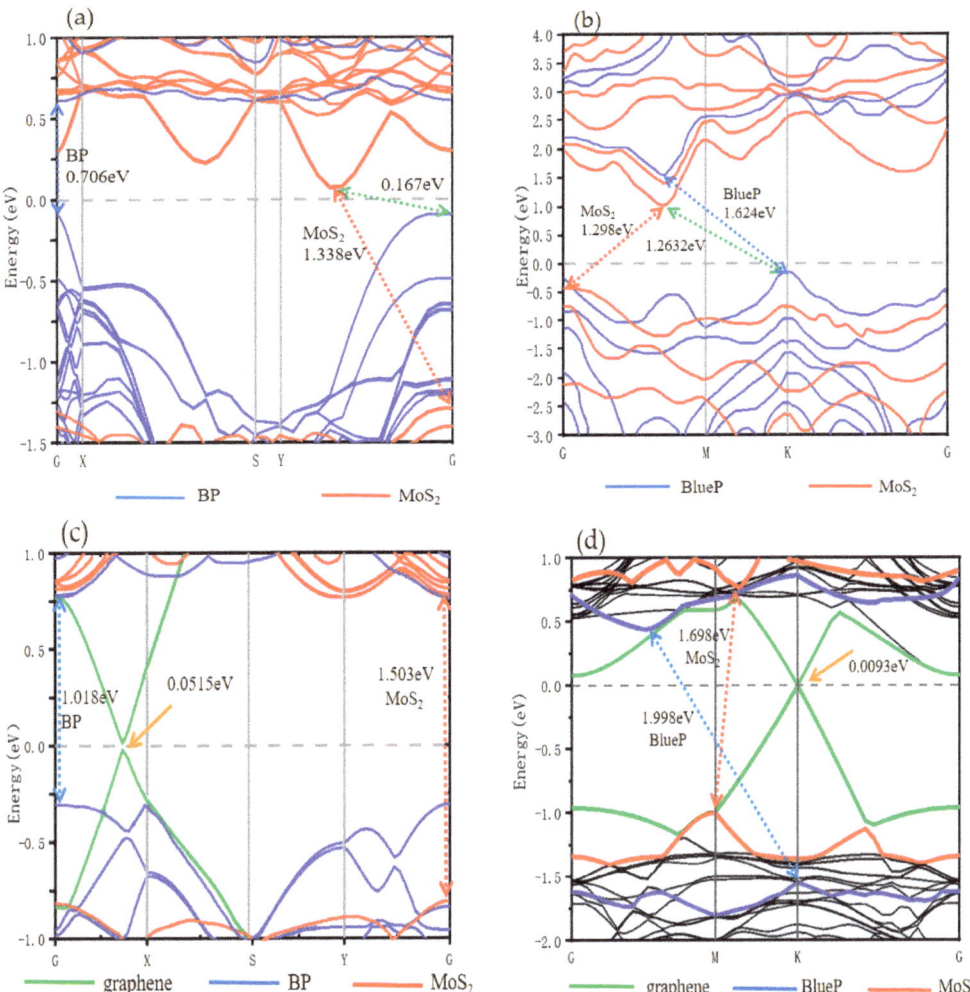

Figure 2. Calculated projected band structures for trilayer vdW heterostructures (**a**) BP/BP/MoS$_2$; (**b**) BlueP/BlueP/MoS$_2$; (**c**) BP/graphene/MoS$_2$; (**d**) BlueP/graphene/MoS$_2$. The Fermi level E_f is set as zero energy.

Figure 2c,d show the projected band structure of BP/graphene/MoS$_2$ and BlueP/graphene/MoS$_2$ trilayer vdW heterostructures calculated using the PBE method. What is immediately significant is that the BP/graphene/MoS$_2$ structure opens the bandgap of graphene to 0.051 eV, which is over one order of magnitude larger than the opening by the reported bilayer heterostructures such as graphene/BP (~0.0013 eV) [24], graphene/ZnO (~0.0057 eV) [30], and graphene/MoS$_2$ (~0.0007 eV) [49], and remarkably larger than those by other trilayer heterostructures such as graphene/ZnO/MoS$_2$ (~0.0048 eV) [30], graphene/nitrogene/graphene (~0.04 eV) [40], and graphene/BN/graphene

(~0.020 eV) [47]. BlueP/graphene/MoS$_2$ opens the bandgap of graphene by 0.0093 eV, similar to many other heterostructures. The effectiveness of the graphene bandgap opening in BP/graphene/MoS$_2$ can hardly be attributed to strain, because the strains of graphene in both structures are close to each other (3.2% and 2.8%) and close to those of the reported structures such as graphene/BP (3%) [24]. We believe it originates from the united vdW coupling forces among BP and MoS$_2$ monolayers and possible built-in electric fields. The remarkable graphene bandgap opening suggests potential applications in temperature-sensitive devices [62]. On the basis of guaranteeing the excellent electronic properties of graphene, it can also be applied to field effect transistors with a high switching ratio. It is worth noting that these band gap values are grossly underestimated owing to adopting PBE functional calculation, so the real bandgaps of graphene in such heterostructures may be larger to match more extended application prospects. Furthermore, the band gap values of BP (~1.018 eV) and BlueP (~1.998 eV) in these two trilayer heterostructures are, qualitatively as graphene does, slightly larger than the free monolayers (BP of 0.9 eV [36] and BlueP of 1.908 [37]), which is in contrast to the first two structures BP/BP/MoS$_2$ and BlueP/BlueP/MoS$_2$. There seem to be two types of bandgap engineering mechanisms in trilayer vdW heterostructures, providing wider space for novel device design.

We also calculated the local density of state (LDOS) and total density of states (TDOS) of BP/BP/MoS$_2$ and BlueP/BlueP/MoS$_2$, as shown in Figure 3. The CBM state is mainly contributed by the Mo element and the VBM state relies on the P element, but the small difference between LDOS and TDOS may suggest that the interlayer vdW interaction also has a role in tuning the energy bands. What is quite significant is that only one more monolayer of BP(BlueP) increases the DOS to more than twice compared with the corresponding bilayer heterostructures. A similar phenomenon occurs in BP/graphene/MoS$_2$ and BlueP/graphene/MoS$_2$, where the DOS also increases due to the insertion of a single graphene monolayer with respect to BP/MoS$_2$ and BlueP/MoS$_2$ bilayer heterostructures. It evidences the effectiveness of the proposed trilayer vdW heterostructures in optimizing the performance of 2D material systems.

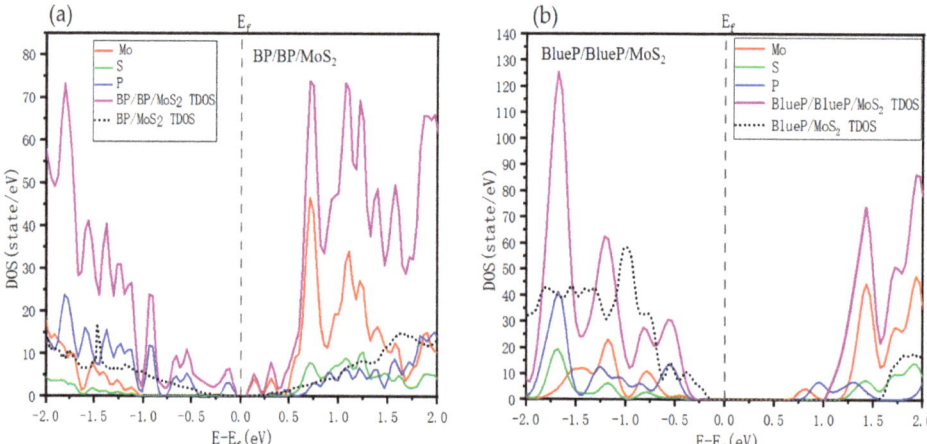

Figure 3. Calculated LDOS and TDOS for (**a**) BP/BP/MoS$_2$ and (**b**) BlueP/BlueP/MoS$_2$. The different colors represent different elements. The gray dashed line indicates Fermi energy. That of bilayer counterpart is also shown for comparison.

3.2.2. Band Alignment

The band alignment of the vdW heterostructure is important in the design of new 2D-material structures. It is thus necessary to specifically pick up the band alignment data from the proceeding band structures with the help of calculating the work function. In Figure 4, for the BP/BP/MoS$_2$ and BlueP/BlueP/MoS$_2$ heterostructures, p-type bilayer-BP(BlueP) (E_f closer to VBM) and n-type MoS$_2$ (E_f closer to CBM) form type-II band alignment, as is expected. This band alignment has carrier confinement effects [63], favoring the electron-hole separation and then increasing the internal gain and responsivity of photodetectors. In brief, BP/BP/MoS$_2$ shows a more obvious type-II band alignment with a larger offset on the valence band and will produce a stronger carrier confinement effect compared with BP/MoS$_2$ [37,40]. With graphene inserted, BP/graphene/MoS$_2$ and BlueP/graphene/MoS$_2$ heterostructures naturally show a type-I quantum-well-like band alignment, providing the potential to construct nano-scaled bipolar transistors, field-effect transistors and photon transistors, which may need three different energy structures. In contrast to the Schottky contact of graphene/MoS$_2$ and graphene/BP(BlueP), here, semiconductor heterojunctions are formed. By applying an electric field across this semiconductor heterojunction, the graphene bandgap can be further flexibly tuned to exhibit different band alignments, which is of significance for the construction of new types of field-effect photodetectors [49].

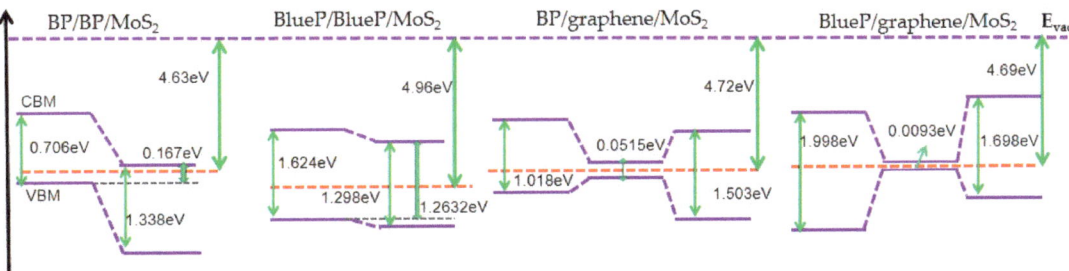

Figure 4. Band alignments of BP/BP/MoS$_2$, BlueP/BlueP/MoS$_2$, BP/graphene/MoS$_2$ and BlueP/graphene/MoS$_2$. The red dashed lines represent the Fermi levels. The vacuum level (E_{vac}) is shown by a purple dashed line.

3.3. Optical Absorption Spectra

On the optical absorption properties of 2D materials, many studies have been devoted to monolayer and bilayer vdW heterostructures, but little research has been carried out on trilayer ones, let alone a comparison between the bilayer and trilayer heterostructures [48]. The absorption spectra of BP/BP/MoS$_2$ and BlueP/BlueP/MoS$_2$ trilayer heterostructures are shown in Figure 5a,b. The absorption edge redshifts effectively expand the application ranges from covering the short-wave infrared band to covering the mid-wave infrared band for BP-based material, and from covering the visible band to covering the short-wave infrared band for BlueP-based material. This change is directly associated with the shrinkage of the bandgaps observed in Figure 3, but there seems to be something more incorporated as we see the following. Obviously, with only one more phosphorene monolayer, i.e., ~1/3 increase in material quantity, the spectral absorption intensity increases to be more or less twice with respect to monolayer and bilayer counterparts. This is more than expected, meaning that these two trilayer heterostructures greatly improve the quantum efficiency in optoelectronic devices [64] in wide spectral ranges. This sort of enhancement might be the result of complex interlayer vdW interactions. In more detail, the absorption of BP/BP/MoS$_2$ is much stronger (reaching 10^5 cm^{-1}) than that of BlueP/BlueP/MoS$_2$ in the near-infrared range, but the latter displays more significant absorption in the ultraviolet range. They can thus be applied effectively in different scenes. As a whole, these results make up for the shortage of monolayer 2D materials in optical characteristics and are beneficial for optoelectronic devices with the need for light absorption enhancement and absorption range broadening.

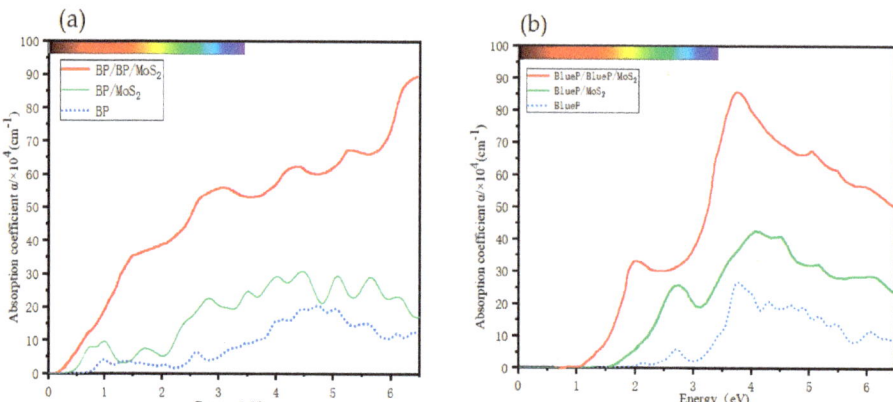

Figure 5. Optical absorption coefficient of (**a**) monolayer BP and MoS_2, bilayer BP, BP/MoS_2 and $BP/BP/MoS_2$; and (**b**) monolayer BlueP and MoS_2, bilayer BlueP, BP/MoS_2 and $BlueP/BlueP/MoS_2$.

Figure 6 displays the absorption coefficients of bilayer heterostructures graphene/BP, graphene/BlueP and graphene/MoS_2, and the trilayer heterostructures BP/graphene/MoS_2 and BlueP/graphene/MoS_2. The inset shows that the two kinds of trilayer heterostructures have stronger light absorption in the range below 0.8 eV than the related bilayer heterostructures, so they are better choices for graphene-based 2D materials to realize photodetectors widely responding to short-, mid- and long-wave infrared light. In the range of 2.2–6 eV, BP/graphene/MoS_2 shows remarkably higher light absorption than the related bilayer heterostructures, making it able to realize better graphene-based optoelectronic devices working in a wide range from visible to deep ultraviolet. BlueP/graphene/MoS_2 shows better light absorption characteristics than the related bilayer heterostructures in the range of 4–6 eV, so it is more suitable for use as a graphene-based solar-blind ultraviolet photodetector. These asymmetric graphene-sandwiched vdW heterostructrues are thus proved to be effective in tuning and improving the performance of graphene-based 2D materials on the basis of guaranteeing the excellent electronic properties of graphene.

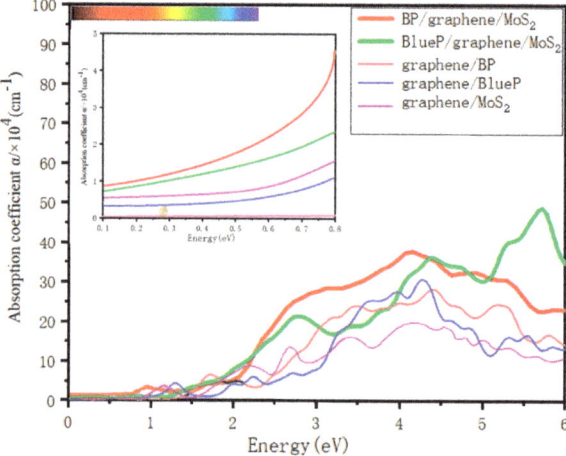

Figure 6. Optical absorption coefficient of BP/graphene/MoS_2 and BlueP/graphene/MoS_2 heterostructures, and the optical absorption coefficient of graphene/BP, graphene/BlueP and graphene/MoS_2 are also shown for comparison.

4. Conclusions

Using the first-principles calculation based on DFT theory, we proposed and simulated trilayer 2D vdW heterostructures: BP/BP/MoS$_2$, BlueP/BlueP/MoS$_2$, BP/graphene/MoS$_2$ and BlueP/graphene/MoS$_2$. The atomic structures of these heterostructures are all stable and look feasible to prepare. For BP/BP/MoS$_2$ and BlueP/BlueP/MoS$_2$, all the bandgaps were tuned to be smaller, to effectively compensate for the shortage of corresponding monolayers and to satisfy a wider spectrum of applications with respect to the bilayer structures BP/MoS$_2$ and BlueP/MoS$_2$, respectively. Their higher DOS, much stronger light absorption and more obvious type-II band alignment will realize better performance of optoelectronic devices working in a wide spectrum from mid-wave (short-wave) infrared to violet. In contrast, for BP/graphene/MoS$_2$ and BlueP/graphene/MoS$_2$, most of the bandgaps were enlarged, implying different tuning mechanisms related to vdW interaction. Specifically in BP/graphene/MoS$_2$, the graphene bandgap was well opened to 0.051 eV, more than one order of magnitude larger than usual, indicating more space for graphene tuning in trilayer vdW heterostructures. Accompanying the bandgap increase, both BP/graphene/MoS$_2$ and BlueP/graphene/MoS$_2$ exhibit absorption enhancement in wide spectral ranges, suggesting that these trilayer vdW heterostructures show better prospects as graphene-based 2D material optoelectronic devices with broad responses in the short- to long-wave infrared, visible to deep ultraviolet and solar-blind ultraviolet bands. It is concluded that the proposed 2D trilayer vdW heterostructures are potentially applicable as novel optoelectronic devices.

Author Contributions: B.C.: model building, simulation, data processing, and writing—original draft; Y.Z., R.J., X.W., S.H., X.H., W.Z., Q.D. and L.Z.: investigation help and discussion; P.L.: simulation instruction, discussion and revision guidance; H.-Z.S.: writing and revision guidance, project administration, resources, and supervision. All authors have read and agreed to the published version of the manuscript.

Funding: This research was supported by the National Key Research and Development Program of China under grants No. 2019YFB2203400 and No. 2021YFA0718803, and by the Natural Science Foundation of Sichuan Province under grant No. 2022NSFSC1817. We also acknowledge the support of the Special Subject of Significant Science and Technology of Sichuan Province under grant 2018TZDZX0001 and the Special Subject of Significant Innovation of Chengdu City under grant 2021-YF08-00159-GX.

Data Availability Statement: The data presented in this study are available on request from the corresponding author.

Conflicts of Interest: The authors declare no conflict of interest.

References

1. Miao, J.; Wang, C. Avalanche photodetectors based on two-dimensional layered materials. *Nano Res.* **2021**, *14*, 1878–1888. [CrossRef]
2. Liu, Y.; Weiss, N.O.; Duan, X.; Cheng, H.C.; Huang, Y.; Duan, X. Van der Waals heterostructures and devices. *Nat. Rev. Mater.* **2016**, *1*, 16042. [CrossRef]
3. Das, S.; Gulotty, R.; Sumant, A.V.; Roelofs, A. All two-dimensional, flexible, transparent, and thinnest thin film transistor. *Nano Lett.* **2014**, *14*, 2861–2866. [CrossRef] [PubMed]
4. Long, M.; Wang, P.; Fang, H.; Hu, W. Progress, challenges, and opportunities for 2D material based photodetectors. *Adv. Funct. Mater.* **2019**, *29*, 1803807. [CrossRef]
5. Akinwande, D.; Huyghebaert, C.; Wang, C.H.; Serna, M.I.; Goossens, S.; Li, L.J.; Koppens, F.H. Graphene and two-dimensional materials for silicon technology. *Nature* **2019**, *573*, 507–518. [CrossRef] [PubMed]
6. Geim, A.K.; Grigorieva, I.V. Van der Waals heterostructures. *Nature* **2013**, *499*, 419–425. [CrossRef]
7. Novoselov, K.S.; Mishchenko, O.A.; Carvalho, O.A.; Castro Neto, A.H. 2D materials and van der Waals heterostructures. *Science* **2016**, *353*, aac9439. [CrossRef]
8. Jariwala, D.; Marks, T.J.; Hersam, M.C. Mixed-dimensional van der Waals heterostructures. *Nat. Mater.* **2017**, *16*, 170–181. [CrossRef]
9. Wang, H.; Li, Z.; Li, D.; Chen, P.; Pi, L.; Zhou, X.; Zhai, T. Van der Waals Integration Based on Two-Dimensional Materials for High-Performance Infrared Photodetectors. *Adv. Funct. Mater.* **2021**, *31*, 2103106. [CrossRef]

10. Wang, P.; Xia, H.; Li, Q.; Wang, F.; Zhang, L.; Li, T.; Hu, W. Sensing infrared photons at room temperature: From bulk materials to atomic layers. *Small* **2019**, *15*, 1904396. [CrossRef]
11. Guan, X.; Yu, X.; Periyanagounder, D.; Benzigar, M.R.; Huang, J.K.; Lin, C.H.; Wu, T. Recent progress in short-to long-wave infrared photodetection using 2D materials and heterostructures. *Adv. Opt. Mater.* **2021**, *9*, 2001708. [CrossRef]
12. Rogalski, A.; Antoszewski, J.; Faraone, L. Third-generation infrared photodetector arrays. *J. Appl. Phys.* **2009**, *105*, 344–348. [CrossRef]
13. Rhodes, D.; Chae, S.H.; Ribeiro-Palau, R.; Hone, J. Disorder in van der Waals heterostructures of 2D materials. *Nat. Mater.* **2019**, *18*, 541–549. [CrossRef] [PubMed]
14. Geim, A.K.; Novoselov, K.S. The rise of graphene. *Nat. Mater.* **2007**, *6*, 183–191. [CrossRef]
15. Bhimanapati, G.R.; Lin, Z.; Meunier, V.; Jung, Y.; Cha, J.; Das, S.; Robinson, J.A. Recent advances in two-dimensional materials beyond graphene. *ACS Nano* **2015**, *9*, 11509–11539. [CrossRef]
16. Ebnonnasir, A.; Narayanan, B.; Kodambaka, S.; Ciobanu, C.V. Tunable MoS_2 bandgap in MoS_2-graphene heterostructures. *Appl. Phys. Lett.* **2014**, *105*, 031603. [CrossRef]
17. Radisavljevic, B.; Radenovic, A.; Brivio, J.; Giacometti, V.; Kis, A. Single-layer MoS_2 transistors. *Nat. Nanotechnol.* **2011**, *6*, 147–150. [CrossRef]
18. Zong, X.; Hu, H.; Ouyang, G.; Wang, J.; Shi, R.; Zhang, L.; Chen, X. Black phosphorus-based van der Waals heterostructures for mid-infrared light-emission applications. *Light Sci. Appl.* **2020**, *9*, 114. [CrossRef]
19. Chen, P.; Li, N.; Chen, X.; Ong, W.J.; Zhao, X. The rising star of 2D black phosphorus beyond graphene: Synthesis, properties and electronic applications. *2d Mater.* **2017**, *5*, 014002. [CrossRef]
20. Castellanos-Gomez, A. Black phosphorus: Narrow gap, wide applications. *J. Phys. Chem. Lett.* **2015**, *6*, 4280–4291. [CrossRef]
21. Zhao, X.; Bo, M.; Huang, Z.; Zhou, J.; Peng, C.; Li, L. Heterojunction bond relaxation and electronic reconfiguration of WS_2-and MoS_2-based 2D materials using BOLS and DFT. *Appl. Surf. Sci.* **2018**, *462*, 508–516. [CrossRef]
22. Hu, W.; Wang, T.; Zhang, R.; Yang, J. Effects of interlayer coupling and electric fields on the electronic structures of graphene and MoS_2 heterobilayers. *J. Mater. Chem. C* **2016**, *4*, 1776–1781. [CrossRef]
23. Le, N.B.; Huan, T.D.; Woods, L.M. Interlayer interactions in van der Waals heterostructures: Electron and phonon properties. *ACS Appl. Mater. Interfaces* **2016**, *8*, 6286–6292. [CrossRef] [PubMed]
24. Cai, Y.; Zhang, G.; Zhang, Y.W. Electronic properties of phosphorene/graphene and phosphorene/hexagonal boron nitride heterostructures. *J. Phys. Chem. C* **2015**, *119*, 13929–13936. [CrossRef]
25. Su, J.; Xiao, B.; Jia, Z. A first principle study of black phosphorene/N-doped graphene heterostructure: Electronic, mechanical and interface properties. *Appl. Surf. Sci.* **2020**, *528*, 146962. [CrossRef]
26. Behera, S.K.; Deb, P. Controlling the bandgap in graphene/h-BN heterostructures to realize electron mobility for high performing FETs. *RSC Adv.* **2017**, *7*, 31393–31400. [CrossRef]
27. Komsa, H.P.; Krasheninnikov, A.V. Electronic structures and optical properties of realistic transition metal dichalcogenide heterostructures from first principles. *Phys. Rev. B* **2013**, *88*, 085318. [CrossRef]
28. Lu, N.; Guo, H.; Li, L.; Dai, J.; Wang, L.; Mei, W.N.; Zeng, X.C. MoS_2/MX_2 heterobilayers: Bandgap engineering via tensile strain or external electrical field. *Nanoscale* **2014**, *6*, 2879–2886. [CrossRef]
29. Ren, K.; Sun, M.; Luo, Y.; Wang, S.; Yu, J.; Tang, W. First-principle study of electronic and optical properties of two-dimensional materials-based heterostructures based on transition metal dichalcogenides and boron phosphide. *Appl. Surf. Sci.* **2019**, *476*, 70–75. [CrossRef]
30. Wang, S.; Tian, H.; Ren, C.; Yu, J.; Sun, M. Electronic and optical properties of heterostructures based on transition metal dichalcogenides and graphene-like zinc oxide. *Sci. Rep.* **2018**, *8*, 12009. [CrossRef]
31. Huang, L.; Li, J. Tunable electronic structure of black phosphorus/blue phosphorus van der Waals pn heterostructure. *Appl. Phys. Lett.* **2016**, *108*, 083101. [CrossRef]
32. Sun, M.; Chou, J.-P.; Yu, J.; Tang, W. Electronic properties of blue phosphorene/graphene and blue phosphorene/graphene-like gallium nitride heterostructures. *Phys. Chem. Chem. Phys.* **2017**, *19*, 17324–17330. [CrossRef] [PubMed]
33. Liao, C.; Zhao, Y.; Ouyang, G. Strain-modulated band engineering in two-dimensional black phosphorus/MoS_2 van der Waals heterojunction. *ACS Omega* **2018**, *3*, 14641–14649. [CrossRef] [PubMed]
34. Huang, L.; Huo, N.; Li, Y.; Chen, H.; Yang, J.; Wei, Z.; Li, S.S. Electric-field tunable band offsets in black phosphorus and MoS_2 van der Waals pn heterostructure. *J. Phys. Chem. Lett.* **2015**, *6*, 2483–2488. [CrossRef] [PubMed]
35. Tang, K.; Qi, W.; Li, Y.; Wang, T. Electronic properties of van der Waals heterostructure of black phosphorus and MoS_2. *J. Phys. Chem. C* **2018**, *122*, 7027–7032. [CrossRef]
36. You, B.; Wang, X.; Zheng, Z.; Mi, W. Black phosphorene/monolayer transition-metal dichalcogenides as two dimensional van der Waals heterostructures: A first-principles study. *Phys. Chem. Chem. Phys.* **2016**, *18*, 7381–7388. [CrossRef]
37. Yang, F.; Han, J.; Zhang, L.; Tang, X.; Zhuo, Z.; Tao, Y.; Dai, Y. Adjustable electronic and optical properties of BlueP/MoS_2 van der Waals heterostructure by external strain: A First-principles study. *Nanotechnology* **2020**, *31*, 375706. [CrossRef]
38. Ye, L.; Li, H.; Chen, Z.; Xu, J. Near-infrared photodetector based on MoS_2/black phosphorus heterojunction. *ACS Photonics* **2016**, *3*, 692–699. [CrossRef]
39. Liu, Y.; Shivananju, B.N.; Wang, Y.; Zhang, Y.; Yu, W.; Xiao, S.; Bao, Q. Highly efficient and air-stable infrared photodetector based on 2D layered graphene–black phosphorus heterostructure. *ACS Appl. Mater. Interfaces* **2017**, *9*, 36137–36145. [CrossRef]

40. Li, R. Electronic properties of hybrid graphene/nitrogene/graphene hetero-trilayers. *Phys. E Low-Dimens. Syst. Nanostructures* **2020**, *123*, 114166. [CrossRef]
41. Liu, B.; Chen, Y.; You, C.; Liu, Y.; Kong, X.; Li, J.; Zhang, Y. High performance photodetector based on graphene/MoS$_2$/graphene lateral heterostrurcture with Schottky junctions. *J. Alloys Compd.* **2019**, *779*, 140–146. [CrossRef]
42. Datta, K.; Khosru, Q.D. Electronic properties of MoS$_2$/MX$_2$/MoS$_2$ trilayer heterostructures: A first principle study. *ECS J. Solid State Sci. Technol.* **2016**, *5*, Q3001–Q3007. [CrossRef]
43. Datta, K.; Shadman, A.; Rahman, E.; Khosru, Q.D.M. Trilayer TMDC Heterostructures for MOSFETs and Nanobiosensors. *J. Electron. Mater.* **2017**, *46*, 1248–1260. [CrossRef]
44. Bafekry, A.; Yagmurcukardes, M.; Akgenc, B.; Ghergherehchi, M.; Nguyen, C.V. Van der Waals heterostructures of MoS$_2$ and Janus MoSSe monolayers on graphitic boron-carbon-nitride (BC$_3$, C$_3$N, C$_3$N$_4$ and C$_4$N$_3$) nanosheets: A first-principles study. *J. Phys. D: Appl. Phys.* **2020**, *53*, 355106. [CrossRef]
45. Liu, S.; Li, X.; Meng, D.; Li, S.; Chen, X.; Hu, T. Tunable electronic properties of MoS$_2$/SiC heterostructures: A First-Principles study. *J. Electron. Mater.* **2022**, *51*, 3714–3726. [CrossRef]
46. Han, J.; Yang, F.; Xu, L.; Zhuo, Z.; Cao, X.; Tao, Y.; Liu, W. Modulated electronic and optical properties of bilayer/trilayer Blue Phosphorene/MoX$_2$ (X= S, Se) van der Waals heterostructures. *Surf. Interfaces* **2021**, *25*, 101228. [CrossRef]
47. Kim, D.; Hashmi, A.; Hwang, C.; Hong, J. Thickness dependent band gap and effective mass of BN/graphene/BN and graphene/BN/graphene heterostructures. *Surf. Sci.* **2013**, *610*, 27–32. [CrossRef]
48. Farooq, M.U.; Hashmi, A.; Hong, J. Thickness dependent optical properties of multilayer BN/graphene/BN. *Surf. Sci.* **2015**, *634*, 25–30. [CrossRef]
49. Xu, L.; Huang, W.Q.; Hu, W.; Yang, K.; Zhou, B.X.; Pan, A.; Huang, G.F. Two-dimensional MoS2-graphene-based multilayer van der Waals heterostructures: Enhanced charge transfer and optical absorption, and electric-field tunable Dirac point and band gap. *Chem. Mater.* **2017**, *29*, 5504–5512. [CrossRef]
50. Xia, C.; Du, J.; Fang, L.; Li, X.; Zhao, X.; Song, X.; Wang, T.; Li, J. PtSe$_2$/graphene hetero-multilayer: Gate-tunable Schottky barrier height and contact type. *Nanotechnology* **2018**, *29*, 465707. [CrossRef]
51. Long, M.; Liu, E.; Wang, P.; Gao, A.; Xia, H.; Luo, W.; Miao, F. Broadband photovoltaic detectors based on an atomically thin heterostructure. *Nano Lett.* **2016**, *16*, 2254–2259. [CrossRef] [PubMed]
52. Li, A.; Chen, Q.; Wang, P.; Gan, Y.; Qi, T.; Wang, P.; Gong, Y. Ultrahigh-sensitive broadband photodetectors based on dielectric shielded MoTe$_2$/Graphene/SnS$_2$ p–g–n junctions. *Adv. Mater.* **2019**, *31*, 1805656. [CrossRef] [PubMed]
53. Sun, G.; Kürti, J.; Rajczy, P.; Kertesz, M.; Hafner, J.; Kresse, G. Performance of the Vienna ab initio simulation package (VASP) in chemical applications. *J. Mol. Struct. THEOCHEM* **2003**, *624*, 37–45. [CrossRef]
54. Hafner, J. Ab-initio simulations of materials using VASP: Density-functional theory and beyond. *J. Comput. Chem.* **2008**, *29*, 2044–2078. [CrossRef] [PubMed]
55. Bucko, T.; Hafner, J.; Lebegue, S.; Angyan, J.G. Improved description of the structure of molecular and layered crystals: Ab initio DFT calculations with van der Waals corrections. *J. Phys. Chem. A* **2010**, *114*, 11814–11824. [CrossRef]
56. Klimeš, J.; Bowler, D.R.; Michaelides, A. Chemical accuracy for the van der Waals density functional. *J. Phys. Condens. Matter* **2010**, *22*, 022201. [CrossRef]
57. Dong, H.M.; Huang, L.S.; Liu, J.L.; Huang, F.; Zhao, C.X. Layer-dependent optoelectronic properties of black phosphorus. *Int. J. Mod. Phys. C* **2020**, *31*, 2050177. [CrossRef]
58. Sun, M.; Chou, J.P.; Gao, J.; Cheng, Y.; Hu, A.; Tang, W.; Zhang, G. Exceptional optical absorption of buckled arsenene covering a broad spectral range by molecular doping. *ACS Omega* **2018**, *3*, 8514–8520. [CrossRef]
59. Yu, L.; Lee, Y.H.; Ling, X.; Santos, E.J.; Shin, Y.C.; Lin, Y.; Palacios, T. Graphene/MoS$_2$ hybrid technology for large-scale two-dimensional electronics. *Nano Lett.* **2014**, *14*, 3055–3063. [CrossRef]
60. Zhang, J.L.; Zhao, S.; Han, C.; Wang, Z.; Zhong, S.; Sun, S.; Guo, R.; Zhou, X.; Gu, C.D.; Yuan, K.D.; et al. Epitaxial growth of single layer blue phosphorus: A new phase of two-dimensional phosphorus. *Nano Lett.* **2016**, *16*, 4903–4908. [CrossRef]
61. Qiao, J.; Kong, X.; Hu, Z.X.; Yang, F.; Ji, W. High-mobility transport anisotropy and linear dichroism in few-layer black phosphorus. *Nat. Commun.* **2014**, *5*, 4475. [CrossRef] [PubMed]
62. Yan, J.; Kim, M.H.; Elle, J.A.; Sushkov, A.B.; Jenkins, G.S.; Milchberg, H.M.; Fuhrer, M.S.; Drew, H.D. Dual-gated bilayer graphene hot-electron bolometer. *Nat. Nanotechnol.* **2012**, *7*, 472–478. [CrossRef] [PubMed]
63. Kim, J.D.; Chen, X.; Li, X.; Coleman, J.J. Photocurrent density enhancement of a III-V inverse quantum dot intermediate band gap photovoltaic device. In Proceedings of the 2015 Conference on Lasers and Electro-Optics (CLEO), San Jose, CA, USA, 10–15 May 2015.
64. Joseph, I.; Wan, K.; Hussain, S.; Guo, L.; Xie, L.; Shi, X. Interlayer angle-dependent electronic structure and optoelectronic properties of BP-MoS2 heterostructure: A first principle study. *Comput. Mater. Sci.* **2021**, *186*, 110056. [CrossRef]

Disclaimer/Publisher's Note: The statements, opinions and data contained in all publications are solely those of the individual author(s) and contributor(s) and not of MDPI and/or the editor(s). MDPI and/or the editor(s) disclaim responsibility for any injury to people or property resulting from any ideas, methods, instructions or products referred to in the content.

Article

A Stable Rechargeable Aqueous Zn–Air Battery Enabled by Heterogeneous MoS₂ Cathode Catalysts

Min Wang [1], Xiaoxiao Huang [2], Zhiqian Yu [1], Pei Zhang [3], Chunyang Zhai [2,*], Hucheng Song [1,*], Jun Xu [1] and Kunji Chen [1]

[1] National Laboratory of Solid State Microstructures, School of Electronics Science and Engineering, College of Engineering and Applied Sciences, Nanjing University, Nanjing 210093, China
[2] School of Materials Science and Chemical Engineering, Ningbo University, Ningbo 315211, China
[3] College of Electrical and Information Engineering, Zhengzhou University of Light Industry, Zhengzhou 450002, China
* Correspondence: zhaichunyang@nbu.edu.cn (C.Z.); hcsong@nju.edu.cn (H.S.)

Citation: Wang, M.; Huang, X.; Yu, Z.; Zhang, P.; Zhai, C.; Song, H.; Xu, J.; Chen, K. A Stable Rechargeable Aqueous Zn–Air Battery Enabled by Heterogeneous MoS₂ Cathode Catalysts. *Nanomaterials* 2022, *12*, 4069. https://doi.org/10.3390/nano12224069

Academic Editor: Sónia Carabineiro

Received: 28 October 2022
Accepted: 16 November 2022
Published: 18 November 2022

Publisher's Note: MDPI stays neutral with regard to jurisdictional claims in published maps and institutional affiliations.

Copyright: © 2022 by the authors. Licensee MDPI, Basel, Switzerland. This article is an open access article distributed under the terms and conditions of the Creative Commons Attribution (CC BY) license (https://creativecommons.org/licenses/by/4.0/).

Abstract: Aqueous rechargeable zinc (Zn)–air batteries have recently attracted extensive research interest due to their low cost, environmental benignity, safety, and high energy density. However, the sluggish kinetics of oxygen (O_2) evolution reaction (OER) and the oxygen reduction reaction (ORR) of cathode catalysts in the batteries result in the high over-potential that impedes the practical application of Zn–air batteries. Here, we report a stable rechargeable aqueous Zn–air battery by use of a heterogeneous two-dimensional molybdenum sulfide (2D MoS₂) cathode catalyst that consists of a heterogeneous interface and defects-embedded active edge sites. Compared to commercial Pt/C-RuO₂, the low cost MoS₂ cathode catalyst shows decent oxygen evolution and acceptable oxygen reduction catalytic activity. The assembled aqueous Zn–air battery using hybrid MoS₂ catalysts demonstrates a specific capacity of 330 mAh g^{-1} and a durability of 500 cycles (~180 h) at 0.5 mA cm^{-2}. In particular, the hybrid MoS₂ catalysts outperform commercial Pt/C in the practically meaningful high-current region (>5 mA cm^{-2}). This work paves the way for research on improving the performance of aqueous Zn–air batteries by constructing their own heterogeneous surfaces or interfaces instead of constructing bifunctional catalysts by compounding other materials.

Keywords: Zn–air batteries; 2D MoS₂; defects-embedded; OER; ORR

1. Introduction

Zinc–air (Zn–air) batteries using oxygen (O_2) as the active medium have recently attracted extensive research interest as a promising energy storage device for the next-generation energy storage technology due to their high theoretical energy density of 1086 Wh kg^{-1}, safety and environmental friendliness, cost-effectiveness, and readily available raw materials [1–6]. The reported Zn–air battery usually consists of a Zn metal anode (Zn-anode), an alkaline aqueous electrolyte, and a porous carbonaceous cathode, where the O_2 can be absorbed and react with the electron and H_2O to convert OH^{-1} via an oxygen reduction reaction (ORR, see Equation (1)) on discharge, showing a theoretical specific energy of ~1086 Wh kg^{-1}, being at least four times higher than that of Li-ion batteries of ~265 Wh kg^{-1} [7–9].

$$\text{ORR on discharge: } O_2 \text{ (g)} + 2H_2O \text{ (l)} + 4e^{-1} \rightarrow 4OH^{-} \text{ (aq)} \qquad (1)$$

$$\text{OER on charge: } 4OH \text{ (aq)} \rightarrow O_2 \text{ (g)} + 2H_2O \text{ (l)} + 4e^{-} \qquad (2)$$

On charge, the OH^{-1} can be converted into O_2 gas and H_2O to store electric energy via the oxygen evolution reaction (OER, see Equation (2)) [8]. However, the OER and ORR of the cathodes are slow in kinetics due to the proton-coupled multistep electron transfer process for the reversible charge/discharge reactions [7–13]. In addition, the alkaline aqueous electrolyte can capture carbon dioxide in the air and generate insoluble and insulating carbonate [14], such as the typical Li_2CO_3 byproduct that has a ~5.09 eV band gap defined by the highest occupied molecular orbital (HOMO) and the lowest unoccupied molecular orbital (LUMO) levels [15], which result in extremely sluggish kinetics for OER and ORR processes [14]. Various advanced strategies have been used to improve the kinetics of the OER/ORR on charge/discharge [16]. Among them, precious metal and metal oxide catalysts, including platinum (Pt), ruthenium (Ru), iridium (Ir), ruthenium dioxide (RuO_2), iridium dioxide (IrO_2), etc., usually show good catalytic activity for ORR or OER. As the most representative cooperation, Pt is used as an ORR electrocatalyst in alkaline media, and the ruthenium dioxide (RuO_2)/iridium dioxide (IrO_2) is used as the OER catalyst [17–21]. However, these precious metals and metal oxides are difficult to use in large-scale industrial applications due to their poor chemical stability, relative scarcity, and high cost [21]. Therefore, the exploitation of cheap and efficient electrocatalysts to boost the OER and ORR processes is highly necessary for developing high-performance aqueous Zn–air batteries.

As one of the most classical 2D transition-metal chalcogenides (TMDs), molybdenum disulfide (MoS_2) has abundant active edge sites and good crystallinity [22,23], and it recently has been considered to be an effective hydrogen evolution reaction (HER) catalyst [24]. In addition, research suggests that crystal-amorphous interface and grain boundaries of MoS_2 can catalyze electrochemical reactions [22]. Nonetheless, Amiinu et al. recently reported multifunctional Mo–N/C@MoS_2 electrocatalysts for HER, OER, ORR, and Zn–air batteries that show a high power density of ≈196.4 mW cm^{-2} and a voltaic efficiency of ≈63% at 5 mA cm^{-2}, as well as excellent cycling stability, even after 48 h at 25 mA cm^{-2} [10]. Bai et al. reported heterostructure Co_9S_8@MoS_2 core–shell structures that exhibited robust OER performance and a 20 h cycle lifespan for Zn–air batteries with low high discharge voltages/low discharge voltages (~1.28 V/2.03 V) [25]. Plulia et al. reported layered MoS_2/graphene nanosheet catalysts that showed enhanced oxygen reduction activity, nearly double that of graphene nanosheets or ~25-fold that of MoS_2 nanosheets, and high open circuit voltages (1.4 V) and high specific energy of up to 130 W h kg^{-1} for assembled Zn–air batteries [26]. Although great efforts have thus been focused on improving the electrocatalytic performance of MoS_2, the OER activity of MoS_2 is largely limited, which makes it difficult to use for OER and Zn–air batteries [27]. Compared with the reported bifunctional catalysts consisting of two typical catalytic function materials, the development of a single and efficient non-precious metal catalyst with a good oxygen evolution reaction and oxygen reduction reaction, by a simple preparation process, is of great significance for the development and application of aqueous zinc–air batteries.

Here, we report an aqueous Zn–air battery that shows advantages in cycle performance and economic cost. The Zn–air battery consists of a Zn-anode, an alkaline aqueous solution electrolyte, and the air cathode (see Figure S1 in Supplementary Materials). The air cathode materials, consisting of electron-conducting carbon nanotube (CNT) and MoS_2 catalysts, are prepared on the carbon paper gas diffusion layer (GDL) that can provide an efficient gas transport for the assembled battery (see Figure 1a). A stable cycling Zn–air battery can be obtained through several mechanisms: (a) An air cathode consisting of heterogeneous MoS_2 catalysts and CNT on the GDL can provide good conductivity for charge transport, and an adequate three-phase interface for ORR and OER for the assembled aqueous Zn–air battery on charge/discharge (see Figure 1b) [21]. (b) A heterogeneous interface consisting of a super-hydrophobic carbon paper GDL and a hydrophilic MoS_2 catalyst enables efficient gas utilization, where the oxygen (O_2) can be directly absorbed and reacted at the interface without being dissolved in aqueous solution on the charge/discharge process (see Figure 1c,d) [28]. (c) The MoS_2 with heterogeneous and abundant active edge sites can enable an efficient electrochemical reaction, especially for the hydrogen evolution reaction (HER) and the oxygen evolution reaction [29]. Notably, the heterogeneous catalyst is only composed of MoS_2 that can form heterogeneous interfaces and defects-embedded active edge sites only by using a relatively low synthesis temperature (see Section 2).

The oxygen evolution activity of the heterogeneous MoS₂ catalyst, by constructing its own heterogeneous surface or interface instead of constructing bifunctional catalysts by compounding another OER catalytically active material, exceeds commercial Pt/C-RuO₂ catalytic activity for the assembled aqueous Zn–air battery.

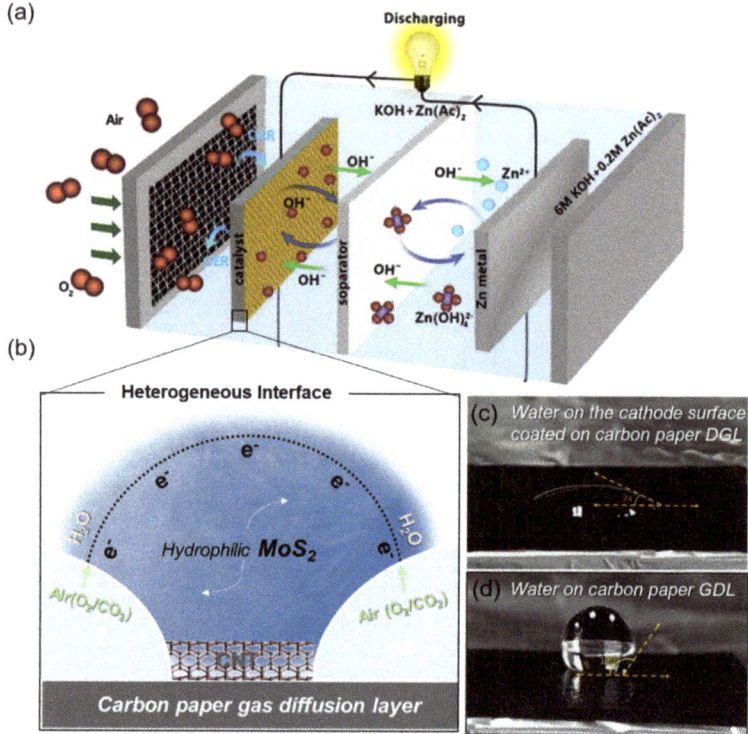

Figure 1. Configuration of the aqueous Zn–air battery with a heterogeneous design. (**a**) Configuration of the aqueous Zn–air battery and the enlarged diagram of the heterogeneous interface consisting of hydrophilic MoS₂ catalysts, electron-conductive CNT, and a super-hydrophobic carbon paper gas diffusion layer. (**b**) Diagram of charging and discharging of an assembled water-zinc-air battery with a three-phase interface ORR and OER. (**c**) Optical photo of a drop of water on the cathode surface coated on carbon paper GDL that shows a contact angle of ~24 degrees. (**d**) Optical photo of a drop of water on carbon paper GDL that shows a contact angle of ~130 degrees.

2. Materials and Methods

2.1. Chemicals

(NH$_4$)$_2$MoS$_4$, dimethylformamide (DMF), ethanol, carbon nanotubes, RuO$_2$, and KOH were purchased from Sigma Aldrich. Nafion D520 (5 wt%), and carbon paper (GDL340) were received from SCI-Materials-Hub. A commercial Pt/C catalyst with 20 wt% Pt was purchased from Shanghai Macklin Biochemical Technology Co., Ltd (Shanghai, China). All the reagents were of analytical grade and used as received without further purification. Deionized water was used throughout the experimental processes.

2.2. Material Preparation

MoS$_2$ was prepared by hydrothermal synthesis. Briefly, 0.2 g (NH$_4$)$_2$MoS$_4$ was dissolved in 15 mL DMF by ultrasonication for 15 min to form a homogeneous solution. Then, the solution was transferred into a 25 mL Teflon-lined stainless steel autoclave maintained at 180 °C for 10 h and cooled to room temperature naturally. The final product was collected by centrifugation, washed with water and ethanol three times each, and subsequently dried in a vacuum oven at 60 °C overnight, resulting in black powder.

2.3. Material Characterization

The crystal structures of samples were identified on a BRUKER D8 ADVANCE X-ray diffractometer with Cu-Kα radiation (λ = 1.54178 Å) over the 2θ range from 5 to 90°. The morphology and microstructure were characterized by a scanning electron microscope (SEM, Hitachi SU8010) and a transmission electron microscope (TEM, TecnaiG2F20).

2.4. Electrode Preparation and Battery Assembly

Firstly, 0.5 mL Nafion D520 solution, 1 mg MOS_2, 1 mg carbon nanotubes, and 10 mL ethanol were measured into a beaker and ultrasonicated for 30 min. After the mixture was evenly mixed, 1 mL of the mixture was taken and sprayed evenly onto 2 × 2 cm carbon paper with a spray gun. While spraying, the surface of the carbon paper was dried with a baking lamp. After drying at 60 °C, the loading capacity of MoS_2 catalyst was estimated to be 1 mg cm^{-2}, which is the air electrode. When making the contrast electrode, it was only necessary to change the MoS_2 catalyst with the same amount of Pt/C-RuO_2 powder (m(Pt/C):m(RuO_2) = 1:1), and the other steps were exactly the same as above. A polished zinc foil (thickness: 0.25 mm) was used as the anode, and the electrolyte was 6.0 M KOH for the primary Zn–air battery and 6.0 M KOH with 0.20 M Zn $(CH_3COO)_2$ for the rechargeable Zn–air battery. Measurements were carried out at 25 °C with an electrochemical workstation (CHI 660E, CH Instrument, Austin, TX, USA).

2.5. Electrochemical Test

The discharge/charge cycling performance of the ASS Zn–air battery was obtained by using an electrochemical test system (Hokuto Denko Corporation, HJ1001SD8, Meguro-ku, Tokyo). Alternating-current impedance spectroscopy of the lithium–air batteries was investigated using an electrochemical workstation (CHI 660E, CH Instrument, Austin, TX, USA).

3. Results

Typically, heterogeneous MoS_2 catalysts are prepared by a simple hydrothermal process (see Figure S1 in Supplementary Materials). The structural and morphological properties of the prepared MoS_2 catalysts were characterized by a transmission electron microscope (TEM), X-ray diffraction (XRD), and an aberration-corrected high-angle annular dark-field scanning TEM (HAADF-STEM). As shown in Figure 2, the prepared MoS_2 catalysts were typical layered structures (see Figure 2a) [30] consisting of Mo and S elements, where the S and Mo elements were evenly distributed on the prepared MoS_2 catalyst (see Figure 2a–d). XRD patterns further suggested the fact that the prepared catalysts were MoS_2, where XRD diffraction peaks located at 14.1°, 33.2°, 39.4° and 58.2° corresponded to the (002), (100), (103), and (110) planes of the 2H-MoS_2 nanosheet (Powder Diffraction File no. 37-1492, Joint Committee on Powder Diffraction Standards), respectively (see Figure 2e) [31]. Other diffraction peaks located at 11.4°, 22.5°, and 29.6° could be attributed to the 2m-1T phase transition of MoS_2 driven by in situ intercalation of ammonium or alkyl amine cations [32]. High-resolution HAADF-STEM images further indicated that the prepared MoS_2 were heterogeneous, where amorphous phases were distributed on the crystalline MoS_2 phases (see Figure 2f) [33]. Further, the STEM of the MoS_2 catalyst showed a heterogeneous surface consisting of active edge sites and abundant defects/disordered phases (see Figure 2g) that could enhance the electrochemical reaction, where the interplanar lattice spacings of 0.678 nm (see Figure 2g,h) were consistent with the crystal reflection of MoS_2 [34]. This result was also consistent with the XRD result of the prepared MoS_2 catalyst. The high-resolution HAADF-STEM and TEM also indicated the formation of heterogeneous crystalline-amorphous MoS_2 (see Figure 2i,j). The above results suggest that the formation of heterogeneous MoS$_2$ consists of crystalline-amorphous interfaces and defects-embedded active edge sites [35].

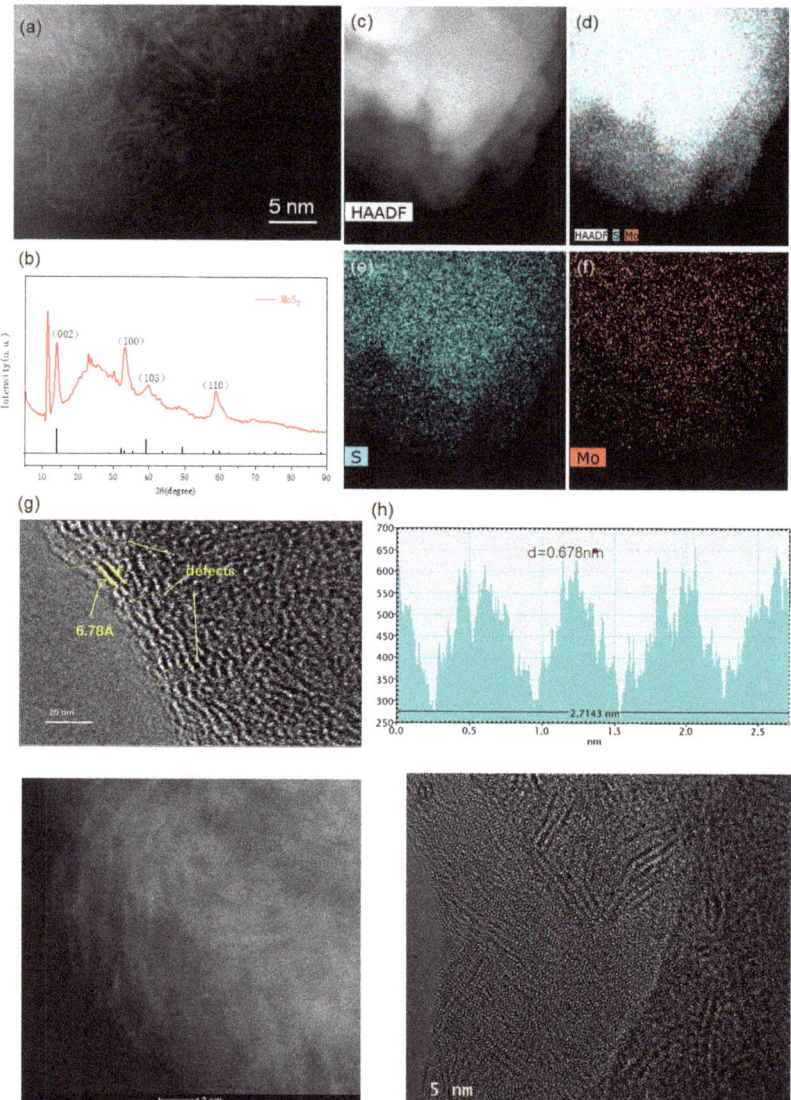

Figure 2. Characterization of the MoS$_2$ catalyst for the Zn–air battery. (**a**) STEM of the prepared MoS$_2$ catalyst. (**b**–**d**) EDS mappings of elementals (Mo and S) of the MoS$_2$ catalyst. (**e**) XRD pattern of the prepared MoS$_2$ catalyst. (**f**) STEM image and (**g**) TEM image of the MoS$_2$ catalyst. (**h**) Interplanar spacings of the MoS$_2$ catalyst, where the spacing of the nanosheets is found to be ∼6.78 Å. (**i**) High-resolution STEM and (**j**) TEM image of the prepared MoS$_2$ catalysts.

Befitting the electrochemical performance of the prepared MoS$_2$, the aqueous Zn–air battery using the MoS$_2$ catalyst showed a lower discharge potential (∼1.17 V) and a higher charge potential (∼2.39 V) than the Zn–air battery using commercial Pt/C-RuO$_2$ catalysts (∼1.35 V and 1.89 V, see Figure 3a) at the first cycle. Notably, the Zn–air battery using the

MoS$_2$ catalyst showed a lower discharge potential (~1.3 V) and a slightly lower charge potential (~1.88 V) than the Zn–air battery using commercial Pt/C-RuO$_2$ catalysts (~1.26 V and 1.94 V, see Figure 3a) after 500 cycles. This indicates that the heterogeneous MoS$_2$ catalysts can enable a more stable and better electrochemical performance than the expensive Pt/C-RuO$_2$ catalysts for the assembled aqueous Zn–air battery [36]. In addition, the Zn–air batteries using the MoS$_2$ catalysts could stably discharge a specific capacity of ~330 mAh g^{-1}, ~660 mAh g^{-1}, ~3300 mAh g^{-1}, and ~6600 mAh g^{-1} at 0.5 mA cm^{-2}, 1 mA cm^{-2}, 5 mA cm^{-2}, and 10 mA cm^{-2}, respectively (see Figure 3b). This suggests that the Zn–air battery using MoS$_2$ can be stably operated at high current density. Note that by using heterogeneous MoS$_2$ catalysts, the assembled aqueous Zn–air battery demonstrated good cycle performance with lower potential than that of the Zn–air battery using commercial Pt/C-RuO$_2$ catalysts (see Figure 3c and Figure S2 in Supplementary Materials). The long-term electrochemical test of the Zn–air battery using the MoS$_2$ catalyst showed an increasingly enhanced electrochemistry performance where the discharge potential decreased from ~2.38 V at the first cycle to ~1.98 V at the 500th cycle with a limited specific capacity of ~330 mAh g^{-1} (see Figure 3d). The enhanced electrochemical performance could also be observed by the reduction of charge potential during cycling (see Figure 3e). This indicates the fact that the MoS$_2$ catalysts with heterogeneous interfaces and defects-embedded active edge sites can demonstrate better OER performance for assembled Zn–air batteries than expensive Pt/C-RuO$_2$ catalysts during long-term limited capacity cycling [37]. Such good cycle performance can be attributed to the heterogeneous interface and defects-embedded active edge sites of the prepared MoS$_2$ catalyst, where the edge sites of the MoS$_2$ nanosheets can enable stronger adsorption toward oxygen (O$_2$), and other intermediates [38,39], defects [40], and amorphous phases can make the MoS$_2$ catalyst maintain high electrochemical activity [41], thus resulting in excellent cycle performance for the assembled aqueous Zn–air batteries.

Further, the heterogeneous MoS$_2$ were employed as cathode catalysts of a typical aqueous Zn-O$_2$ battery and a Zn-CO$_2$ battery. As showed in Figure 4a, all the batteries using the MoS$_2$ catalysts could operate at 0.5 mA cm^{-2} with a limited capacity of ~330 mAh g^{-1}. Notably, the aqueous Zn–air battery showed the highest discharge potential (~1.17 V) and the lowest charge potential (~2.39 V) at the first cycle compared to the Zn-O$_2$ battery (with a 1.3 V discharge potential and 2.2 V charge potential, see Figure 4a) and the Zn-CO$_2$ battery (with a 1.18 V discharge potential and 2.16 V charge potential, see Figure 4a). Additionally, the Zn–air battery using the heterogeneous MoS$_2$ catalysts demonstrated a long cycle lifespan (500 cycles, see Figure 4b), being at least four times more than the Zn-O$_2$ battery (~100 cycles, see Figure S2) and Zn-CO$_2$ battery (~100 cycles, see Figure S2). Clearly, the Zn–air battery using the heterogeneous MoS$_2$ catalyst showed the best cycle performance and the lowest potential gap of ~0.8 V compared to the Zn-O$_2$ battery and the Zn-CO$_2$ battery of more than 1.0 V (see Figure 4c and Figure S3 in Supplementary Materials). In addition, the aqueous Zn–air battery showed a lower charge potential (~1.35 V) and a higher discharge potential (~1.89 V) than the Zn–air battery using commercial Pt/C-RuO$_2$ catalysts (~1.35 V and 1.89 V, see Figure 3a) at the first cycle. Notably, the Zn–air battery using the MoS$_2$ catalyst showed a lower discharge potential (~1.3 V) and a slightly lower charge potential (~1.88 V) than the Zn–air battery using commercial Pt/C-RuO$_2$ catalysts (~1.26 V and 1.94 V, see Figure 3a) after 500 cycles. Furthermore, the assembled Zn–air battery showed a long cycle lifespan of over 500 cycles, being at least five times higher than that of the assembled Zn-O$_2$ battery and the Zn-CO$_2$ battery (~100 cycles, see Figure 4d). Notably, the Zn–air battery showed enhanced electrochemical cycle performance during long-term cycling that could be attributed to the activation of the cathode materials [42]. This indicates that the heterogeneous MoS$_2$ can be successfully used as cathode catalysts for both the Zn-O$_2$ battery and the Zn-CO$_2$ battery.

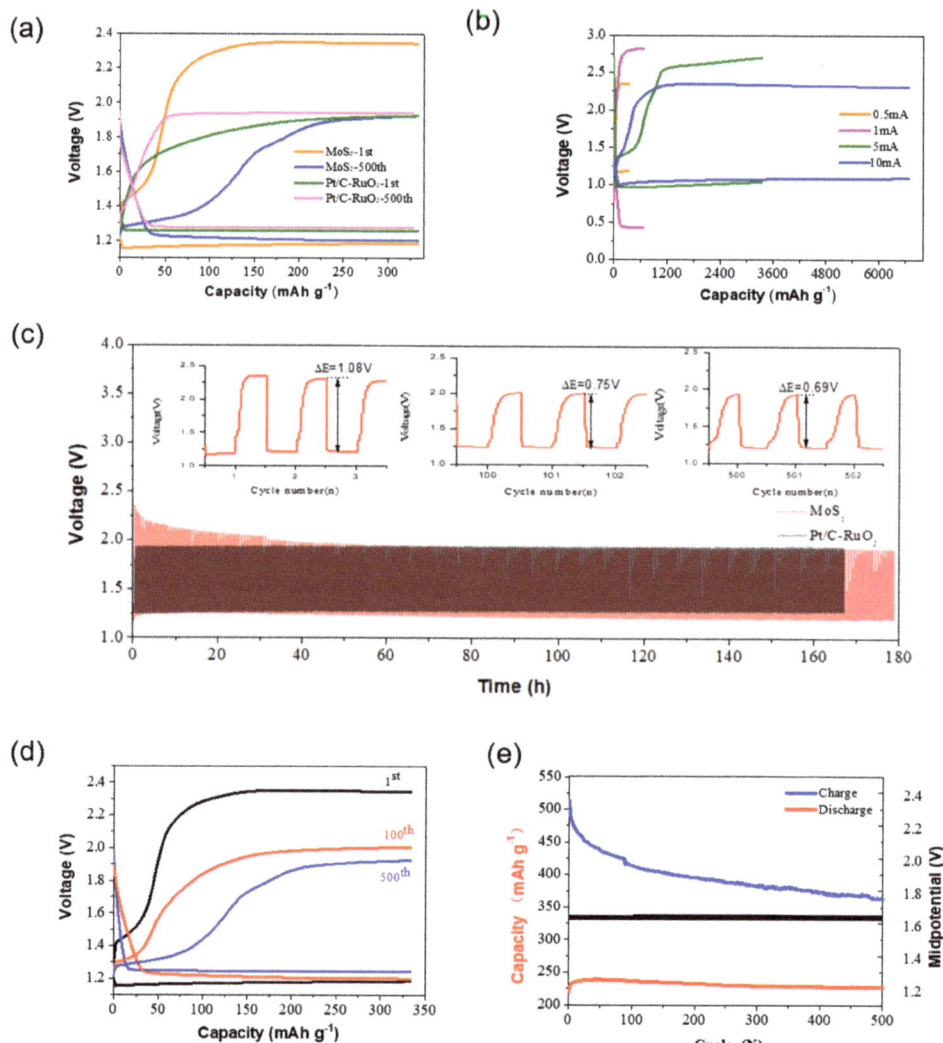

Figure 3. Electrochemical performance of the Zn–air batteries. (**a**) The discharge and charge curves of Zn–air batteries using the prepared MoS$_2$ catalysts and commercial Pt/C-RuO$_2$ catalysts at the 1st and 500th cycles, with a limited capacity of 330 mAh g^{-1}, respectively. (**b**) The discharge and charge curves of the assembled Zn–air batteries using the MoS$_2$ catalysts at various current densities of 0.5 A cm^{-2}, 1 A cm^{-2}, 5 A cm^{-2}, and 10 A cm^{-2}, respectively. (**c**) Electrochemical cycle performance of the Zn–air batteries using the prepared MoS$_2$ catalysts and commercial Pt/C-RuO$_2$ catalysts operating at 0.5 A cm^{-2}, respectively. (**d**) The discharge and charge curves of Zn–air batteries using the prepared MoS$_2$ catalysts at the 1st, 100th and 500th, respectively, and (**e**) corresponding to the voltage and capacity change of the Zn–air battery during the cycle.

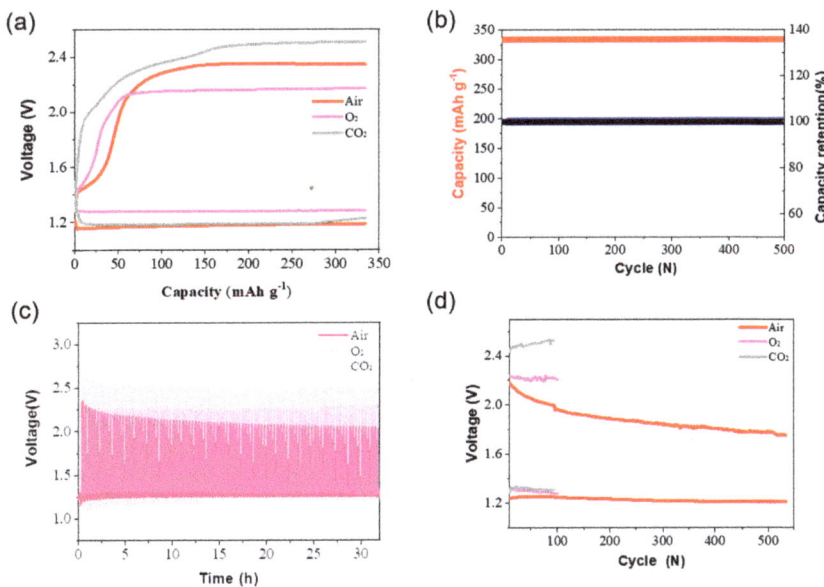

Figure 4. Electrochemical performance of the Zn–air battery, Zn-O_2 battery, and Zn-CO_2 battery using a heterogeneous MoS_2 catalyst. (**a**) The discharge and charge curves of the Zn–air battery, Zn-O_2 battery, and Zn-CO_2 battery using a heterogeneous MoS_2 catalyst at the 1st cycle with a limited capacity of 330 mAh g^{-1} and a current density of 0.5 mA cm^{-2}. (**b**) The specific capacity and corresponding capacity retention of the aqueous Zn–air battery, Zn-O_2 battery, and Zn-CO_2 battery during cycling with a current density of 0.5 mA cm^{-2}. (**c**) Voltage–time curves of the Zn–air battery, Zn-O_2 battery, and Zn-CO_2 battery, respectively, for cycling over 30 h with a limited capacity of ~330 mAh g^{-1} at 0.5 A cm^{-2}. (**d**) The discharge and charge voltage change of the Zn–air battery, Zn-O_2 battery, and Zn-CO_2 battery for long-term cycling with a limited capacity of ~330 mAh g^{-1}.

4. Conclusions

In summary, by simple hydrothermal synthesis, we prepared hydrophilic and heterogeneous MoS_2 catalysts consisting of crystalline-amorphous interfaces and defects-embedded active edge sites that enables a good three phase interface on carbon paper GDL, and efficient O_2 and CO_2 utilization by their hydrophilic characteristics (with a 24 degree water contact angle) and the superhydrophobic characteristics of carbon paper GDL (with a 130 degree water contact angle). Such MoS_2 catalysts showed decent oxygen evolution and acceptable oxygen reduction catalytic activity compared to commercial Pt/C and RuO_2, which enabled a cycling durability of 500 cycles (~180 h) for an assembled aqueous Zn–air battery at 0.5 mA cm^{-2} with a limited capacity of 330 mAh g^{-1}, and lower charge potentials (~1.88 V after 500 cycles) than the Zn–air battery using expensive Pt/C and RuO_2 after cycles. Notably, the Zn–air battery using the prepared MoS_2 catalysts could operate stably even at a large current density of 10 mA cm^{-2}. The Zn–Air battery using single MoS_2 catalyst also shows comparable performance among the Zn–Air batteries using MoS_2-based catalysts (see Table S1). In addition, the heterogeneous MoS_2, as an effective cathode catalyst, could catalyze the reversible circulation of the Zn-O_2 battery and the Zn-CO_2 battery, demonstrating that the heterogeneous MoS_2 catalyst can potentially replace Pt/C and RuO_2 catalysts in aqueous rechargeable Zn–air batteries, Zn-O_2 batteries, and Zn-CO_2 batteries.

Supplementary Materials: The following supporting information can be downloaded at: https://www.mdpi.com/article/10.3390/nano12224069/s1, Figure S1. Auqeous Zn–Air battery system.

(**A**) The photo of aquesous Zn–Air battery system that consists of aquesous electrolyte, pump and Zn–Air battery. (**B**) Photo of aqueous Zn–Air battery that consists of air cathode, aqueous and Zn-anode; Figure S2. Electrochemical cycle performance of aqueous Zn–Air batteries using the heterogeneous MoS_2 cathode catalyst and commercial Pt/C-RuO_2 cathode catalyst where the Zn–Air battery using the prepared MoS_2 cathode catalysts shows an excellent cycle stability than the Zn–Air battery using expensive Pt/C-RuO_2 cathode catalyst at a large current density of 5 mA cm^{-2}. It indicates that the heterogeneous MoS_2 catalyst has better oxygen evolution and oxygen reduction catalytic activity, especailly at operating current desnsity, than commercial Pt/C-RuO_2 catalyst; Figure S3. Electrochemical cycle performance of aqueous Zn-O_2 battery and Zn-CO_2 battery. (**a**) Voltage-time curves of the assembled aqueous Zn-O_2 battery using the heterogeneous MoS_2 catalyst that shows a stable cycle and less than 1.0 V potential gap (with a ~1.3V discharge potential and less than 2.3 V charge potential). (**b**) Voltage-time curves of the assembled aqueous Zn-O_2 battery using the heterogeneous MoS_2 catalyst that shows a stable cycle and ~1.35 V potential gap (with a ~1.25 V discharge potential and ~2.6 V charge potential); Table S1. Comparison of electrochemical properties of Zn–Air batteries using MoS_2-based catalysts. References [10,25,43] are cited in Supplementary Materials.

Author Contributions: M.W.: Device preparation, performance testing, data curation, formal analysis, writing—original draft; M.W., X.H., Z.Y. and P.Z.: Investigation; H.S.: Writing and revising guidance; H.S. and C.Z.: Project administration, resources, supervision; J.X. and K.C.: Supervision. All authors have read and agreed to the published version of the manuscript.

Funding: This research was partially supported financially by NSFC (nos. 61921005, 62104099, 62105048, 61735008, 11774155 and 62004078), the National Key R&D Program of China (2018YFB2200101), and the Natural Science Foundation of Jiangsu province (BK20190313 and BK20201073), Key Scientific Research Project in Colleges and Universities of Henan Province of China (Grant No. 21A416001) and Open project of Nanjing University Solid State Microstructure Laboratory (Grant No. M34057, M35033, M35062, M35059).

Data Availability Statement: The data presented in this study are available on request from the corresponding author.

Conflicts of Interest: The authors declare no conflict of interest.

References

1. Yi, Z.; Chen, G.; Hou, F.; Wang, L.; Liang, J. Zinc-Ion Batteries: Strategies for the Stabilization of Zn Metal Anodes for Zn-Ion Batteries. *Adv. Energy Mater.* **2021**, *11*, 2170001. [CrossRef]
2. Cao, R.; Lee, J.-S.; Liu, M.; Cho, J. Recent progress in non-precious catalysts for metal-air batteries. *Adv. Energy Mater.* **2012**, *2*, 816–829. [CrossRef]
3. Li, Y.; Dai, H. Recent advances in zinc-air batteries. *Chem. Soc. Rev.* **2014**, *43*, 5257–5275. [CrossRef] [PubMed]
4. Yan, L.; Zhang, Y.; Ni, Z.; Zhang, Y.; Xu, J.; Kong, T.Y.; Huang, J.H.; Li, W.; Ma, J.; Wang, Y.G. Chemically self-charging aqueous zinc-organic battery. *J. Am. Chem. Soc.* **2021**, *143*, 15369–15377. [CrossRef]
5. Liu, W.; Feng, J.; Wei, T.; Liu, Q.; Zhang, S.; Luo, Y.; Luo, J.; Liu, X. Active-site and interface engineering of cathode materials for aqueous Zn—gas batteries. *Nano Res.* **2022**, 1–22. [CrossRef]
6. Liu, J.N.; Zhao, C.X.; Ren, D.; Wang, J.; Zhang, R.; Wang, S.H.; Zhao, C.; Li, B.Q.; Zhang, Q. Preconstructing asymmetric interface in air cathodes for high-performance rechargeable Zn—air batteries. *Adv. Mater.* **2022**, *34*, 2109407. [CrossRef]
7. Lee, J.S.; Tai Kim, S.; Cao, R.; Choi, N.S.; Liu, M.; Lee, K.T.; Cho, J. Metal–air batteries with high energy density: Li–air versus Zn–air. *Adv. Energy Mater.* **2011**, *1*, 34–50. [CrossRef]
8. Venezia, E.; Salimi, P.; Chauque, S.; Proietti Zaccaria, R. Sustainable Synthesis of Sulfur-Single Walled Carbon Nanohorns Composite for Long Cycle Life Lithium-Sulfur Battery. *Nanomaterials* **2022**, *12*, 3933. [CrossRef]
9. Glaydson, S.D.R.; Chandrasekar, M.S.; Angélica, D.C.; Sylvia, H.L.; Mikael, T.; Ulla, L.; Flaviano, G.A. Facile Synthesis of Sustainable Activated Biochars with Different Pore Structures as Efficient Additive-Carbon-Free Anodes for Lithium- and Sodium-Ion Batteries. *ACS Omega* **2022**. [CrossRef]
10. Amiinu, I.S.; Pu, Z.; Liu, X.; Owusu, K.A.; Monestel, H.G.R.; Boakye, F.O.; Zhang, H.; Mu, S. Multifunctional Mo–N/C@MoS2 Electrocatalysts for HER, OER, ORR, and Zn–Air Batteries. *Adv. Funct. Mater.* **2017**, *27*, 1702300. [CrossRef]
11. Liu, Q.; Jin, J.; Zhang, J. NiCo2S4@graphene as a Bifunctional Electrocatalyst for Oxygen Reduction and Evolution Reactions. *ACS Appl. Mater. Interfaces* **2013**, *5*, 5002–5008. [CrossRef] [PubMed]
12. Xia, B.Y.; Yan, Y.; Li, N.; Wu, H.B.; Lou, X.W.; Wang, X. A metal–organic framework-derived bifunctional oxygen electrocatalyst. *Nat. Energy* **2016**, *1*, 15006. [CrossRef]
13. Li, L.; Tsang, Y.C.A.; Xiao, D.; Zhu, G.; Zhi, C.; Chen, Q. Phase-transition tailored nanoporous zinc metal electrodes for rechargeable alkaline zinc-nickel oxide hydroxide and zinc-air batteries. *Nat. Commun.* **2022**, *13*, 2870. [CrossRef] [PubMed]

14. Wang, T.; Kunimoto, M.; Takanori, M.; Masahiro, Y.; Niikura, J.; Takahashi, I.; Morita, M.; Abe, T.; Homma, M. Carbonate formation on carbon electrode in rechargeable zinc-air battery revealed by in-situ Raman measurements. *J. Power Sources* **2022**, *533*, 231237. [CrossRef]
15. Gauthier, M.; Carney, T.J.; Grimaud, A.; Giordano, L.; Pour, N.; Chang, H.H.; Fenning, D.P.; Lux, S.F.; Paschos, O.; Bauer, C.; et al. Electrode-electrolyte interface in Li-ion batteries: Current understanding and new insights. *J. Phys. Chem. Lett.* **2015**, *6*, 4653–4672. [CrossRef]
16. Xu, Y.; Sumboja, A.; Zong, Y.; Darr, J. Bifunctionally active nanosized spinel cobalt nickel sulfides for sustainable secondary zinc–air batteries: Examining the effects of compositional tuning on OER and ORR activity. *Catal. Sci. Technol.* **2020**, *10*, 2173–2182. [CrossRef]
17. Wu, M.; Zhang, G.; Qiao, J.; Chen, N.; Chen, W.; Sun, S. Ultra-Long Life Rechargeable Zinc–Air Battery Based on High-Performance Trimetallic Nitride and NCNT Hybrid Bifunctional Electrocatalysts. *Nano Energy* **2019**, *61*, 86–95. [CrossRef]
18. Turney, D.; Gallaway, J.; Yadav, G.; Ramirez, R.; Nyce, M.; Banerjee, S.; Chen-Wiegart, Y.; Wang, J.; Ambrose, M.; Kolhekar, S.; et al. Rechargeable Zinc Alkaline Anodes for Long-Cycle Energy Storage. *Chem. Mater.* **2017**, *29*, 4819–4832. [CrossRef]
19. Li, X.; Lin, Z.; Cheng, L.; Chen, X. Layered $MoSi_2N_4$ as Electrode Material of Zn–Air Battery. *Phys. Status Solidi RRL* **2022**, *16*, 2200007. [CrossRef]
20. Sivanantham, A.; Ganesan, P.; Shanmugam, S. Hierarchical $NiCo_2S_4$ Nanowire Arrays Supported on Ni Foam: An Efficient and Durable Bifunctional Electrocatalyst for Oxygen and Hydrogen Evolution Reactions. *Adv. Funct. Mater.* **2016**, *26*, 4661–4672. [CrossRef]
21. Wagh, N.; Lee, C.; Kim, D.; Kim, S.; Shinde, S.; Lee, J. Heuristic Iron–Cobalt-Mediated Robust pH-Universal Oxygen Bifunctional Lusters for Reversible Aqueous and Flexible Solid-State Zn–Air Cells. *ACS Nano* **2021**, *15*, 14683–14696. [CrossRef] [PubMed]
22. Puglia, M.K.; Malhotra, M.; Kumar, C.V. Engineering functional inorganic nanobiomaterials: Controlling interactions between 2D-nanosheets and enzymes. *Dalton Trans.* **2020**, *49*, 3917–3933. [CrossRef] [PubMed]
23. Tang, Y.J.; Wang, Y.; Wang, X.L.; Li, S.L.; Huang, W.; Dong, L.Z.; Liu, C.H.; Li, Y.F.; Lan, Y.Q. Molybdenum Disulfide/Nitrogen-Doped Reduced Graphene Oxide Nanocomposite with Enlarged Interlayer Spacing for Electrocatalytic Hydrogen Evolution. *Adv. Energy Mater.* **2016**, *6*, 1600116. [CrossRef]
24. Prins, R.; Debeer, V.H.J.G.; Somorjai, A. Structure and Function of the Catalyst and the Promoter in Co—Mo Hydrodesulfurization Catalysts. *Catal. Rev. Sci. Eng.* **1989**, *31*, 1. [CrossRef]
25. Bai, J.; Meng, T.; Guo, D.L.; Wang, S.G.; Mao, B.G.; Cao, M.H. $Co_9S_8@MoS_2$ Core–Shell Heterostructures as Trifunctional Electrocatalysts for Overall Water Splitting and Zn–Air Batteries. *ACS Appl. Mater. Interfaces* **2018**, *10*, 1678–1689. [CrossRef]
26. Puglia, M.K.; Malhotra, M.; Chivukula, A.; Kumar, C.V. "Simple-Stir" Heterolayered MoS_2/Graphene Nanosheets for Zn–Air Batteries. *ACS Appl. Nano Mater.* **2021**, *4*, 10389–10398. [CrossRef]
27. Drew, A.A.; Yi, S.; Douglas, G. Zn-Based Oxides Anchored to Nitrogen-Doped Carbon Nanotubes as Efficient Bifunctional Catalysts for Zn–Air Batteries. *ChemElectroChem* **2020**, *7*, 2283.
28. Hu, H.; Xu, J.C.; Zheng, Y.H.; Yao Yao, Z.; Rong, J.; Zhang, T.; Yang, D.Y.; Qiu, F.X. NiS_2-Coated Carbon Fiber Paper Decorated with MoS_2 Nanosheets for Hydrogen Evolution. *ACS Appl. Nano Mater.* **2022**, *5*, 10933–10940. [CrossRef]
29. Hou, X.B.; Zhou, H.M.; Zhao, M.; Cai, Y.B.; Wei, Q.F. MoS_2 Nanoplates Embedded in Co-N-Doped Carbon Nanocages as Efficient Catalyst for HER and OER. *ACS Sustain. Chem. Eng.* **2020**, *8*, 5724–5733. [CrossRef]
30. Li, J.C.; Zhang, C.; Ma, H.J.; Wang, T.H.; Guo, Z.Q.; Yang, Y.; Wang, Y.Y.; Ma, H.X. Modulating interfacial charge distribution of single atoms confined in molybdenum phosphosulfide heterostructures for high efficiency hydrogen evolution. *Chem. Eng. J.* **2021**, *414*, 128834. [CrossRef]
31. Wen, F.S.; Li, Y.J.; Liu, W.L.; Liu, G.D.; Zhang, T.; Xu, Y.; Zhang, X.Y. Comprehensive experiment on simple morphology regulation and photocatalytic degradation of Rhodamine B by molybdenum disulfide. *Exp. Technol. Manag.* **2022**, *39*, 157–161. (In Chinese)
32. Tan, J.W.; Zhang, W.B.; Shu, Y.J.; Lu, H.Y.; Tang, Y.; Gao, Q.S. Interlayer engineering of molybdenum disulfide toward efficient electrocatalytic hydrogenation. *Sci. Bull.* **2021**, *66*, 1003–1012. [CrossRef]
33. Chen, B.; Lu, H.; Zhou, J.; Ye, C.; Shi, C.; Zhao, N.; Qiao, S.-Z. Porous MoS_2/Carbon Spheres Anchored on 3D Interconnected Multiwall Carbon Nanotube Networks for Ultrafast Na Storage. *Adv. Energy Mater.* **2018**, *8*, 1702909. [CrossRef]
34. Bau, J.A.; Emwas, A.H.; Nikolaienko, P.; Aljarb, A.A.; Tung, V.; Rueping, M. Mo^{3+} hydride as the commonorigin of H2 evolution and selective NADH regeneration in molybdenum sulfide electrocatalysts. *Nat. Catal.* **2022**, *5*, 397–404. [CrossRef]
35. Li, L.; Qin, Z.; Ries, L.; Hong, S.; Michel, T.; Yang, J.; Salameh, C.; Bechelany, M.; Miele, P.; Kaplan, D.; et al. Role of Sulfur Vacancies and Undercoordinated Mo Regions in MoS_2 Nanosheets toward the Evolution of Hydrogen. *ACS Nano* **2019**, *13*, 6824–6834. [CrossRef] [PubMed]
36. Chen, B.; He, X.; Yin, F.; Wang, H.; Liu, D.-J.; Shi, R.; Chen, J.; Yin, H. MO-Co@N-Doped Carbon (M = Zn or Co): Vital Roles of Inactive Zn and Highly Efficient Activity toward Oxygen Reduction/Evolution Reactions for Rechargeable Zn–Air Battery. *Adv. Funct. Mater.* **2017**, *27*, 1700795. [CrossRef]
37. Liu, Y.; Wu, X. Recent Advances of Transition Metal Chalcogenides as Cathode Materials for Aqueous Zinc-Ion Batteries. *Nanomaterials* **2022**, *12*, 3298. [CrossRef]
38. Mohammad, A.; Bijandra, K.; Liu, C.; Patrick, P.; Poya, Y.; Amirhossein, B.; Peter, Z.; Robert, F.K.; Larry, A.C.; Amin, S.K. Cathode Based on Molybdenum Disulfide Nanoflakes for Lithium–Oxygen Batteries. *ACS Nano* **2016**, *10*, 2167–2175.

39. Zoya, S.; Liu, J.P.; Zhao, L.; Francesco, C.; Jang-Kyo, K. Metallic MoS$_2$ nanosheets: Multifunctional electrocatalyst for the ORR, OER and Li–O$_2$ batteries. *Nanoscale* **2018**, *10*, 22549–22559.
40. Liu, Y.; Zang, Y.P.; Liu, X.M.; Cai, J.Y.; Lu, Z.; Niu, S.W.; Pei, Z.B.; Zhai, T.; Wang, G.M. Three-Dimensional Carbon-Supported MoS2 with Sulfur Defects as Oxygen Electrodes for Li-O$_2$ Batteries. *Nano Energy* **2019**, *65*, 103996. [CrossRef]
41. Wu, L.F.; Alessandro, L.; Nelson, Y.D.; Akhil, S.; Marco, M.R.M.M.; Ageeth, A.B.; Nora, H.; Emiel, J.M.; Jan, P.H. The Origin of High Activity of Amorphous MoS2 in the Hydrogen Evolution Reaction. *ChemSusChem* **2019**, *12*, 4383. [CrossRef] [PubMed]
42. Gong, J.; Zhu, B.; Zhang, Z.; Xiang, Y.; Tang, C.; Ding, Q.; Wu, X. The Synthesis of Manganese Hydroxide Nanowire Arrays for a High-Performance Zinc-Ion Battery. *Nanomaterials* **2022**, *12*, 2514. [CrossRef] [PubMed]
43. Liu, X.; Yin, Z.H.; Cui, M.; Gao, L.G.; Liu, A.M.; Su, W.N.; Chen, S.R.; Ma, T.L.; Li, Y.Q. Double shelled hollow CoS$_2$@MoS$_2$@NiS$_2$ polyhedron as advanced trifunctional electrocatalyst for zinc-air battery and self-powered overall water splitting. *J. Colloid Interface Sci.* **2022**, *610*, 653–662. [CrossRef] [PubMed]

Article

A Wide-Range-Response Piezoresistive–Capacitive Dual-Sensing Breathable Sensor with Spherical-Shell Network of MWCNTs for Motion Detection and Language Assistance

Shuming Zhang [†], Xidi Sun *,[†], Xin Guo [†], Jing Zhang, Hao Li, Luyao Chen, Jing Wu, Yi Shi and Lijia Pan *

Collaborative Innovation Center of Advanced Microstructures, School of Electronic Science and Engineering, Nanjing University, Nanjing 210093, China
* Correspondence: xidisun@smail.nju.edu.cn (X.S.); ljpan@nju.edu.cn (L.P.)
† These authors contributed equally to this work.

Citation: Zhang, S.; Sun, X.; Guo, X.; Zhang, J.; Li, H.; Chen, L.; Wu, J.; Shi, Y.; Pan, L. A Wide-Range-Response Piezoresistive–Capacitive Dual-Sensing Breathable Sensor with Spherical-Shell Network of MWCNTs for Motion Detection and Language Assistance. *Nanomaterials* 2023, 13, 843. https://doi.org/10.3390/nano13050843

Academic Editor: Antonino Gulino

Received: 15 January 2023
Revised: 11 February 2023
Accepted: 22 February 2023
Published: 24 February 2023

Copyright: © 2023 by the authors. Licensee MDPI, Basel, Switzerland. This article is an open access article distributed under the terms and conditions of the Creative Commons Attribution (CC BY) license (https://creativecommons.org/licenses/by/4.0/).

Abstract: It is still a challenge for flexible electronic materials to realize integrated strain sensors with a large linear working range, high sensitivity, good response durability, good skin affinity and good air permeability. In this paper, we present a simple and scalable porous piezoresistive/capacitive dual-mode sensor with a porous structure in polydimethylsiloxane (PDMS) and with multi-walled carbon nanotubes (MWCNTs) embedded on its internal surface to form a three-dimensional spherical-shell-structured conductive network. Thanks to the unique spherical-shell conductive network of MWCNTs and the uniform elastic deformation of the cross-linked PDMS porous structure under compression, our sensor offers a dual piezoresistive/capacitive strain-sensing capability, a wide pressure response range (1–520 kPa), a very large linear response region (95%), excellent response stability and durability (98% of initial performance after 1000 compression cycles). Multi-walled carbon nanotubes were coated on the surface of refined sugar particles by continuous agitation. Ultrasonic PDMS solidified with crystals was attached to the multi-walled carbon nanotubes. After the crystals were dissolved, the multi-walled carbon nanotubes were attached to the porous surface of the PDMS, forming a three-dimensional spherical-shell-structure network. The porosity of the porous PDMS was 53.9%. The large linear induction range was mainly related to the good conductive network of the MWCNTs in the porous structure of the crosslinked PDMS and the elasticity of the material, which ensured the uniform deformation of the porous structure under compression. The porous conductive polymer flexible sensor prepared by us can be assembled into a wearable sensor with good human motion detection ability. For example, human movement can be detected by responding to stress in the joints of the fingers, elbows, knees, plantar, etc., during movement. Finally, our sensors can also be used for simple gesture and sign language recognition, as well as speech recognition by monitoring facial muscle activity. This can play a role in improving communication and the transfer of information between people, especially in facilitating the lives of people with disabilities.

Keywords: porous conducting polymer; strain sensing; dual sensing mode; breathable

1. Introduction

With the progress of science and technology, various electronic products have been invented to provide help for human life and work. In this process, portable computers, smart watches, smart glasses and other wearable devices have been created and developed rapidly [1–5]. In the research of wearable devices, we focused on the skin, one of the largest and most important organs of the human body. Human skin is not only the first barrier against the outside world but also has many tactile receptors for sensing pressure, temperature, strain and other external stimuli [6,7]. By mimicking the sensory capabilities and characteristics of natural skin, a great deal of research has been conducted to develop bio-inspired electronic skin [3,8–11], which has important applications in wearable healthcare devices [12,13], smart robots [14,15], implantable medical devices [16], motion monitoring [17,18] and environmental perception [19,20]. As an important part of electronic skin, stress sensors based on different principles have been widely studied, including transistor sensing [21], capacitive sensing [22], piezoresistive sensing [23], piezoelectric sensing [24] and triboelectric sensing [25]. Piezoresistive and capacitive strain sensors are widely studied by researchers because of their simple preparation process, low cost and excellent flexibility [18,23,26–32]. On the other hand, compared with metal-based and semiconductor-based stress sensors, conductive polymer composite stress sensors based on percolation thresholds are one of the best choices for the next generation of electronic skin stress sensors due to their superior strain-sensing capability and their unique flexibility and light weight [8,33]. Breathability sensors provide comfortable contact with the human body by balancing the temperature and humidity of the human skin through the exchange of gases between the skin and the outside world [34]. However, conventional polymeric strain sensors have a single sensing mode, and it is difficult to provide a wide range of detection, high sensitivity and excellent response stability while maintaining soft and breathable properties using these devices. This greatly limits their application in various situations. In order to solve these problems, researchers have designed a variety of different microstructures in recent years, such as pyramid array structures [22,35], fiber structures [4,30,31,36], microcrack structures [33,37,38], sphere structures [39] and porous structures [26,32,40,41]. Polymer foam with a porous structure has been widely used in the sensor field due to its low density, large specific surface area, good compression recovery performance, simple preparation process and other advantages [23]. Meanwhile, due to their high electrical conductivity, large surface area and chemical stability, MWCNTs have been investigated in a variety of frontier sensing applications, such as optoelectronic sensors [42,43], medical sensors [44,45], chemical sensors [46,47], mechanical sensors [48] and nano semiconductor devices [49]. Among them, conducting polymer sensors doped with MWCNTs have received much attention in the field of mechanical sensing due to their fast response and high sensitivity. In addition, devices with multiple sensing mechanisms are important for artificial limbs and robots that need to be able to sense fine motion and environmental information. A variety of single-signal haptic sensors have been developed, piezoresistive and capacitive-based sensors can independently enable the sensing of pressure or temperature information. There are also e-skin systems that integrate multiple sensors to further improve the perception of the environment and motion [20,50]. In addition, special functional properties such as self-healing, self-powering, biodegradability, biocompatibility and breathability have been gradually integrated into these devices to obtain e-skins with comprehensive performance for practical applications [51–53]. In particular, breathability is an important way to improve the comfort of wearable devices by balancing the thermal humidity between the human body and the external environment. Therefore, haptic sensors with breathability are in high demand and of great relevance in wearable health monitoring, biomedical monitoring and implantable device applications [54].

In this paper, a novel strain sensor with a porous structure and a three-dimensional spherical shell network composed of MWCNTs is proposed, in which a dissolvable crystal was used as a sacrificial template. Firstly, the embedded three-dimensional spherical-shell network of MWCNTs enabled the sensor to have both piezoresistive and pressure-capacity-sensing capabilities, resulting in more accurate sensing data. In addition, the porous structure prepared by the dissolvable crystal as a sacrificial template gave the sensor a high sensitivity of $-1.2/\text{kPa}$ (piezoresistive) and $0.38/\text{kPa}$ (pressure capacitive). With a wide pressure sensing range (1–520 kPa), it fully met the requirements of human health monitoring applications. In addition, the porous structure gave the sensor good comfort and skin affinity for adapting to different human application scenarios. Finally, a porous-structure sensor based on doped MWCNTs was applied to the human body to monitor physiological signals and joint movements, including muscle movement, finger flexion, elbow movement, knee movement and plantar pressure.

2. Materials and Methods

2.1. Materials

Commercia PDMS (Sylgard 184, Dow Corning Co., Midland, MI, USA) and a curing agent (Dow Corning Co., USA) were purchased from Dow Materials Sciences and were used as the polymer matrix in this study. Hydroxylated multi-walled carbon nanotubes (10–20 μm) were purchased from Jiangsu Xianfeng Nano Company (Nanjing, China) and were used as conductive fillers. Refined sucrose crystals were purchased from Taikoo to construct the cavity structure.

2.2. Fabrication of Sphere-Shell Three-Dimensional Structure of MWCNTs

A three-dimensional spherical-shell network of MWCNTs was prepared using refined sucrose crystal as a removable skeleton. MWCNTs with a mass ratio of 1% were first mixed with refined sucrose particles, which were stirred in a blender for 30 min to make the MWCNTs evenly coated on the surface of the sucrose crystals. Finally, an MWCNT network with a three-dimensional spherical-shell structure was formed by allowing the sample to stand at room temperature for 60 min.

2.3. Fabrication of CNT–PDMS Sponges

Highly compressible CNT–PDMS sponges were prepared by a simple sacrificial template method. First, 20 g of Dow Corning PDMS (Sylgard 184, Dow Corning Co. USA) was mixed with a curing agent in a ratio of 10:1 and mixed with carbon nanotubes (200 mg, Xianfeng, China). After removing the bubbles captured during agitation under mild vacuum conditions, the refined sucrose particles, which were evenly wrapped in the three-dimensional spherical-shell structure of the MWCNTs, were placed into a PDMS mixture that could penetrate the porous structure through a vacuum suction process and capillary forces. The pore size was determined by the grain size of the sugar. Then, the sample was cured at 100 °C under atmospheric pressure for 1 h. After curing, the sugar was dissolved in an ultrasonic bath in hot water for 1 h, and finally the sample was dried overnight. Then, two conductive copper tapes were assembled on the upper and lower surfaces of the prepared porous PDMS sponge with MWCNTs as electrodes, and the capacitance and resistance responses under corresponding compression were recorded.

2.4. Characterization

The surface morphology of the sensor was observed using an on-site scanning electron microscope (Nova Nano SEM 230, FEI. Hillsboro, Oregon, USA) at an operating voltage of 5 kV. All specimen thicknesses were confirmed by measuring the cross-section of the specimens. Considering the PDMS insulator, the samples were sputtered with a thin Au layer on the cross-section of the samples before SEM observation. The internal shape of the device was observed by an Ultra-Depth Three-Dimensional Microscope (KEYENCE VHX-6000. Osaka, Japan) with a 40× magnification of the cross-section. A series of strain and pressure tests were carried out on the microcomputer, a strain cycle was applied to the sensor by a ZHIQU ZQ-990 (ZHIQU. Dongguan, Guangdong, China) pressure stress tester and the capacitance and resistance signals of the sensor were detected by a KEYSIGHT E4980AL LCR (KEYSIGHT. Santa Rosa, CA, USA) tester at an operating voltage of 20 V and an operating frequency of 20 kHz. The information was collected and recorded by the software developed by LABVIEW.

3. Results

3.1. Structural Design and Sensing Principle of Dual-Mechanism Pressure Sensor

As shown in Figure 1a, MWCNTs and polydimethylsiloxane (PDMS) were used as pressure-sensitive materials to fabricate the pressure sensor. After the structural optimization processing and design integration of the MWCNTs, a wearable flexible pressure sensor with excellent breathability and flexibility was realized. The pressure sensor can be directly attached to human skin for the continuous monitoring of various stress–strain signals without causing any damage to the skin, and its schematic diagram is shown in Figure 1b. The entire pressure sensor does not require an encapsulation layer and substrate typical of conventional pressure sensors.

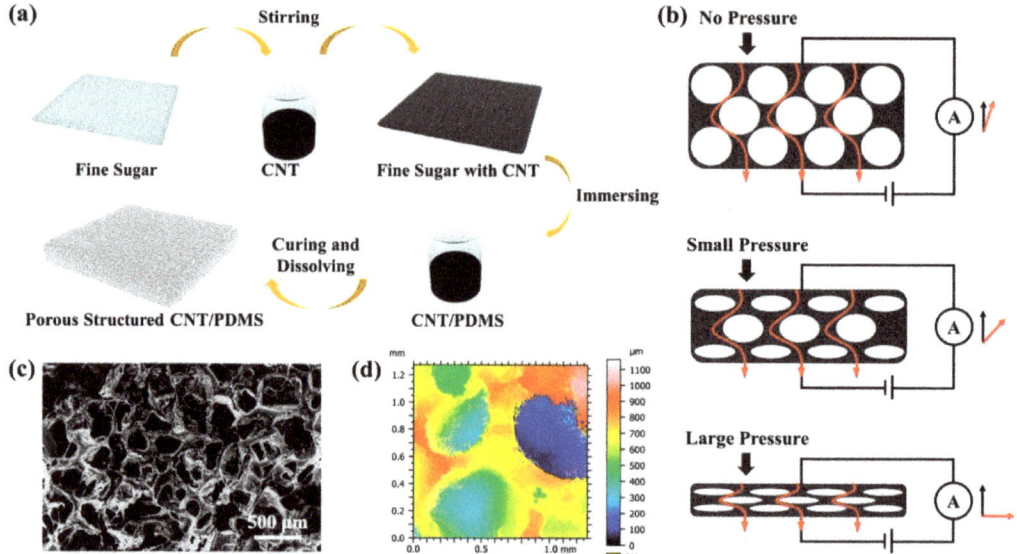

Figure 1. Preparation of multifunctional flexible and breathable electronic skin system with porous structure using PDMS and MWCNTs. (**a**) A diagram of the manufacturing process. (**b**) Schematic diagram of how the stress sensor works. (**c**) Porous-structure strain sensor SEM photo. (**d**) Porous-structure strain sensor 3D depth-of-field photo.

The porous structure of the pressure-sensitive layer was made of fine sugar wrapped in MWCNTs, and finally the fine sugars contained in the pressure-sensitive layer were treated with deionized water. The preparation methods and processes detail of the specific materials and devices are shown in Figure S1 (Supporting Information). Figure 1c shows the surface SEM images of the porous structure of the pressure-sensitive layer. The surface morphology of the sensor was observed using an on-site scanning electron microscope (Nova Nano SEM 230, FEI) at an operating voltage of 5 kV. The thicknesses of all the samples were determined by measuring the cross-sections of the samples. Due to the insulating properties of PDMS, a thin gold layer was sputtered onto cross-sections of the samples prior to the SEM observation. As can be seen in Figure 1c, the pores formed by the fine sugar were uniformly distributed throughout the PDMS layer. The internal morphology was observed using an Ultra-Depth Three-Dimensional Microscope (KEYENCE VHX-6000) at a 40× magnification, as shown in Figure 1d. The presence of the pores greatly increased the specific surface area of the entire pressure-sensitive layer, increased the contact area of the MWCNTs in the pressure-sensitive layer and provided the change in the conductive path inside the piezoresistive sensor, which is an effective way to achieve a high sensitivity over a wide sensing range.

3.2. Breathability and Comfort

The breathability and flexibility of wearable pressure sensors have remarkable effects in health monitoring. As shown in Figure 2a, the MWCNT-based pressure sensors could be directly attached to human skin. This was due to the overall porous structure of the pressure-sensitive layer, and when attached to human skin, the epidermal temperature and sweat from the human body could be quickly dissipated into the surrounding environment through the pressure sensor. In addition, as seen in Figure 1c, the sensor had a large number of cracks and porous structures, which have the advantage of optimizing their surface coverage while maintaining breathability. The properties of this structure could provide great advantages in terms of the design and construction of medical devices for physiological signal monitoring.

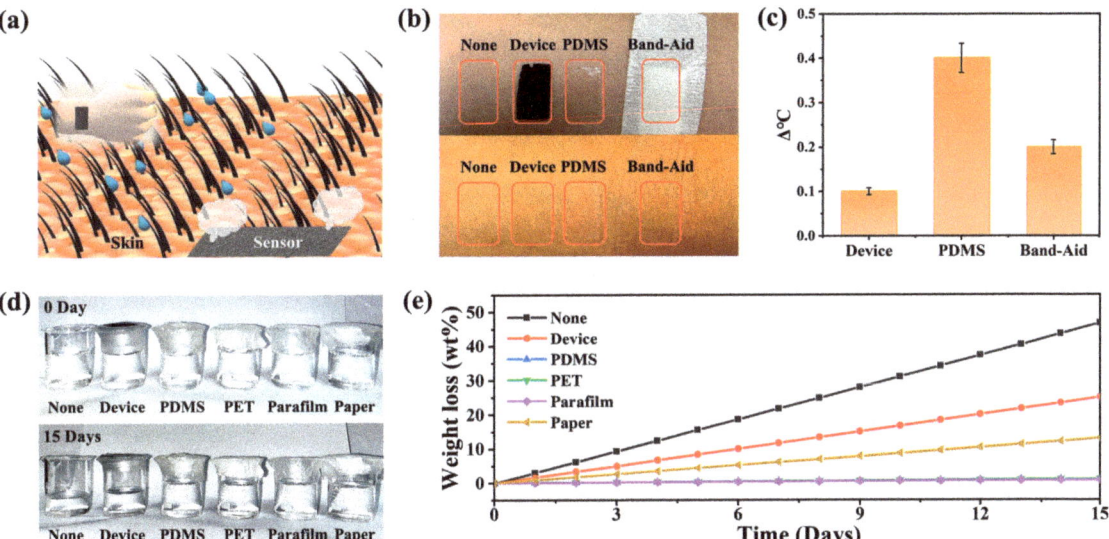

Figure 2. The breathability and skin-friendly nature of the pressure sensor based on the porous structure of the three-dimensional spherical-shell conductive network. (a) Schematic diagram showing that sensors based on porous structures could be directly connected to human skin and had skin affinity and breathability. (b) Changes in skin surface after different materials were adhered to arm skin for 6 h and (c) skin temperature changes. (d) Optical photos of the breathability test process at room temperature. (e) Water loss in glass bottles covered with films of different materials every 24 h.

To demonstrate the permeability of the sensor, the permeability performance of the pressure sensor was evaluated using the heat dissipation rate of the human epidermis, and the optical photographs of the process are shown in Figure 2b. The skin epidermis temperature of bare leaky skin, the skin temperature of the porous pressure sensor covered with MWCNTs, the skin epidermis temperature covered with a PDMS film and the epidermis temperature covered with band-aids were tested separately. After 360 min, the epidermal temperature of the skin covered with the MWCNT-based porous pressure sensor was almost the same as that of the bare epidermal skin temperature. In contrast, the skin surface covered by PDMS showed a significant increase in temperature and was accompanied by redness of the skin surface. The skin epidermal temperature was recorded by IR thermometer (Figure S2) and is shown in Figure 2c, and comparative photographs of the skin are shown in Figure 2b. In addition, we used the rate of water vapor molecules penetrating the film to evaluate the permeability of the porous pressure sensor, and an optical photograph of the process is shown in Figure 2d. Glass bottles holding colored water were sealed by various functional films including para film, PDMS film, PET film, paper and porous MWCNT film. Afterwards, they were compared to open glass bottles that also held the same mass of water at room temperature. After 15 days, the level of water evaporation in the glass bottle sealed with the porous MWCNT film was second only to that of the unsealed control group. In contrast, the mass of water in the glass bottles encapsulated by the other films remained almost constant. The weight loss of water was recorded every 24 h throughout the experiment (Figure 2e), and photographs of the water weight loss process are shown in Figure S3 (Supporting Information). Based on these results, the device exhibited excellent permeability.

3.3. Electrical Output Performance

Stress sensors for monitoring physiological signals and movements of human joints require their collected data to have a high accuracy and stability. In order to improve the accuracy of the data, the sensitivity and linear sensing range of the pressure sensor were tested using the dual sensing mode. The test instrument shown in Figure 3a was used to test the change in the resistance and capacitance of the composite porous-structure strain sensor under different strains at an operating voltage of 20 V and an operating frequency of 20 kHz. In order to optimize the range and sensitivity of the sensor, different MWCNT-doping concentrations were tested (Figure S4). It was found that when the mass ratio of the MWCNTs inside the PDMS skeleton was 1% and the mass ratio of the MWCNTs attached to the surface of the porous structure was 1%, the piezoresistive performance of the sensor provided excellent uniformity under pressure application and release while providing an optimal detection range and sensitivity (Figure S5). The capacitance and resistance values of the sensor under a load pressure ranging from 1 kPa to 520kPa, are shown in Figure 3b. The pressure-sensitive layer of the sensor had opposite resistance and pressure changes throughout a pressure cycle (1–520 kPa), while its capacitance and pressure changed with the same trend. As shown in Figure 3c, the sensor demonstrated a good consistency in terms of boosting and lowering the pressure within the range of 1~520 kPa, and it had good elasticity and stability. We introduced the sensitivity (S) and gauge factor (GF) in order to characterize and discuss the sensitivity of the sensor. In detail, S is defined as $S = \delta(\Delta R/R0)/\delta P$, where $\Delta R/R0$ is the relative capacitance change, and P is the applied pressure, which corresponds to the slope of the pressure–response curve in Figure 3c. At the first stage (1–20 kPa), the porous structure began to deform, the number of conductive paths increased rapidly and the distance of the conductive paths shortened. At this time, the maximum sensitivity was $-1.2\%/kPa$. At the second stage (20 kPa–130 kPa), the deformation rate of the porous structure decreased, and the sensitivity was $-0.42\%/kPa$. At the third stage (130 kPa–520 kPa), the sensitivity was $-0.053\%/kPa$, and the porous structure no longer generated deformation. At this time, the deformation of the sensor was mainly caused by the deformation of the PDMS itself, the number of conductive paths was basically unchanged and the conductive paths continued to shorten. For the capacitive response, the sensor capacitance increased with the increase in the sensor strain degree. When pressure was applied to the surface of the device, the device underwent vertical deformation, and the spherical holes inside the device gradually closed, increasing the conductive path and reducing the resistance. The capacitance of the device was provided by both the PDMS skeleton structure and the air inside the spherical cavity, while the PDMS doped with MWCNTs inside it had a higher dielectric constant than air. When the sensor was compressed, the cavity was compressed, the air inside the cavity was reduced and the PDMS ratio was increased, resulting in an increase in the sensor's capacitance [32]. When the applied pressure was removed, the vertical deformation of the device disappeared, and the spherical cavity inside the device gradually recovered, leading to a reduction in the conductive path and an increase in the resistance. At this point, the proportion of air increases, resulting in a decrease in the capacitance. Due to the capacitance structure formed by the microcavities inside the sensor and the MWCNT layer on its surface, the critical points at the different stages of capacitance change were almost the same as the critical points at the different stages of the strain of the sensor, resulting in a three-stage process similar to the resistance response. As shown in Figure 3d, we explored the sensitivity of the sensor to strain. $\Delta R/R0$ was estimated by the formula $\Delta R/R0 = -3.43\varepsilon + 0.03\varepsilon^2 - 0.08(\%)$ (Figure S5a), and $\Delta C/C0$ was estimated by the formula $\Delta C/C0 = 1.97\varepsilon - 0.009\varepsilon^2 + 0.08(\%)$ (Figure S5b), where $\Delta R/R0$ is the change in the relative resistance, $\Delta C/C0$ is the change in the relative capacitance and ε is the strain applied, as shown in Figure 3d. Therefore, the corresponding gauge factor (GF) showed a linear trend (GF (resistance)= $0.06\varepsilon - 3.43$ (%),

GF (capacitance) = 1.97 − 0.018ε (%)). It turned out that the strain sensor was sensitive and comparable to some other piezoresistive sensors (Table S1). The strain of the sensor under pressure is shown in Figure 3e. At 1–130 kPa, the strain was mainly generated by the porous structure, and at 130kPa–520kPa, the strain was mainly generated by the PDMS matrix.

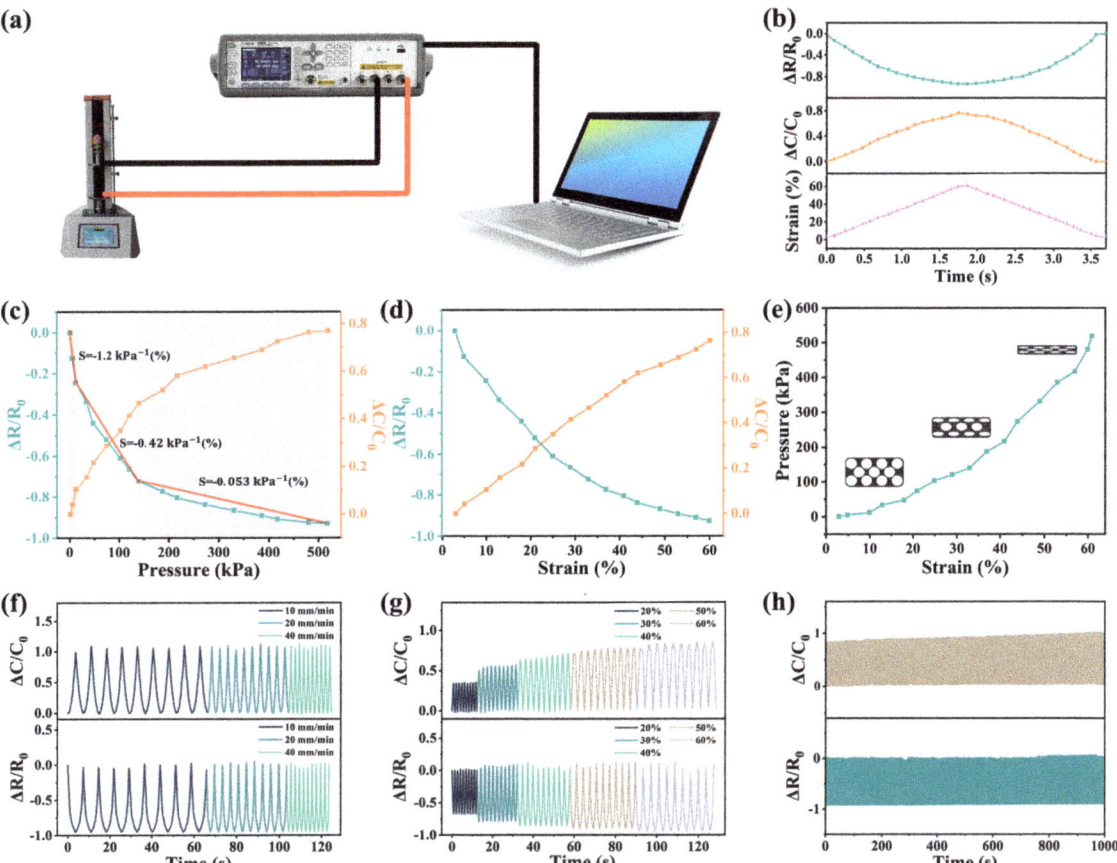

Figure 3. (**a**) A testing instrument used to study the piezoresistive and capacitive behavior. (**b**) The sensor had a dual-response-sensing mode that produced resistance/capacitance changes under strain. Normalized resistance and capacitance curves of the sensor in relation to (**c**) pressure and (**d**) strain. (**e**) Three strain stages of the sensor. The corresponding resistance and capacitance behaviors of sensor under (**f**) different strain frequencies, (**g**) different strains and (**h**) 1000−cycle stability test.

We compared the pressure-sensing performance of the prepared porous MWCNT-based pressure sensors with that of some other devices mentioned in the literature, and Table S1 (Supporting Information) lists more specific comparisons of the performance between these pressure sensors mentioned in the literature. Some pressure sensors have a large stress monitoring range but a low sensitivity. Some sensors have a high sensitivity but a narrow and nonlinear detection range. Moreover, most conventional sensors are single-physical-quantity sensing, whereas the dual-sensing-mode pressure sensor produced in this work had a relatively high sensitivity in the high-stress detection range, which is a great advantage compared to conventional wearable electronic devices. As shown in Figure 3c, the highest sensitivity of this pressure sensor was 0.38/kPa (capacitance) and

−1.2/kPa (resistance) over a wide detection range of 1–520 kPa. It is noteworthy that the MWCNT-based pressure sensor with a porous structure had excellent linearity over a wide detection range up to 520 kPa, which provides a good foundation for later planar pressure mapping.

Finally, the durability and stability of pressure sensors are also important in terms of practical applications. In order to study the stability of piezoresistive and capacitance response of the sensor under different strain rates and different strain degrees, nine strain cycles under different strain rates were studied and are shown in Figure 3f, and the stable sensing behavior of the sensor under different strain rates was found. The sensing performance was independent of the strain rate, which is conducive to the detection of human motion in complex motion scenes. Figure 3g shows the resistance response under a step strain. The strain ranged from 20% to 60% (increasing by 10% at each step). The initial resistance and capacitance values fluctuated slightly with each applied strain release. It was understood that when a strain removal was applied, the contact between the MWCNTs was damaged due to the recovery of the porous structure deformation, and the conductive network tended to return to its initial state. However, the applied strain could damage the microstructure of the sensor, resulting in a slight change in the resistance during compression. To demonstrate the excellent stability of the sensor, 1000 cycles of the sensor were measured at 1–520 kPa (Figure 3h) and details of the sensor response signal during the compression cycle test are shown in Figure S6. The resistance and capacitance values of the sensor remained almost unchanged after 1000 cycles, indicating that the pressure sensor had excellent durability and stability. The above data showed that the porous PDMS sensor doped with MWCNTs had a good elasticity and compressibility. $\Delta R/R0$ and $\Delta C/C0$ also had a good recovery rate. The results showed that the stress sensor had a high repeatability and stability.

3.4. Behavior Monitoring and Pressure-Sensing Arrays

As technology continues to advance and develop, more and more convenience is provided to people, but this also leaves more people in a sub-healthy state due to a lack of exercise. Therefore, in today's wearable medical monitoring devices, it is important to monitor the human body's behavioral signals and movement conditions at all times. Our sensors show excellent performance in this regard. The motion and human behavior of some major joints, including plantar pressure, knee motion, finger flexion, elbow motion, etc., were monitored to demonstrate the real-time motion health detection capability of these sensors as wearable electronic devices.

Due to the improvement of information technology worldwide, more and more people are using computers in their work, and the risk of cervical spondylosis caused by poor sitting posture for a long time is increasing. Posture and motion detection devices are urgently needed in people's daily life. They can be connected to the joints of the human body and can record the movement of the flexion cycle. By detecting these signals, people's activity status at work or in life in real time can be analyzed, and occupational diseases such as cervical spondylosis caused by sitting for too long can be prevented.

In addition, our pressure sensors could be used to track human movement. The device can be fitted to the elbow, fingers, knees and ankles and can record the body's movements. We recorded the plantar pressure during elbow flexure (30°, 45°, 60°, 90° and 135°), finger flexure (30°, 60°, 90° and 120°) and knee joint bending (15°, 45°, 90° and 135°), as shown in Figure 4a–d. Similarly, the output signal had a good real-time response and repeatability under the different bending angles, indicating that the sensor could provide a stable real-time monitoring function in practical applications. With the increase in the bending angle of the sensor, the change in the amplitude of the output signal was significantly enhanced, and the output signal under the different bending angles could be clearly distinguished. Videos of the body-worn sensor performing the motion tests are shown in Videos S1–S5. These results showed that the porous sensor with a 3D spherical-shell-structure MWCNT network could monitor the human motion state in real time and has great application potential in terms of medical and health monitoring and motion monitoring.

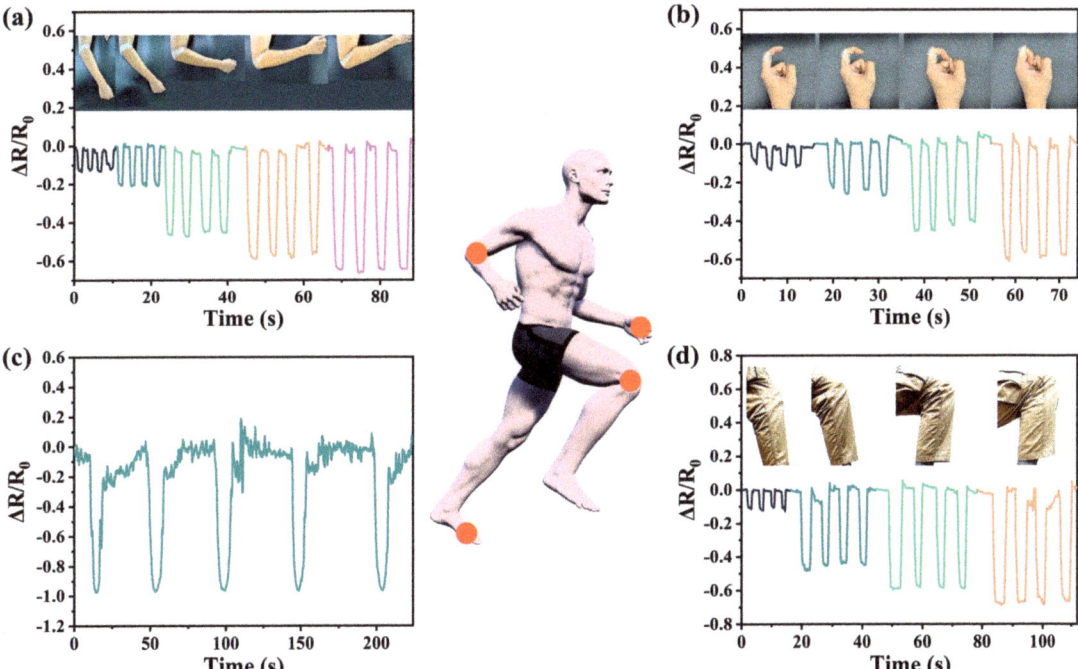

Figure 4. Monitoring and detection of human physiological signals and joint movements. (**a**) Detecting elbow movement. The tested motion angles of elbow joints were 30°, 60°, 90°, 120° and 150°, respectively. (**b**) Detecting the bending movement of the index finger. The tested motion angles of fingers were 30°, 60°, 90° and 120°, respectively. (**c**) Plantar pressure detection. The figure shows the significant change in plantar pressure during exercise. (**d**) Detecting knee movement. The tested knee joint motion angles were 10°, 45°, 90° and 135°, respectively.

Finally, we constructed a sensor array consisting of five sensors and placed the five sensors at each of the five finger joints to detect the movement of the finger joints in order to recognize simple sign language. When different gestures were made, the sensor array received different signals due to the different changes in the finger joints. The type of sign language or gesture was detected by detecting the position and intensity of the signal changes, as shown in Figure 5b. This showed the great potential of the devices in the field of gesture recording and sign language translation. In addition, we applied the sensors to speech recognition. The strain sensor placed in the interlayer of a mask monitored the dynamics of facial muscles when people spoke and generated signal responses to the changes in the mouth shape, thus realizing speech recognition, as shown in Figure 5d.

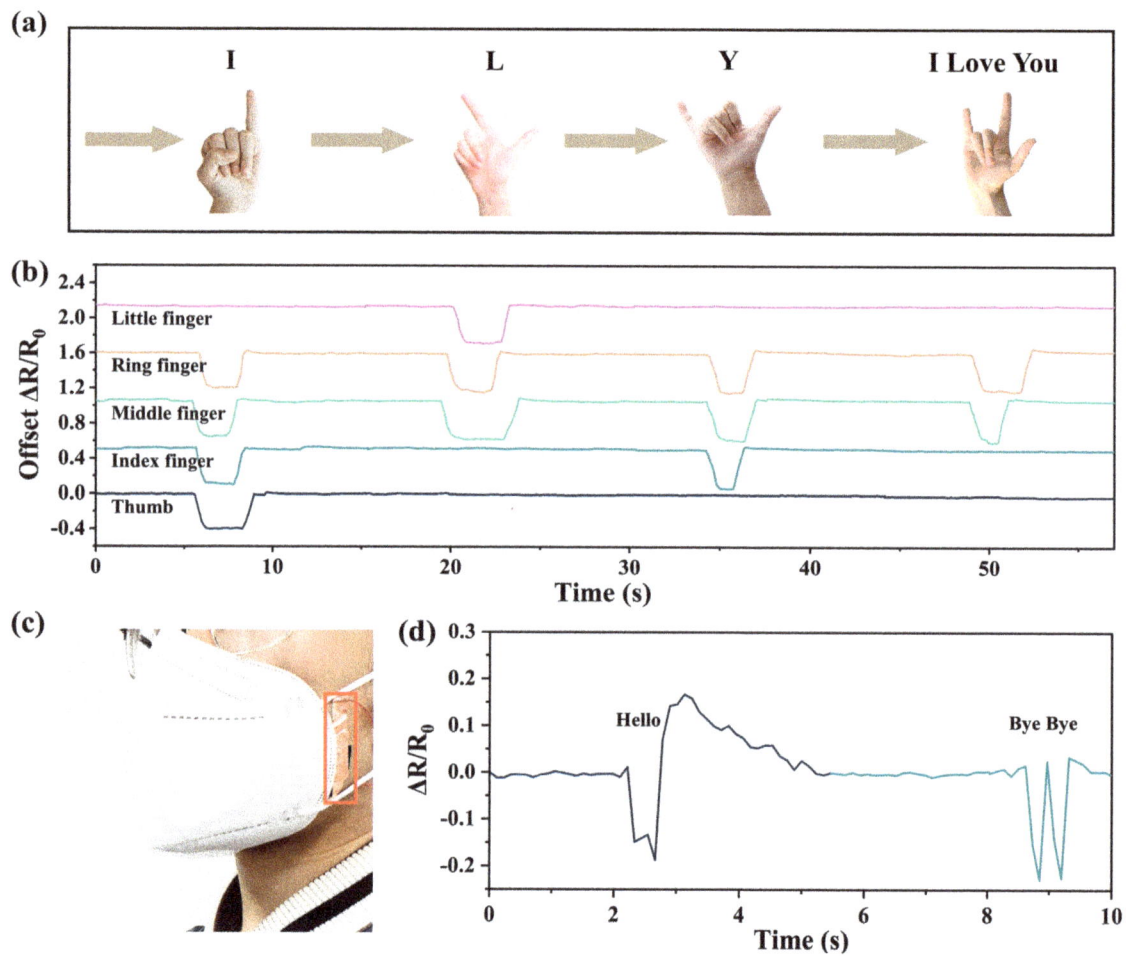

Figure 5. The recognition of human sign language and gestures as well as pronunciation was realized through the combination of multiple sensors. (**a**) Schematic diagram of different gestures and (**b**) resistance response signals generated by corresponding finger joint movements. (**c**) The sensor was placed on the face to recognize different articulation movements by monitoring the resistance response signal (**d**) generated by the stress change in the sensor caused by the movement of the facial muscles during articulation.

4. Conclusions

In summary, we reported a strain sensor with a wide response range based on MWC-NTs. To further improve the sensitivity of the sensor, we transferred a three-dimensional MWCNT structure to the internal pore surface of a porous PMDS flexible structure by wrapping MWCNTs on the surface of refined sucrose. In addition, some MWCNTs were dispersed within the PDMS structure using ultrasound. Through the above process, a porous PDMS flexible sensor with piezoresistive/capacitance dual-sensing performance was obtained, and stress changes were more accurately reflected through the changes in the capacitance and resistance. The sensor had an excellent sensing performance with a high sensitivity of 1.2/kPa and an ultra-wide response range (1–520 kPa). In addition, the sensor had a durable stability of more than 1000 loading and unloading cycles. The high sensitivity and wide detection range were explained by the sensor's structure, which was observed by SEM and three-dimensional depth-of-field experiments. In addition, the flexibility and air permeability of the porous PDMS flexible sensor made it suitable for attaching to the surface of the human body in order to realize human posture detection, such as gesture and sign language detection. Looking ahead, these porous flexible PDMS strain sensors show potential for application in skin sensors, artificial intelligence and bionic robots.

Supplementary Materials: The following supporting information can be downloaded at: https://www.mdpi.com/article/10.3390/nano13050843/s1, Figure S1: Sensor preparation process; Figure S2: Arm skin temperature was 30.6 °C before the material was placed, 30.7 °C after the porous structure strain sensor was placed for 3 h, 31.0 °C after the PDMS was placed for 3 h, and 30.8 °C after the Band-aid was placed for 3 h; Figure S3: The air permeability test used six sets of materials sealed with water beakers placed at room temperature to record water evaporation from (**a**) day 0 to (**b**) day 15; Figure S4: The (**a**) resistance, (**b**) resistance sensitivity, (**c**) capacitance and (**d**) capacitance sensitivity of the sensor under pressure cycling were tested when the mass ratio of MWCNTs in the PDMS matrix was different from that on the surface of the porous structure; Figure S5: Equation fitting of (**a**) $\Delta R/R0$ and (**b**) $\Delta C/C0$ properties of sensors; Figure S6: The response curve of sensor (**a**) resistance and (**b**) capacitance under 1000 compression cycles and its local magnification diagram; Table S1: Summary of the performance of the porous stress sensors; Video S1: Finger motion detection; Video S2: Speech detection; Video S3: Arm motion detection; Video S4: Leg motion detection; Video S5: Foot motion detection. References [55–63] are cited in the supplementary materials.

Author Contributions: S.Z., X.S. and X.G. contributed equally to this work. Methodology, experiment, writing—original draft and writing—review and editing, S.Z.; methodology and writing—review and editing, X.S.; writing—review and editing, X.G.; resources, J.Z.; validation, H.L.; formal analysis, L.C.; investigation, J.W.; conceptualization and supervision, Y.S. and L.P. All authors have read and agreed to the published version of the manuscript.

Funding: This work was supported by the National Key Research and Development program of China under Grant No. 2021YFA1401103 and the National Natural Science Foundation of China under Grants 61825403, 61921005 and 61674078.

Institutional Review Board Statement: Not applicable.

Informed Consent Statement: Informed consent was obtained from all subjects involved in the study.

Data Availability Statement: No new data were created or analyzed in this study. Data sharing is not applicable to this article.

Conflicts of Interest: The authors declare no conflict of interest.

References

1. Kaltenbrunner, M.; Sekitani, T.; Reeder, J.; Yokota, T.; Kuribara, K.; Tokuhara, T.; Drack, M.; Schwödiauer, R.; Graz, I.; Bauer-Gogonea, S.; et al. An ultra-lightweight design for imperceptible plastic electronics. *Nature* **2013**, *499*, 458–463. [CrossRef] [PubMed]
2. Kim, J.; Gutruf, P.; Chiarelli, A.M.; Heo, S.Y.; Cho, K.; Xie, Z.; Banks, A.; Han, S.; Jang, K.I.; Lee, J.W.; et al. Miniaturized Battery-Free Wireless Systems for Wearable Pulse Oximetry. *Adv. Funct. Mater.* **2017**, *27*, 1604373. [CrossRef]
3. Kim, J.; Campbell, A.S.; De Ávila, B.E.-F.; Wang, J. Wearable biosensors for healthcare monitoring. *Nat. Biotechnol.* **2019**, *37*, 389–406. [CrossRef] [PubMed]
4. Shi, J.; Liu, S.; Zhang, L.; Yang, B.; Shu, L.; Yang, Y.; Ren, M.; Wang, Y.; Chen, J.; Chen, W.; et al. Smart Textile-Integrated Microelectronic Systems for Wearable Applications. *Adv. Mater.* **2020**, *32*, 1901958. [CrossRef] [PubMed]
5. Wu, W. Stretchable electronics: Functional materials, fabrication strategies and applications. *Sci. Technol. Adv. Mater.* **2019**, *20*, 187–224. [CrossRef]
6. Nguyen, A.V.; Soulika, A.M. The Dynamics of the Skin's Immune System. *Int. J. Mol. Sci.* **2019**, *20*, 1811. [CrossRef]
7. Abraira, V.E.; Ginty, D.D. The Sensory Neurons of Touch. *Neuron* **2013**, *79*, 618–639. [CrossRef]
8. Yang, J.C.; Mun, J.; Kwon, S.Y.; Park, S.; Bao, Z.; Park, S. Electronic Skin: Recent Progress and Future Prospects for Skin-Attachable Devices for Health Monitoring, Robotics, and Prosthetics. *Adv. Mater.* **2019**, *31*, 1904765. [CrossRef]
9. Hammock, M.L.; Chortos, A.; Tee, B.C.K.; Tok, J.B.H.; Bao, Z. 25th Anniversary Article: The Evolution of Electronic Skin (E-Skin): A Brief History, Design Considerations, and Recent Progress. *Adv. Mater.* **2013**, *25*, 5997–6038. [CrossRef]
10. Fang, Z.; Zhang, H.; Qiu, S.; Kuang, Y.; Zhou, J.; Lan, Y.; Sun, C.; Li, G.; Gong, S.; Ma, Z. Versatile Wood Cellulose for Biodegradable Electronics. *Adv. Mater. Technol.* **2021**, *6*, 2000928. [CrossRef]

11. Ma, Y.; Zhang, Y.; Cai, S.; Han, Z.; Liu, X.; Wang, F.; Cao, Y.; Wang, Z.; Li, H.; Chen, Y.; et al. Flexible Hybrid Electronics for Digital Healthcare. *Adv. Mater.* **2020**, *32*, 1902062. [CrossRef] [PubMed]
12. Arakawa, T.; Kuroki, Y.; Nitta, H.; Chouhan, P.; Toma, K.; Sawada, S.-I.; Takeuchi, S.; Sekita, T.; Akiyoshi, K.; Minakuchi, S.; et al. Mouthguard biosensor with telemetry system for monitoring of saliva glucose: A novel cavitas sensor. *Biosens. Bioelectron.* **2016**, *84*, 106–111. [CrossRef] [PubMed]
13. Lee, Y.; Howe, C.; Mishra, S.; Lee, D.S.; Mahmood, M.; Piper, M.; Kim, Y.; Tieu, K.; Byun, H.-S.; Coffey, J.P.; et al. Wireless, intraoral hybrid electronics for real-time quantification of sodium intake toward hypertension management. *Proc. Natl. Acad. Sci. USA* **2018**, *115*, 5377–5382. [CrossRef] [PubMed]
14. Sun, Z.; Zhu, M.; Zhang, Z.; Chen, Z.; Shi, Q.; Shan, X.; Yeow, R.C.H.; Lee, C. Artificial Intelligence of Things (AIoT) Enabled Virtual Shop Applications Using Self-Powered Sensor Enhanced Soft Robotic Manipulator. *Adv. Sci.* **2021**, *8*, 2100230. [CrossRef] [PubMed]
15. Larson, C.; Peele, B.; Li, S.; Robinson, S.; Totaro, M.; Beccai, L.; Mazzolai, B.; Shepherd, R. Highly stretchable electroluminescent skin for optical signaling and tactile sensing. *Science* **2016**, *351*, 1071–1074. [CrossRef] [PubMed]
16. Liu, Y.; Li, J.; Song, S.; Kang, J.; Tsao, Y.; Chen, S.; Mottini, V.; McConnell, K.; Xu, W.; Zheng, Y.-Q.; et al. Morphing electronics enable neuromodulation in growing tissue. *Nat. Biotechnol.* **2020**, *38*, 1031–1036. [CrossRef]
17. Moin, A.; Zhou, A.; Rahimi, A.; Menon, A.; Benatti, S.; Alexandrov, G.; Tamakloe, S.; Ting, J.; Yamamoto, N.; Khan, Y.; et al. A wearable biosensing system with in-sensor adaptive machine learning for hand gesture recognition. *Nat. Electron.* **2020**, *4*, 54–63. [CrossRef]
18. Wei, J.; Xie, J.; Zhang, P.; Zou, Z.; Ping, H.; Wang, W.; Xie, H.; Shen, J.Z.; Lei, L.; Fu, Z. Bioinspired 3D Printable, Self-Healable, and Stretchable Hydrogels with Multiple Conductivities for Skin-like Wearable Strain Sensors. *ACS Appl. Mater. Interfaces* **2021**, *13*, 2952–2960. [CrossRef]
19. Wang, S.; Xu, J.; Wang, W.; Wang, G.-J.N.; Rastak, R.; Molina-Lopez, F.; Chung, J.W.; Niu, S.; Feig, V.R.; Lopez, J.; et al. Skin electronics from scalable fabrication of an intrinsically stretchable transistor array. *Nature* **2018**, *555*, 83–88. [CrossRef]
20. Li, G.; Liu, S.; Wang, L.; Zhu, R. Skin-inspired quadruple tactile sensors integrated on a robot hand enable object recognition. *Sci. Robot.* **2020**, *5*, eabc8134. [CrossRef]
21. Ruther, P.; Baumann, M.; Gieschke, P.; Herrmann, M.; Lemke, B.; Seidl, K.; Paul, O. *CMOS-Integrated Stress Sensor Systems*; IEEE: Piscataway, NJ, USA, 2010.
22. Ruth, S.R.A.; Beker, L.; Tran, H.; Feig, V.R.; Matsuhisa, N.; Bao, Z. Rational Design of Capacitive Pressure Sensors Based on Pyramidal Microstructures for Specialized Monitoring of Biosignals. *Adv. Funct. Mater.* **2019**, *30*, 1903100. [CrossRef]
23. Cao, M.; Su, J.; Fan, S.; Qiu, H.; Su, D.; Li, L. Wearable piezoresistive pressure sensors based on 3D graphene. *Chem. Eng. J.* **2021**, *406*, 126777. [CrossRef]
24. Cui, H.; Hensleigh, R.; Yao, D.; Maurya, D.; Kumar, P.; Kang, M.G.; Priya, S.; Zheng, X.R. Three-dimensional printing of piezoelectric materials with designed anisotropy and directional response. *Nat. Mater.* **2019**, *18*, 234–241. [CrossRef] [PubMed]
25. Lin, L.; Xie, Y.; Wang, S.; Wu, W.; Niu, S.; Wen, X.; Wang, Z.L. Triboelectric Active Sensor Array for Self-Powered Static and Dynamic Pressure Detection and Tactile Imaging. *ACS Nano* **2013**, *7*, 8266–8274. [CrossRef]
26. Zhai, W.; Xia, Q.; Zhou, K.; Yue, X.; Ren, M.; Zheng, G.; Dai, K.; Liu, C.; Shen, C. Multifunctional flexible carbon black/polydimethylsiloxane piezoresistive sensor with ultrahigh linear range, excellent durability and oil/water separation capability. *Chem. Eng. J.* **2019**, *372*, 373–382. [CrossRef]
27. Song, Y.; Chen, H.; Su, Z.; Chen, X.; Miao, L.; Zhang, J.; Cheng, X.; Zhang, H. Highly Compressible Integrated Supercapacitor-Piezoresistance-Sensor System with CNT-PDMS Sponge for Health Monitoring. *Small* **2017**, *13*, 1702091. [CrossRef]
28. Ha, K.H.; Zhang, W.; Jang, H.; Kang, S.; Wang, L.; Tan, P.; Hwang, H.; Lu, N. Highly Sensitive Capacitive Pressure Sensors over a Wide Pressure Range Enabled by the Hybrid Responses of a Highly Porous Nanocomposite. *Adv. Mater.* **2021**, *33*, 2103320. [CrossRef]
29. Cho, M.-Y.; Lee, J.H.; Kim, S.-H.; Kim, J.S.; Timilsina, S. An Extremely Inexpensive, Simple, and Flexible Carbon Fiber Electrode for Tunable Elastomeric Piezo-Resistive Sensors and Devices Realized by LSTM RNN. *ACS Appl. Mater. Interfaces* **2019**, *11*, 11910–11919. [CrossRef]
30. Wang, Y.; Niu, W.; Lo, C.Y.; Zhao, Y.; He, X.; Zhang, G.; Wu, S.; Ju, B.; Zhang, S. Interactively Full-Color Changeable Electronic Fiber Sensor with High Stretchability and Rapid Response. *Adv. Funct. Mater.* **2020**, *30*, 2000356. [CrossRef]
31. Ge, J.; Sun, L.; Zhang, F.-R.; Zhang, Y.; Shi, L.-A.; Zhao, H.-Y.; Zhu, H.-W.; Jiang, H.-L.; Yu, S.-H. A Stretchable Electronic Fabric Artificial Skin with Pressure-, Lateral Strain-, and Flexion-Sensitive Properties. *Adv. Mater.* **2016**, *28*, 722–728. [CrossRef]
32. Lo, L.-W.; Zhao, J.; Wan, H.; Wang, Y.; Chakrabartty, S.; Wang, C. A Soft Sponge Sensor for Multimodal Sensing and Distinguishing of Pressure, Strain, and Temperature. *ACS Appl. Mater. Interfaces* **2022**, *14*, 9570–9578. [CrossRef] [PubMed]
33. Liu, H.; Chen, X.; Zheng, Y.; Zhang, D.; Zhao, Y.; Wang, C.; Pan, C.; Liu, C.; Shen, C. Lightweight, Superelastic, and Hydrophobic Polyimide Nanofiber/MXene Composite Aerogel for Wearable Piezoresistive Sensor and Oil/Water Separation Applications. *Adv. Funct. Mater.* **2021**, *31*, 2008006. [CrossRef]
34. Chao, M.; He, L.; Gong, M.; Li, N.; Li, X.; Peng, L.; Shi, F.; Zhang, L.; Wan, P. Breathable $Ti_3C_2T_x$ MXene/Protein Nanocomposites for Ultrasensitive Medical Pressure Sensor with Degradability in Solvents. *ACS Nano* **2021**, *15*, 9746–9758. [CrossRef] [PubMed]

35. Qiu, Y.; Tian, Y.; Sun, S.; Hu, J.; Wang, Y.; Zhang, Z.; Liu, A.; Cheng, H.; Gao, W.; Zhang, W.; et al. Bioinspired, multifunctional dual-mode pressure sensors as electronic skin for decoding complex loading processes and human motions. *Nano Energy* **2020**, *78*, 105337. [CrossRef]
36. Zhou, Z.; Chen, K.; Li, X.; Zhang, S.; Wu, Y.; Zhou, Y.; Meng, K.; Sun, C.; He, Q.; Fan, W.; et al. Sign-to-speech translation using machine-learning-assisted stretchable sensor arrays. *Nat. Electron.* **2020**, *3*, 571–578. [CrossRef]
37. Liu, Y.; Xu, H.; Dong, M.; Han, R.; Tao, J.; Bao, R.; Pan, C. Highly Sensitive Wearable Pressure Sensor over a Wide Sensing Range Enabled by the Skin Surface-like 3D Patterned Interwoven Structure. *Adv. Mater. Technol.* **2022**, *7*, 2200504. [CrossRef]
38. Zhou, K.; Xu, W.; Yu, Y.; Zhai, W.; Yuan, Z.; Dai, K.; Zheng, G.; Mi, L.; Pan, C.; Liu, C.; et al. Tunable and Nacre-Mimetic Multifunctional Electronic Skins for Highly Stretchable Contact-Noncontact Sensing. *Small* **2021**, *17*, 2100542. [CrossRef]
39. Xiong, Y.; Shen, Y.; Tian, L.; Hu, Y.; Zhu, P.; Sun, R.; Wong, C.-P. A flexible, ultra-highly sensitive and stable capacitive pressure sensor with convex microarrays for motion and health monitoring. *Nano Energy* **2020**, *70*, 104436. [CrossRef]
40. Wang, Z.; Guan, X.; Huang, H.; Wang, H.; Lin, W.; Peng, Z. Full 3D Printing of Stretchable Piezoresistive Sensor with Hierarchical Porosity and Multimodulus Architecture. *Adv. Funct. Mater.* **2019**, *29*, 1807569. [CrossRef]
41. Yao, W.; Mao, R.; Gao, W.; Chen, W.; Xu, Z.; Gao, C. Piezoresistive effect of superelastic graphene aerogel spheres. *Carbon* **2020**, *158*, 418–425. [CrossRef]
42. Pelella, A.; Capista, D.; Passacantando, M.; Faella, E.; Grillo, A.; Giubileo, F.; Martucciello, N.; Di Bartolomeo, A. A Self-Powered CNT–Si Photodetector with Tuneable Photocurrent. *Adv. Electron. Mater.* **2023**, *9*, 2200919. [CrossRef]
43. Li, J.; Dwivedi, P.; Kumar, K.S.; Roy, T.; Crawford, K.E.; Thomas, J. Growing Perovskite Quantum Dots on Carbon Nanotubes for Neuromorphic Optoelectronic Computing. *Adv. Electron. Mater.* **2021**, *7*, 2000535. [CrossRef]
44. Zamzami, M.A.; Rabbani, A.; Ahmad, A.; Basalah, A.A.; Al-Sabban, W.H.; Nate Ahn, S.; Choudhry, H. Carbon nanotube field-effect transistor (CNT-FET)-based biosensor for rapid detection of SARS-CoV-2 (COVID-19) surface spike protein S1. *Bioelectrochemistry* **2022**, *143*, 107982. [CrossRef]
45. Guillaume, Y.C.; André, C. ACE2 and SARS-CoV-2-Main Protease Capillary Columns for Affinity Chromatography: Testimony of the Binding of Dexamethasone and its Carbon Nanotube Nanovector. *Chromatographia* **2022**, *85*, 773–781. [CrossRef]
46. Kareem, M.H.; Hussein, H.T.; Abdul Hussein, A.M. Study of the effect of CNTs, and (CNTs-ZnO) on the porous silicon as sensor for acetone gas detection. *Optik* **2022**, *259*, 168825. [CrossRef]
47. Bolotov, V.V.; Stenkin, Y.A.; Sokolov, D.V.; Roslikov, V.E.; Knyazev, E.V.; Ivlev, K.E. *Gas Sensing Properties of MWCNT/ZnO and MWCNT/ZnO/In$_2$O$_3$ Nanostructures*; AIP Publishing: Woodbury, NY, USA, 2020.
48. Nag, A.; Alahi, M.E.E.; Mukhopadhyay, S.C.; Liu, Z. Multi-Walled Carbon Nanotubes-Based Sensors for Strain Sensing Applications. *Sensors* **2021**, *21*, 1261. [CrossRef]
49. Shulaker, M.M.; Hills, G.; Patil, N.; Wei, H.; Chen, H.-Y.; Wong, H.S.P.; Mitra, S. Carbon nanotube computer. *Nature* **2013**, *501*, 526–530. [CrossRef]
50. Li, C.; Liu, D.; Xu, C.; Wang, Z.; Shu, S.; Sun, Z.; Tang, W.; Wang, Z.L. Sensing of joint and spinal bending or stretching via a retractable and wearable badge reel. *Nat. Commun.* **2021**, *12*, 2950. [CrossRef] [PubMed]
51. Kim, C.S.; Yang, H.M.; Lee, J.; Lee, G.S.; Choi, H.; Kim, Y.J.; Lim, S.H.; Cho, S.H.; Cho, B.J. Self-Powered Wearable Electrocardiography Using a Wearable Thermoelectric Power Generator. *ACS Energy Lett.* **2018**, *3*, 501–507. [CrossRef]
52. Wen, Z.; Yeh, M.-H.; Guo, H.; Wang, J.; Zi, Y.; Xu, W.; Deng, J.; Zhu, L.; Wang, X.; Hu, C.; et al. Self-powered textile for wearable electronics by hybridizing fiber-shaped nanogenerators, solar cells, and supercapacitors. *Sci. Adv.* **2016**, *2*, e1600097. [CrossRef] [PubMed]
53. Gao, C.; Huang, J.; Xiao, Y.; Zhang, G.; Dai, C.; Li, Z.; Zhao, Y.; Jiang, L.; Qu, L. A seamlessly integrated device of micro-supercapacitor and wireless charging with ultrahigh energy density and capacitance. *Nat. Commun.* **2021**, *12*, 2647. [CrossRef] [PubMed]
54. Liu, Y.; Tao, J.; Yang, W.; Zhang, Y.; Li, J.; Xie, H.; Bao, R.; Gao, W.; Pan, C. Biodegradable, Breathable Leaf Vein-Based Tactile Sensors with Tunable Sensitivity and Sensing Range. *Small* **2022**, *18*, 2106906. [CrossRef] [PubMed]
55. Yin, Y.M.; Li, H.Y.; Xu, J.; Zhang, C.; Liang, F.; Li, X.; Jiang, Y.; Cao, J.W.; Feng, H.F.; Mao, J.N.; et al. Facile Fabrication of Flexible Pressure Sensor with Programmable Lattice Structure. *ACS Appl. Mater. Interfaces* **2021**, *13*, 10388–10396. [CrossRef] [PubMed]
56. Lee, J.; Kim, J.; Shin, Y.; Jung, I. Ultra-robust wide-range pressure sensor with fast response based on polyurethane foam doubly coated with conformal silicone rubber and CNT/TPU nanocomposites islands. *Compos. Part B Eng.* **2019**, *177*, 107364. [CrossRef]
57. Tewari, A.; Gandla, S.; Bohm, S.; McNeill, C.R.; Gupta, D. Highly Exfoliated MWNT–rGO Ink-Wrapped Polyurethane Foam for Piezoresistive Pressure Sensor Applications. *ACS Appl. Mater. Interfaces* **2018**, *10*, 5185–5195. [CrossRef] [PubMed]
58. Hwang, J.; Kim, Y.; Yang, H.; Oh, J.H. Fabrication of hierarchically porous structured PDMS composites and their application as a flexible capacitive pressure sensor. *Compos. Part B Eng.* **2021**, *211*, 108607. [CrossRef]
59. Mu, C.; Guo, X.; Zhu, T.; Lou, S.; Tian, W.; Liu, Z.; Jiao, W.; Wu, B.; Liu, Y.; Yin, L.; et al. Flexible strain/pressure sensor with good sensitivity and broad detection range by coupling PDMS and carbon nanocapsules. *J. Alloy. Compd.* **2022**, *918*, 165696. [CrossRef]
60. Hsieh, G.W.; Shih, L.C.; Chen, P.Y. Porous Polydimethylsiloxane Elastomer Hybrid with Zinc Oxide Nanowire for Wearable, Wide-Range, and Low Detection Limit Capacitive Pressure Sensor. *Nanomaterials* **2022**, *12*, 256. [CrossRef]
61. Zhu, G.; Dai, H.; Yao, Y.; Tang, W.; Shi, J.; Yang, J.; Zhu, L. 3D Printed Skin-Inspired Flexible Pressure Sensor with Gradient Porous Structure for Tunable High Sensitivity and Wide Linearity Range. *Adv. Mater. Technol.* **2022**, *7*, 2101239. [CrossRef]

62. Iglio, R.; Mariani, S.; Robbiano, V.; Strambini, L.; Barillaro, G. Flexible Polydimethylsiloxane Foams Decorated with Multiwalled Carbon Nanotubes Enable Unprecedented Detection of Ultralow Strain and Pressure Coupled with a Large Working Range. *ACS Appl. Mater. Interfaces* **2018**, *10*, 13877–13885. [CrossRef]
63. Li, W.; Jin, X.; Han, X.; Li, Y.; Wang, W.; Lin, T.; Zhu, Z. Synergy of Porous Structure and Microstructure in Piezoresistive Material for High-Performance and Flexible Pressure Sensors. *ACS Appl. Mater. Interfaces* **2021**, *13*, 19211–19220. [CrossRef] [PubMed]

Disclaimer/Publisher's Note: The statements, opinions and data contained in all publications are solely those of the individual author(s) and contributor(s) and not of MDPI and/or the editor(s). MDPI and/or the editor(s) disclaim responsibility for any injury to people or property resulting from any ideas, methods, instructions or products referred to in the content.

Article
A Flexible Piezocapacitive Pressure Sensor with Microsphere-Array Electrodes

Shu Ying †, Jiean Li †, Jinrong Huang *, Jia-Han Zhang, Jing Zhang, Yongchang Jiang, Xidi Sun *, Lijia Pan * and Yi Shi

Collaborative Innovation Center of Advanced Microstructures, School of Electronic Science and Engineering, Nanjing University, Nanjing 210093, China; yingshu@smail.nju.edu.cn (S.Y.); jali@smail.nju.edu.cn (J.L.); jhzhang@smail.nju.edu.cn (J.-H.Z.); zjing@smail.nju.edu.cn (J.Z.); ycjiang@nju.edu.cn (Y.J.); yshi@nju.edu.cn (Y.S.)
* Correspondence: jinronghuang@smail.nju.edu.cn (J.H.); xidisun@smail.nju.edu.cn (X.S.); ljpan@nju.edu.cn (L.P.)
† These authors contributed equally to this work.

Abstract: Flexible pressure sensors that emulate the sensation and characteristics of natural skins are of great importance in wearable medical devices, intelligent robots, and human–machine interfaces. The microstructure of the pressure-sensitive layer plays a significant role in the sensor's overall performance. However, microstructures usually require complex and costly processes such as photolithography or chemical etching for fabrication. This paper proposes a novel approach that combines self-assembled technology to prepare a high-performance flexible capacitive pressure sensor with a microsphere-array gold electrode and a nanofiber nonwoven dielectric material. When subjected to pressure, the microsphere structures of the gold electrode deform via compressing the medium layer, leading to a significant increase in the relative area between the electrodes and a corresponding change in the thickness of the medium layer, as simulated in COMSOL simulations and experiments, which presents high sensitivity (1.807 kPa^{-1}). The developed sensor demonstrates excellent performance in detecting signals such as slight object deformations and human finger bending.

Keywords: microsphere arrays; nanofiber dielectric layers; piezocapacitive sensor; flexible pressure sensors; electronic skins

1. Introduction

In contemporary society, the burgeoning Internet of Things (IoT) technology has led to an upsurge in wearable devices, which continuously monitor the human body while interacting with the terminal, and sensors play a crucial role in this regard [1–5]. Human skin, being the largest and most basic organ of the human body, serves as both the primary barrier to the external environment and a source of sensory information regarding external stimuli such as pressure, deformation, and touch [6,7]. Hence, electronic skin has become essential in wearable medical devices, human–computer interaction, artificial intelligence, and health monitoring [8–15]. An ideal e-skin must possess high flexibility, sensitivity, affordability, and structural simplicity and should mimic the natural skin's ability to sense tactile pressure ranging from light contact to object haptic perception (0–30 kPa) [16–19]. Piezoresistive, piezocapacitive, triboelectric, and piezoelectric sensing mechanisms are commonly employed to convert the applied stress signal into an electrical signal in electronic skin pressure sensors. In particular, piezocapacitive pressure sensors are highly attractive owing to their ease of preparation, structural simplicity, low power consumption, and compact circuit layout [20–30].

The sensing performance of piezocapacitive sensors primarily depends on the mechanical properties of the elastic dielectric layer used. In some respects, the greater the compressibility under the same pressure of the dielectric material used, the higher the sensitivity of the sensor [15,20]. Capacitive pressure sensors usually consist of two parallel electrodes sandwiched on a polymer dielectric layer. Since the sensor's capacitance changes by altering the thickness of the dielectric layer when an external force is applied to the sensor, its sensitivity is directly affected by the mechanical properties of the elastic dielectric layer and the overall thickness of the sensor. Therefore, polydimethylsiloxane (PDMS) with high compressibility is considered a suitable material for pressure-sensitive dielectric films [31–33]. By modifying the microstructure of PDMS, the surface of the pressure-sensitive layer can be elastically deformed, and it can quickly recover its original shape upon the application/release of external pressure, in which the microstructure improves the overall sensor performance and mitigates issues associated with viscoelastic behavior. To further enhance the overall performance of piezocapacitive pressure sensors—measured in terms of linearity, sensitivity, detection range, and stability—microstructures such as pyramidal structures, micropores, and air gaps are often incorporated into the dielectric layer to obtain an appropriate Young's modulus or silver nanowire embedding is introduced to increase the dielectric constant [14,34,35]. However, these preparation methods typically employ porous structures to enhance the sensor's sensitivity, making it challenging to achieve high sensitivity, wide detection range, and good linearity simultaneously [12,33,36–38]. Moreover, these microstructures necessitate complex processes such as photolithography, sacrificial templates, or chemical etching, which are unsuitable for low-cost and large-scale production. Hence, there is an urgent need to develop a simple and reliable preparation method that can customize the microstructures easily, thus enhancing the sensor's performance while reducing the preparation cost.

In this paper, we propose a low-cost, high-sensitivity, and large-area compatible inverse-mold technique for preparing a flexible capacitive pressure sensor with a self-assembled microsphere-type array. We simulate the designed sensor using COMSOL, including the simulation of the nonwoven fabrics with electrostatic spinning and the calculation of the overall sensor performance, to provide a theoretical basis for the experiments. With the help of COMSOL simulations, we prepared a highly sensitive piezocapacitive sensor, which exhibited a relative rate of change $\Delta C/C_0$ maximum of about 2.13, a sensitivity of 1.807 kPa^{-1} maximum, and a hysteresis of 7.69%. These performance improvements were attributed to the interaction between the microsphere-shaped gold electrode array and the rough surface of the dielectric layer, resulting in a significant change in the contact area of the electrodes and the thickness of the dielectric layer. Moreover, the electrostatic spinning method used to prepare the nanofibrous dielectric layer makes it more easily compressible compared to a conventional dielectric layer. The exclusion of air from the dielectric layer also increases its dielectric constant, making the change in capacitance more pronounced. Consequently, this flexible pressure sensor exhibits excellent performance in pressure measurement and has significant potential for electronic skin applications. Overall, the results of our study suggest that the proposed low-cost and high-performance capacitive pressure sensor could find wide applications in the fields of robotics, prosthetics, and healthcare monitoring.

2. Materials and Methods

2.1. COMSOL Multiphysics Simulation

We performed simulations based on the focuses of two modules within COMSOL Multiphysics: electrostatics and solid mechanics. To obtain the potential map of a capacitive pressure sensor, we first used the electrostatics sub-module under the electromagnetics module to calculate the electric field generated by electrostatics. Under free space, the spatial electric field is irrotational, and the relationship between the potential and the spatial charge density can be determined using Maxwell's system of equations:

$$-\nabla \cdot \nabla V = \frac{\rho}{\varepsilon_0} \tag{1}$$

where ρ is the charge density, ε_0 is the free dielectric constant in space, and V is the electric potential. However, the internal electric field of a dielectric material is generated by an electric dipole, which is not consistent with the free electric field. Therefore, this phenomenon needs to be described from a macroscopic point of view:

$$\rho_P = -\nabla \cdot P \tag{2}$$

$$D = \varepsilon_0 E + P \tag{3}$$

where P is the polarization vector field and D is the electrical displacement density. The set of electrostatic equations can be finally associated as a system of equations:

$$-\nabla \cdot (\varepsilon_0 \nabla V - P) = \rho_P \tag{4}$$

When a material is compressed or stretched, the stress distribution within it is not uniform. Therefore, our simulation needed to consider the physical properties of the material, the stress distribution, and the deformation. In solid mechanics, the stress tensor of a material is used to describe the stress distribution in three dimensions:

$$\sigma = \begin{pmatrix} \sigma_{xx} & \sigma_{xy} & \sigma_{xz} \\ \sigma_{yx} & \sigma_{yy} & \sigma_{yz} \\ \sigma_{zx} & \sigma_{zy} & \sigma_{zz} \end{pmatrix} \tag{5}$$

where the first component represents the component of the force and the second represents the direction of the normal to the force. Newton's second law can be rewritten to be expressed using the stress tensor:

$$\nabla \cdot \sigma + f = \rho \frac{\partial^2 u}{\partial t^2} \tag{6}$$

where f is the force per unit volume, ρ is the mass density, and u is the displacement vector. When a material is subjected to stress, it deforms, and stresses are generated within it. When deformation occurs, all parts of the material should be in harmony and reach a state of equilibrium. In order to describe this strain, a stacking tensor element is introduced to represent it:

$$\varepsilon = \begin{pmatrix} \varepsilon_{xx} & \varepsilon_{xy} & \varepsilon_{xz} \\ \varepsilon_{yx} & \varepsilon_{yy} & \varepsilon_{yz} \\ \varepsilon_{zx} & \varepsilon_{zy} & \varepsilon_{zz} \end{pmatrix} \tag{7}$$

Each tensor element is defined as the derivative of the displacement:

$$\begin{bmatrix} \varepsilon_{xx} \\ \varepsilon_{yy} \\ \varepsilon_{zz} \\ \varepsilon_{xy} \\ \varepsilon_{yz} \\ \varepsilon_{xz} \end{bmatrix} = \begin{bmatrix} \frac{\partial u}{\partial x} \\ \frac{\partial v}{\partial y} \\ \frac{\partial w}{\partial z} \\ \frac{1}{2}\left(\frac{\partial u}{\partial y} + \frac{\partial v}{\partial x}\right) \\ \frac{1}{2}\left(\frac{\partial v}{\partial z} + \frac{\partial w}{\partial y}\right) \\ \frac{1}{2}\left(\frac{\partial u}{\partial z} + \frac{\partial w}{\partial x}\right) \end{bmatrix} \tag{8}$$

These tensors are not characterized by an arbitrary spatial distribution of displacements, but rather by fundamental constants that can represent deformation processes and provide coordination relations for the deformation of materials under the action of external forces.

2.2. Chemicals and Materials

PDMS (Dow Corning, Sylgard 184) and polystyrene microspheres (PS, dm 0.6–1.0 μm, 2.5% w/v) were purchased as received. Deionized water (water purifier), ethanol solution (AR, Nanjing Reagent), gold target (ZhongNuo Advanced Material Technology Co., Ltd., Beijing, China), glass slides (10 cm diameter), and acetone solution (Shanghai Reagent) were purchased from the reagent platform of Nanjing University.

To prepare the nanofiber PVDF nonwoven fabrics, 5 mL of DMF and 5 mL of acetone solution were mixed in a beaker at a ratio of 1:1 and placed in a magnetic stirrer at 60 °C for 20 min. A certain amount of PVDF powder was added to the DMF and acetone solution to form a mixture with a mass fraction of 20% wt. The stirred mixture was then placed in a magnetic stirrer at 70 °C for 1 h. The stirred mixture was left at room temperature for 1 h to remove any air bubbles. The mixed PVDF solution was electrostatically spun at 20 kV for 6 h to obtain the nanofiber PVDF nonwoven fabrics with electrostatic spinning.

2.3. Preparation of Microsphere Array Electrodes

The PDMS was obtained by mixing the crosslinker with the prepolymer in the ratio of 1:10 and stirring well. It was then put into a constant-temperature desiccator for the first evacuation operation to remove the air bubbles generated during the mixing process. The PDMS was uniformly poured on the glass sheet with PS microspheres, and then the second evacuation was performed to remove the air bubbles generated during the pouring process. The glass sheet with PDMS was then placed in a constant-temperature drying oven and baked at 100 °C for 2 h to cure and form a smooth PDMS film. The PDMS substrate with inverted microsphere arrays can be obtained by tearing off the PDMS film. The PDMS substrate was washed in acetone solution to remove the residual PS microspheres, then cleaned with ethanol and deionized water, respectively, and placed on a hot table for drying. The PDMS substrates with inverted microsphere arrays obtained from the first inversion were placed in a UV-Ozone cleaning machine for 20 min for hydrophilic treatment to produce a surface with good hydrophilicity. The second PDMS layer was uniformly poured onto the PDMS surface of the first layer, placed in an oven and baked at 80 °C for 2 h, and then removed. The PDMS substrate with microsphere arrays was prepared by the "secondary inversion" method by carefully tearing off the PDMS film of the second layer. Finally, the final PDMS substrate was sputtered with gold for 5 min using a JS-1600 magnetron sputterer to obtain gold electrode arrays of 50 nm thickness.

2.4. Preparation of Pressure Sensors

Copper wires and gold electrodes with a microsphere array structure were bonded together with conductive silver paste. Then, the lower electrode, nonwoven fabrics with electrostatic spinning, and upper electrode were assembled together in a sandwich structure to obtain a flexible capacitive pressure sensor.

3. Results and Discussion

Figure 1a depicts the fabrication process of microsphere array electrodes (content A, Supporting Information) for flexible capacitive pressure sensors. The process begins with the dispersion of PS microspheres in an ethanol solution, followed by the uniform deposition of the solution on a glass sheet to form a monolayer film through self-assembly. Once the ethanol evaporates completely, the PS microspheres are evenly distributed on the monolayer glass sheet, as shown in Figure 1b. Next, the configured PDMS solution is poured onto the PS microsphere film to form a concave microsphere structure after curing. Subsequently, PDMS is poured onto the concave PDMS film formed for the first time by the second inversion method and cured in the same way to form a PDMS substrate with a microsphere array structure, as shown in Figure 1c. Finally, the microsphere gold electrode arrays, with a gold layer of 50 nm in thickness, are obtained through magnetron sputtering.

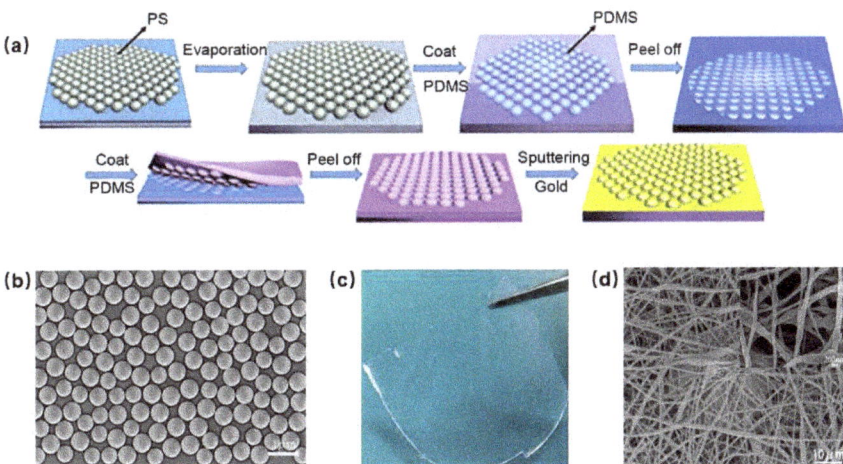

Figure 1. (a) Preparation process of microsphere array electrodes. (b) SEM image of monolayer PS microsphere array. (c) Photograph of PDMS with surface microsphere structure. (d) SEM image of nanofiber PVDF nonwoven fabrics; inset is SEM image at high magnification.

In the process of preparing flexible capacitive pressure sensors, electrostatic spinning plays a crucial role. Due to the complex network structure of the nanofiber nonwoven fabrics, the electrostatic spinning time needs to be controlled for more than six hours to prevent any leakage current or a device short-circuit. The electrostatic spinning method is used to prepare the PVDF film, as shown in Figure 1d. A dielectric layer with fibers exhibits a higher displacement increase compared to a flat dielectric layer without fibers. The nanofibers with voids constitute the inner part of the dielectric layer prepared by the electrostatic spinning technique. The interlaced nanofibers can be easily deformed by applying external forces, which is advantageous in detecting small force changes. From the SEM photos in the inset of Figure 1d, it can be seen that the diameter of PVDF nanofibers of the nonwoven fabrics is about 200 nm, the thickness is relatively uniform and smooth, the boundary is clear and uniformly distributed, and there is practically no agglomeration phenomenon.

The performance of capacitive pressure sensors can be improved through altering the electrode contact area and dielectric layer parameters. We first simulated the pressure sensor model with a microsphere-array-structure gold electrode on the surface and a fibrous dielectric material layer with surface roughness produced via electrostatic spinning technology through COMSOL, as shown in Figure 2a. Since the nanofibrous dielectric layer itself has a very complex fiber structure on the surface and inside, it is not feasible to make a complete fibrous dielectric layer. Thus, only rectangular voids can be added in the middle of the model of the nanofibrous dielectric layer to simulate the complex fiber structure. Figure 2b shows the deformation of the nonwoven fabrics with electrostatic spinning in the pressure range of 0–20 kPa. As the thickness of the dielectric layer decreases, its contact area with the motor increases gradually, resulting in an increase in capacitance value with the increase of pressure.

Subsequently, the simulation model without the nanofibrous structured dielectric layer is selected for comparison experiments. Figure 2c shows that the displacement increase of the dielectric layer with the fibrous form is significantly higher than that without the fibrous form. When P_0 is 2 kPa, the displacement with the fibrous dielectric layer reaches about 0.57 μm, while the displacement without the fibrous dielectric layer is only 0.40 μm.

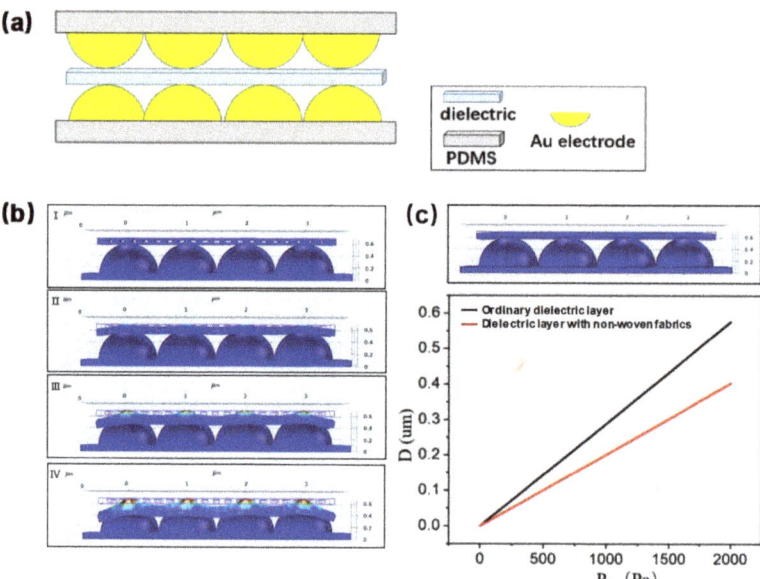

Figure 2. (a) Model drawing of flexible capacitive sensor. (b) The simulated deformation of the dielectric layer under the pressure of 0, 5, 10, and 20 kPa, respectively. (c) The upper figure shows the model of ordinary dielectric layer, and the lower figure shows the comparison of displacement of nonwoven fabrics with electrostatic spinning and ordinary dielectric layer.

In a flexible capacitive pressure sensor utilizing a microsphere array, the electrode portion is formed by sputtering a metal layer onto a PDMS surface with an array of microspheres, and this electrode design can effectively improve the sensitivity of the device (SC, Supporting Information). The application of pressure onto the substrate portion of the sensor results in the deformation of the electrodes as they come into contact with the dielectric layer, as illustrated in Figure 3a. The gold microspheres, with a thickness of only 100 nm, cause the electrodes to be squeezed and their contact area to change, ultimately leading to a change in the capacitance of the sensor. In contrast to the parallel plate electrode, the electrode area of the capacitive pressure sensor undergoes a greater degree of change, resulting in a greater amount of capacitance change. This paper presents a flexible capacitive pressure sensor model designed to account for changes in both the dielectric layer and electrode, as shown in Figure 3b. Based on this model, the theoretical sensitivity achievable by the sensor is calculated. Due to the large calculation volume of the three-dimensional model, a two-dimensional model calculation was utilized to maintain consistent imposed conditions. While an ordinary dielectric layer is used in this paper instead of a nanofibrous dielectric layer, parameters such as the elastic modulus and Poisson's ratio of the nanofibrous dielectric layer are introduced to prevent the experimental results from being affected. Additionally, the microsphere-structure gold electrode is known to deform under applied pressure, which is difficult to simulate; thus, this paper employs a microsphere structure that does not deform instead. Figure 3c displays the potential diagram of the electric field following pressure application. The thickness of the dielectric layer decreases under the action of pressure P0, resulting in an increase in the contact area of the electrode and, consequently, an increase in the capacitance of the entire capacitor. The initial capacitance (C_0) is 1.51 pF when no pressure is applied. As pressure P0 increases, the capacitance also increases and ultimately reaches approximately 6.4 pF, as illustrated in Figure 3d. The graph in Figure 3e illustrates the relative rate of change of capacitance

ΔC/C₀ for the simulation model, with a maximum value of about 3.238 observed when the pressure reaches a maximum of 20 kPa.

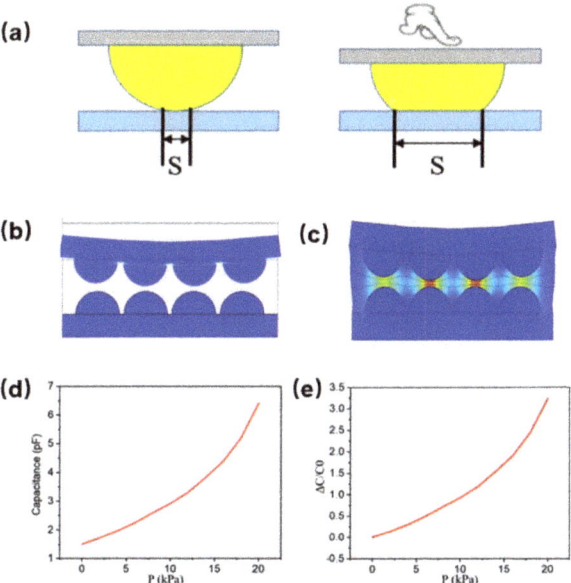

Figure 3. (**a**) Initial model of gold electrode with microsphere structure (**left**) and deformation under pressure (**right**). (**b**) Two−dimensional simulation model of flexible capacitive pressure sensor based on nonwoven fabrics with electrostatic spinning and microsphere−array electrode. (**c**) Simulated potential diagram of a flexible capacitive pressure sensor under external action. (**d**) Simulation curve of capacitance pressure of flexible capacitive pressure sensor. (**e**) Simulation graph of flexible capacitive pressure sensor ΔC/C₀−pressure.

The performance variation curve plotted from the simulation data shows that the nanofibrous dielectric layer has good deformation capability, while the sensor has excellent sensing performance under small stress. We bonded copper wires and gold electrodes with microsphere arrays together using a conductive silver paste/glue. Then, we assembled the lower electrode, the nonwoven fabrics with electrostatic spinning, and the upper electrode according to the "sandwich" structure to produce a flexible capacitive pressure sensor based on the microsphere array. In order to obtain high sensitivity, the dielectric layer thickness of the device needs to be controlled within 5 μm, so the test voltage needs to be controlled at 1 V to prevent the device from breakdown during testing. Figure 4a shows the capacitance pressure curve of the sensor, which shows that the capacitance value of the sensor increases with increasing pressure, increasing to 27.90 pF when the pressure reaches 80 kPa. The sensitivity of the sensor is very large at stresses less than 1 kPa. Similar results can be obtained for the variation curve of the relative rate of change of capacitance's relative intensity, as shown in Figure 4b, which can reach a value of 2.13 when the pressure reaches 80 kPa. The sensitivity curve in Figure 4c shows that the decrease in sensitivity is very sharp when the pressure is less than 1 kPa, and the sensitivity can reach a maximum of 1.807 kPa^{-1} at 0.05 kPa. This is due to the fact that when the electrode is subjected to a small pressure, the surface of the dielectric layer is squeezed against the electrode, resulting in a change in contact area, which can lead to a dramatic change in the capacitance value of the sensor, making the sensor excellent for small force measurements. Moreover, when more pressure is applied to the sensor, the thickness of the dielectric layer decreases further, causing an increase in its capacitance value. This property is related to the internal structure of the dielectric layer prepared by electrostatic spinning, which comprises nanofibers filled with air. When pressure is applied to the sensor, the air within its internal dielectric layer is compressed, causing the dielectric layer's structure to become very dense and the dielectric constant to change accordingly, leading to an increase in the capacitance value of the sensor. This is one of the primary reasons for the change in the capacitance value of flexible capacitive pressure sensors. The capacitance test curves of the sensor are generally consistent with the simulated expectations but show significant deviations in the low-stress range. This is primarily due to the intricate fibers in the nonwoven fabrics with electrostatic spinning, where the nanofibers undergo a significant change in dielectric constant under small stresses, resulting in a sharp increase in capacitance. Treating the microscopic spheres as non-deformable objects in the simulation also leads to significant deviations between the real and simulated results in the low-pressure section. Therefore, we can observe that the performance of flexible capacitive pressure sensors can be improved through changing the microstructure of the elastomer or the surface roughness and internal structure of the dielectric layer. Figure 4d illustrates the hysteresis performance of the sensor. The process of increasing and then decreasing the external pressure from 0–80 kPa back to 0 kPa is represented by two curves. The red curve shows the response when the pressure increases, while the black curve shows the response when the pressure decreases. The hysteresis of the flexible capacitive pressure sensor is approximately 7.69%, which is excellent and generally satisfies the hysteresis performance requirements of general flexible sensors. However, due to viscoelastic deformation within the device, the contact area and dielectric layer of the electrode do not quickly return to their original size and shape when the pressure decreases, leading to a difference in capacitance value compared to when the pressure increases. Furthermore, the air inside the nanofibers does not immediately return, indirectly affecting the dielectric constant of the dielectric layer. Additionally, as the substrate of the sensor is made of PDMS and the dielectric layer is prepared by electrostatic spinning technology, there is some viscosity between the contact surfaces, leading to a small adhesion phenomenon during the process of pressure increase and decrease and resulting in the response lag of the sensor. Figure 4e shows the recovery and response time of the device; the device has a fast response time of 0.048 s and a fast recovery time of 0.072 s. The cycling stability of the device is shown in Figure 4f. We have carried out 1000 cycles of the device under the same 15 N stress and can see that the device has high cycling stability.

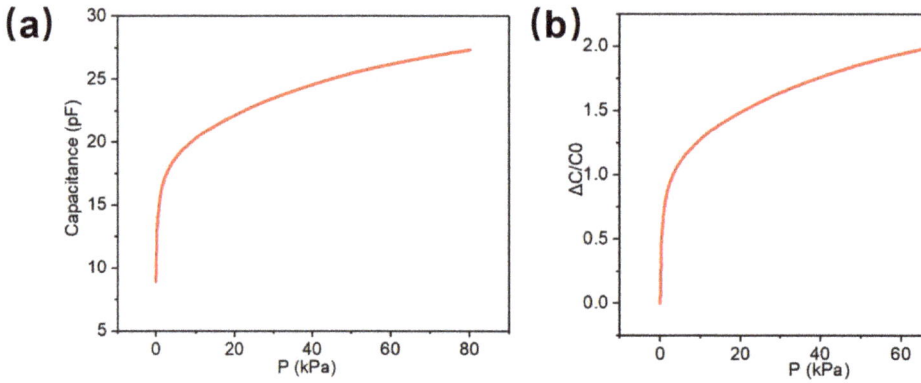

Figure 4. *Cont.*

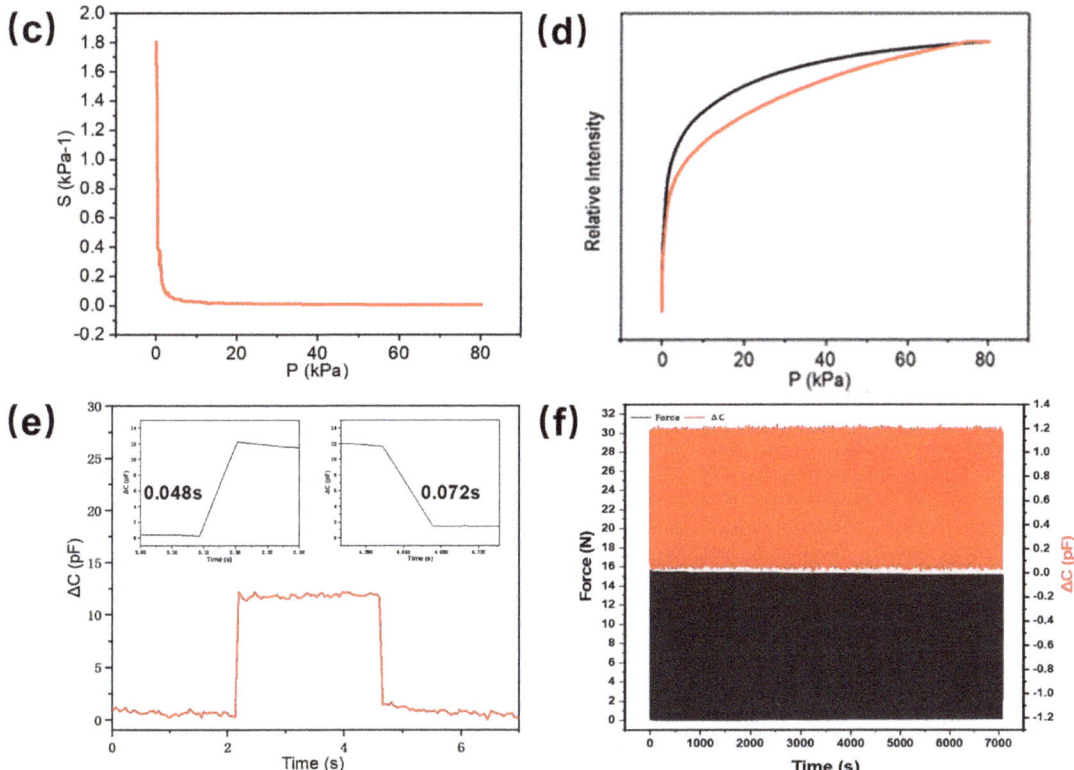

Figure 4. Capacitive pressure sensor based on PVDF nonwoven dielectric layer. (**a**) Capacitance versus pressure curve. (**b**) Relative rate of change of capacitance's relative intensity versus pressure curve. (**c**) Sensitivity pressure response curve. (**d**) Capacitance single–cycle curve. (**e**) Response time. (**f**) Cycle stability.

We designed several experiments to demonstrate the high sensitivity of our prepared microsphere-array electrode capacitive pressure sensor over a wide pressure range. First, we attached the device to a nitrile glove. As shown in Figure 5a, with the passage of nitrogen into the glove, our sensor can recognize the different deformations of the glove. The illustration in Figure 5a shows an example of the sensor on the re-glove, showing that the measured capacitance changes with the magnitude of the glove expansion. Next, we attached the sensor to the finger to distinguish the degree of finger flexion. As shown in Figure 5b, the capacitance value of the device changes as the degree of finger bending is changed, with the greater the bending, the greater the capacitance. The inset shows the capacitance cycling curve of the sensor under repeated finger bending extension, demonstrating the excellent durability and stability of the device (Figure S1, Supporting Information).

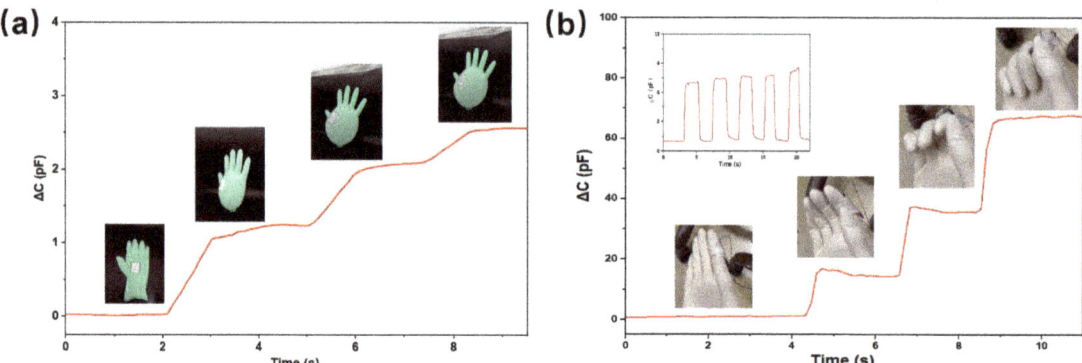

Figure 5. Monitoring of capacitance changes caused by various pressure sources: (**a**) different expansion sizes of the inflated gloves; (**b**) finger bending.

4. Discussion

In this paper, gold electrodes with microsphere arrays on the surface were prepared using the methods of "self-assembly" and "secondary inversion", and the dielectric layer materials with nanofiber nonwoven fabric structure were prepared using electrostatic spinning technology. The sensor performance based on the microsphere-array electrode prepared in this paper shows a high sensitivity under low stress, with a maximum relative change rate $\Delta C/C_0$ of 2.13, a maximum sensitivity of 1.807 kPa^{-1}, and a hysteresis of 7.69%. This is attributed to the microsphere-shaped gold electrode array and the rough-surface dielectric layer as well as the nanofiber-like internal structure. We developed and experimentally validated the COMSOL model, which successfully revealed the sensing mechanism of the microsphere-array electrodes and nanofibrous dielectric. Finally, we demonstrated the sensitivity of this capacitive sensor by detecting subtle pressure changes under various experimental conditions. The excellent sensing range and ultra-high sensitivity make the sensor promising for many other potential applications in human haptic sensing, soft robotics, prosthetics, and surgical e-skin.

Supplementary Materials: The following supporting information can be downloaded at: https://www.mdpi.com/article/10.3390/nano13111702/s1, Contents A: Preparation of piezocapacitive pressure sensor; B: Cyclic stability test; C: Mechanism for enhancing the performance of microsphere electrode-based sensors; Figure S1: Finger bending cycle test. Reference [39] is cited in Supplementary Materials.

Author Contributions: Conceptualization, S.Y. and J.L.; methodology, J.H. and J.Z.; software, J.Z.; validation, J.-H.Z., X.S. and L.P.; data curation, Y.J.; writing—original draft preparation, S.Y.; writing—review and editing, X.S.; supervision, Y.S.; project administration, Y.S. All authors have read and agreed to the published version of the manuscript.

Funding: This work was supported by the National Key Research and Development program of China under Grant No. 2021YFA1401103; the National Natural Science Foundation of China under Grants 61825403, 61921005 and 61674078.

Data Availability Statement: The data presented in this study are available upon request from the corresponding author.

Conflicts of Interest: The authors declare no conflict of interest.

References

1. Shi, J.; Liu, S.; Zhang, L.; Yang, B.; Shu, L.; Yang, Y.; Ren, M.; Wang, Y.; Chen, J.; Chen, W.; et al. Smart Textile-Integrated Microelectronic Systems for Wearable Applications. *Adv. Mater.* **2020**, *32*, 1901958. [CrossRef]
2. Wu, W. Stretchable electronics: Functional materials, fabrication strategies and applications. *Sci. Technol. Adv. Mater.* **2019**, *20*, 187–224. [CrossRef]

3. Sun, Z.; Zhu, M.; Zhang, Z.; Chen, Z.; Shi, Q.; Shan, X.; Yeow, R.C.H.; Lee, C. Artificial Intelligence of Things (AIoT) Enabled Virtual Shop Applications Using Self-Powered Sensor Enhanced Soft Robotic Manipulator. *Adv. Sci.* **2021**, *8*, 2100230. [CrossRef]
4. Yang, J.C.; Mun, J.; Kwon, S.Y.; Park, S.; Bao, Z.; Park, S. Electronic Skin: Recent Progress and Future Prospects for Skin-Attachable Devices for Health Monitoring, Robotics, and Prosthetics. *Adv. Mater.* **2019**, *31*, 1904765. [CrossRef] [PubMed]
5. Sundaram, S.; Kellnhofer, P.; Li, Y.; Zhu, J.-Y.; Torralba, A.; Matusik, W. Learning the signatures of the human grasp using a scalable tactile glove. *Nature* **2019**, *569*, 698–702. [CrossRef] [PubMed]
6. Abraira, V.E.; Ginty, D.D. The Sensory Neurons of Touch. *Neuron* **2013**, *79*, 618–639. [CrossRef]
7. Derler, S.; Gerhardt, L.C. Tribology of Skin: Review and Analysis of Experimental Results for the Friction Coefficient of Human Skin. *Tribol. Lett.* **2012**, *45*, 1–27. [CrossRef]
8. Arakawa, T.; Kuroki, Y.; Nitta, H.; Chouhan, P.; Toma, K.; Sawada, S.-I.; Takeuchi, S.; Sekita, T.; Akiyoshi, K.; Minakuchi, S.; et al. Mouthguard biosensor with telemetry system for monitoring of saliva glucose: A novel cavitas sensor. *Biosens. Bioelectron.* **2016**, *84*, 106–111. [CrossRef]
9. Lee, Y.; Howe, C.; Mishra, S.; Lee, D.S.; Mahmood, M.; Piper, M.; Kim, Y.; Tieu, K.; Byun, H.-S.; Coffey, J.P.; et al. Wireless, intraoral hybrid electronics for real-time quantification of sodium intake toward hypertension management. *Proc. Natl. Acad. Sci. USA* **2018**, *115*, 5377–5382. [CrossRef]
10. Kaltenbrunner, M.; Sekitani, T.; Reeder, J.; Yokota, T.; Kuribara, K.; Tokuhara, T.; Drack, M.; Schwödiauer, R.; Graz, I.; Bauer-Gogonea, S.; et al. An ultra-lightweight design for imperceptible plastic electronics. *Nature* **2013**, *499*, 458–463. [CrossRef] [PubMed]
11. Wang, S.; Xu, J.; Wang, W.; Wang, G.-J.N.; Rastak, R.; Molina-Lopez, F.; Chung, J.W.; Niu, S.; Feig, V.R.; Lopez, J.; et al. Skin electronics from scalable fabrication of an intrinsically stretchable transistor array. *Nature* **2018**, *555*, 83–88. [CrossRef]
12. Zhang, Y.; Zhang, T.; Huang, Z.; Yang, J. A New Class of Electronic Devices Based on Flexible Porous Substrates. *Adv. Sci.* **2022**, *9*, 2105084. [CrossRef] [PubMed]
13. Meng, K.; Xiao, X.; Wei, W.; Chen, G.; Nashalian, A.; Shen, S.; Xiao, X.; Chen, J. Wearable Pressure Sensors for Pulse Wave Monitoring. *Adv. Mater.* **2022**, *34*, 2109357. [CrossRef]
14. Luo, Y.; Abidian, M.R.; Ahn, J.-H.; Akinwande, D.; Andrews, A.M.; Antonietti, M.; Bao, Z.; Berggren, M.; Berkey, C.A.; Bettinger, C.J.; et al. Technology Roadmap for Flexible Sensors. *ACS Nano* **2023**, *17*, 5211–5295. [CrossRef] [PubMed]
15. Kim, J.; Campbell, A.S.; De Ávila, B.E.-F.; Wang, J. Wearable biosensors for healthcare monitoring. *Nat. Biotechnol.* **2019**, *37*, 389–406. [CrossRef] [PubMed]
16. Lee, G.H.; Moon, H.; Kim, H.; Lee, G.H.; Kwon, W.; Yoo, S.; Myung, D.; Yun, S.H.; Bao, Z.; Hahn, S.K. Multifunctional materials for implantable and wearable photonic healthcare devices. *Nat. Rev. Mater.* **2020**, *5*, 149–165. [CrossRef] [PubMed]
17. Chen, J.; Zhu, Y.; Chang, X.; Pan, D.; Song, G.; Guo, Z.; Naik, N. Recent Progress in Essential Functions of Soft Electronic Skin. *Adv. Funct. Mater.* **2021**, *31*, 2104686. [CrossRef]
18. Wang, X.; Liu, Y.; Cheng, H.; Ouyang, X. Surface Wettability for Skin-Interfaced Sensors and Devices. *Adv. Funct. Mater.* **2022**, *32*, 2200260. [CrossRef]
19. Heng, W.; Solomon, S.; Gao, W. Flexible Electronics and Devices as Human–Machine Interfaces for Medical Robotics. *Adv. Mater.* **2022**, *34*, 2107902. [CrossRef]
20. Liu, H.; Xu, T.; Cai, C.; Liu, K.; Liu, W.; Zhang, M.; Du, H.; Si, C.; Zhang, K. Multifunctional Superelastic, Superhydrophilic, and Ultralight Nanocellulose-Based Composite Carbon Aerogels for Compressive Supercapacitor and Strain Sensor. *Adv. Funct. Mater.* **2022**, *32*, 2113082. [CrossRef]
21. Lin, M.; Zheng, Z.; Yang, L.; Luo, M.; Fu, L.; Lin, B.; Xu, C. A High-Performance, Sensitive, Wearable Multifunctional Sensor Based on Rubber/CNT for Human Motion and Skin Temperature Detection. *Adv. Mater.* **2022**, *34*, 2107309. [CrossRef]
22. Gao, J.; Fan, Y.; Zhang, Q.; Luo, L.; Hu, X.; Li, Y.; Song, J.; Jiang, H.; Gao, X.; Zheng, L.; et al. Ultra-Robust and Extensible Fibrous Mechanical Sensors for Wearable Smart Healthcare. *Adv. Mater.* **2022**, *34*, 2107511. [CrossRef]
23. Wang, X.; Gu, Y.; Xiong, Z.; Cui, Z.; Zhang, T. Silk-Molded Flexible, Ultrasensitive, and Highly Stable Electronic Skin for Monitoring Human Physiological Signals. *Adv. Mater.* **2014**, *26*, 1336–1342. [CrossRef] [PubMed]
24. Wei, J.; Xie, J.; Zhang, P.; Zou, Z.; Ping, H.; Wang, W.; Xie, H.; Shen, J.Z.; Lei, L.; Fu, Z. Bioinspired 3D Printable, Self-Healable, and Stretchable Hydrogels with Multiple Conductivities for Skin-like Wearable Strain Sensors. *ACS Appl. Mater. Interfaces* **2021**, *13*, 2952–2960. [CrossRef] [PubMed]
25. Lo, L.-W.; Zhao, J.; Wan, H.; Wang, Y.; Chakrabartty, S.; Wang, C. A Soft Sponge Sensor for Multimodal Sensing and Distinguishing of Pressure, Strain, and Temperature. *ACS Appl. Mater. Interfaces* **2022**, *14*, 9570–9578. [CrossRef] [PubMed]
26. Tolvanen, J.; Hannu, J.; Jantunen, H. Hybrid Foam Pressure Sensor Utilizing Piezoresistive and Capacitive Sensing Mechanisms. *IEEE Sens. J.* **2017**, *17*, 4735–4746. [CrossRef]
27. Zhou, Y.; Shen, M.; Cui, X.; Shao, Y.; Li, L.; Zhang, Y. Triboelectric nanogenerator based self-powered sensor for artificial intelligence. *Nano Energy* **2021**, *84*, 105887. [CrossRef]
28. Chen, Y.; Pu, X.; Xu, X.; Shi, M.; Li, H.-J.; Wang, D. PET/ZnO@MXene-Based Flexible Fabrics with Dual Piezoelectric Functions of Compression and Tension. *Sensors* **2022**, *23*, 91. [CrossRef]
29. Ruth, S.R.A.; Beker, L.; Tran, H.; Feig, V.R.; Matsuhisa, N.; Bao, Z. Rational Design of Capacitive Pressure Sensors Based on Pyramidal Microstructures for Specialized Monitoring of Biosignals. *Adv. Funct. Mater.* **2019**, *30*, 1903100. [CrossRef]

30. Cui, H.; Hensleigh, R.; Yao, D.; Maurya, D.; Kumar, P.; Kang, M.G.; Priya, S.; Zheng, X.R. Three-dimensional printing of piezoelectric materials with designed anisotropy and directional response. *Nat. Mater.* **2019**, *18*, 234–241. [CrossRef] [PubMed]
31. Ma, Y.; Zhang, Y.; Cai, S.; Han, Z.; Liu, X.; Wang, F.; Cao, Y.; Wang, Z.; Li, H.; Chen, Y.; et al. Flexible Hybrid Electronics for Digital Healthcare. *Adv. Mater.* **2020**, *32*, 1902062. [CrossRef] [PubMed]
32. Amjadi, M.; Kyung, K.-U.; Park, I.; Sitti, M. Stretchable, Skin-Mountable, and Wearable Strain Sensors and Their Potential Applications: A Review. *Adv. Funct. Mater.* **2016**, *26*, 1678–1698. [CrossRef]
33. Ruth, S.R.A.; Feig, V.R.; Tran, H.; Bao, Z. Microengineering Pressure Sensor Active Layers for Improved Performance. *Adv. Funct. Mater.* **2020**, *30*, 2003491. [CrossRef]
34. Mannsfeld, S.C.B.; Tee, B.C.K.; Stoltenberg, R.M.; Chen, C.V.H.H.; Barman, S.; Muir, B.V.O.; Sokolov, A.N.; Reese, C.; Bao, Z. Highly sensitive flexible pressure sensors with microstructured rubber dielectric layers. *Nat. Mater.* **2010**, *9*, 859–864. [CrossRef]
35. Ma, Z.; Kong, D.; Pan, L.; Bao, Z. Skin-inspired electronics: Emerging semiconductor devices and systems. *J. Semicond.* **2020**, *41*, 041601. [CrossRef]
36. Zhang, S.; Sun, X.; Guo, X.; Zhang, J.; Li, H.; Chen, L.; Wu, J.; Shi, Y.; Pan, L. A Wide-Range-Response Piezoresistive–Capacitive Dual-Sensing Breathable Sensor with Spherical-Shell Network of MWCNTs for Motion Detection and Language Assistance. *Nanomaterials* **2023**, *13*, 843. [CrossRef] [PubMed]
37. Wen, N.; Zhang, L.; Jiang, D.; Wu, Z.; Li, B.; Sun, C.; Guo, Z. Emerging flexible sensors based on nanomaterials: Recent status and applications. *J. Mater. Chem. A* **2020**, *8*, 25499–25527. [CrossRef]
38. Zhang, J.-H.; Li, Z.; Xu, J.; Li, J.; Yan, K.; Cheng, W.; Xin, M.; Zhu, T.; Du, J.; Chen, S.; et al. Versatile self-assembled electrospun micropyramid arrays for high-performance on-skin devices with minimal sensory interference. *Nat. Commun.* **2022**, *13*, 5839. [CrossRef] [PubMed]
39. Huang, Y.-C.; Liu, Y.; Ma, C.; Cheng, H.-C.; He, Q.; Wu, H.; Wang, C.; Lin, C.-Y.; Huang, Y.; Duan, X. Sensitive pressure sensors based on conductive microstructured air-gap gates and two-dimensional semiconductor transistors. *Nat. Electron.* **2020**, *3*, 59–69. [CrossRef]

Disclaimer/Publisher's Note: The statements, opinions and data contained in all publications are solely those of the individual author(s) and contributor(s) and not of MDPI and/or the editor(s). MDPI and/or the editor(s) disclaim responsibility for any injury to people or property resulting from any ideas, methods, instructions or products referred to in the content.

MDPI
St. Alban-Anlage 66
4052 Basel
Switzerland
www.mdpi.com

Nanomaterials Editorial Office
E-mail: nanomaterials@mdpi.com
www.mdpi.com/journal/nanomaterials

Disclaimer/Publisher's Note: The statements, opinions and data contained in all publications are solely those of the individual author(s) and contributor(s) and not of MDPI and/or the editor(s). MDPI and/or the editor(s) disclaim responsibility for any injury to people or property resulting from any ideas, methods, instructions or products referred to in the content.

www.ingramcontent.com/pod-product-compliance
Lightning Source LLC
LaVergne TN
LVHW070436100526
838202LV00014B/1611